MW00761160

Brilliant Light in Life and Material Sciences

NATO Security through Science Series

This Series presents the results of scientific meetings supported under the NATO Programme for Security through Science (STS).

Meetings supported by the NATO STS Programme are in security-related priority areas of Defence Against Terrorism or Countering Other Threats to Security. The types of meeting supported are generally "Advanced Study Institutes" and "Advanced Research Workshops". The NATO STS Series collects together the results of these meetings. The meetings are co-organized by scientists from NATO countries and scientists from NATO's "Partner" or "Mediterranean Dialogue" countries. The observations and recommendations made at the meetings, as well as the contents of the volumes in the Series, reflect those of participants and contributors only; they should not necessarily be regarded as reflecting NATO views or policy.

Advanced Study Institutes (ASI) are high-level tutorial courses to convey the latest developments in a subject to an advanced-level audience

Advanced Research Workshops (ARW) are expert meetings where an intense but informal exchange of views at the frontiers of a subject aims at identifying directions for future action

Following a transformation of the programme in 2004 the Series has been re-named and re-organised. Recent volumes on topics not related to security, which result from meetings supported under the programme earlier, may be found in the NATO Science Series.

The Series is published by IOS Press, Amsterdam, and Springer, Dordrecht, in conjunction with the NATO Public Diplomacy Division.

Sub-Series

A. Chemistry and Biology	Springer
B. Physics and Biophysics	Springer
C. Environmental Security	Springer
D. Information and Communication Security	IOS Press
E. Human and Societal Dynamics	IOS Press

http://www.nato.int/science
http://www.springer.com
http://www.iospress.nl

Series B: Physics and Biophysics

Brilliant Light in Life and Material Sciences

edited by

Vasili Tsakanov

Center for the Advancement of Natural Discoveries Using Light Emission,
Yerevan, Armenia

and

Helmut Wiedemann

Advanced Physics Department and SSRL/SLAC
Stanford University, Stanford,
CA, U.S.A.

 Springer

Published in cooperation with NATO Public Diplomacy Division

Proceedings of the NATO Advanced Research Workshop on
Brilliant Light Facilities and Research in Life and Material Sciences
Yerevan, Armenia
17-21 July 2006

A C.I.P. Catalogue record for this book is available from the Library of Congress.

ISBN-10 1-4020-5723-7 (PB)
ISBN-13 978-1-4020-5723-6 (PB)
ISBN-10 1-4020-5722-9 (HB)
ISBN-13 978-1-4020-5722-9 (HB)
ISBN-10 1-4020-5724-5 (e-book)
ISBN-13 978-1-4020-5724-3 (e-book)

Published by Springer,
P.O. Box 17, 3300 AA Dordrecht, The Netherlands.

www.springer.com

Printed on acid-free paper

All Rights Reserved
© 2007 Springer
No part of this work may be reproduced, stored in a retrieval system, or transmitted in
any form or by any means, electronic, mechanical, photocopying, microfilming,
recording or otherwise, without written permission from the Publisher, with the exception
of any material supplied specifically for the purpose of being entered and executed on a
computer system, for exclusive use by the purchaser of the work.

PREFACE

A NATO Advanced Research Workshop on "Brilliant Light Facilities and Research in Life and Material Sciences" was held from July 17 to July 21, 2006. The workshop was hosted by the Center for the Advancement of Natural Discoveries using Light Emission, Yerevan - the newly established institute in Armenia with the aim to create a synchrotron radiation facility, CANDLE, as an international laboratory for advanced research in life and material sciences. About 50 researchers from NATO, partner countries and Armenia gathered at Yerevan to discuss modern trends in developments of advanced light sources with high spectral brilliance and applications in basic and applied research in a wide range of fields.

Research with high brilliant photon beams are used, for example for practical applications in pharmacy, electronics and nanotechnology. Such practical relevance promoted the design and construction of now more than 50 such facilities worldwide. Overview and specialized talks on the status and highlights of newly constructed light sources (ALBA, SPEAR3, European XFEL Facility, Siberian Synchrotron Radiation Center, CANDLE), on instrumentation and development of experimental techniques, and frontier research in life and material sciences using synchrotron radiation have been presented. More than 60% of the program was devoted to application of synchrotron radiation in biophysics, biochemistry, biomedicine, material and environmental investigations. The workshop brought together scientists from a wide spectrum of research fields emphasizing the wide application and demand of synchrotron radiation and underlining the necessity of user involvement in the early design stages of a new project.

This workshop proceeding includes the presentations by participants from Armenia, Bulgaria, Canada, France, Germany, Greece, Hungary, Italy, Poland, Russia, Spain, Turkey, Ukraine and USA. A total of 50 presentations in form of invited or contributed talks and posters were. Ample time was provided for informal communication among the participants in an open and friendly atmosphere ensuring the desired intense contact

This meeting became possible through the generous funding provided by the NATO Scientific and Affairs Division. Special thanks goes to Yerevan State University and the Armenian National Academy of for support in organizing the opening session of the workshop. The workshop directors would like to thank the full CANDLE staff and mostly Dr. Gayane Amatuni and Mrs. Marine Baghiryan for their complete dedication towards a smooth and expert organization and administration of the workshop.

October 1, 2006   Vasili Tsakanov
                                                         Helmut Wiedemann

TABLE OF CONTENTS

3. Biomedical Research

4. Material and Environmental Sciences

5. Instrumentation and Experimental Technique

PART 1

Brilliant Light Facilities and New Projects

SYNCHROTRON LIGHT SOURCES, STATUS AND NEW PROJECTS

Dieter Einfeld
CELLS, P.O. Box 68, Campus UAB, 08193, Bellaterra (Spain)

Abstract: The first synchrotron radiation was used in a so-called parasitic mode from high energy machine. At the end of the 1970s and at the beginning of the 1980s accelerators dedicated to the production of synchrotron light (second generation sources) were built (SRS in Daresbury, NSLS in Brookhaven, and BESSY in Berlin). With the investigation of wigglers and undulators the design and construction of the so-called third generation sources (high-brilliance light sources) started (ESRF in Grenoble, ALS in Berkeley, and ELETTRA in Trieste). At present there exist roughly 50 synchrotron light sources around the world. All machines have reached their target specifications (emittance, current, life time, stability, etc) without any problems and in a short time. From the energy point of view the synchrotron light sources are divided in three categories: low- (<1.5 GeV), intermediate- (1.5 to 3.5 GeV), and high-energy (>3.5 GeV). The most attractive light sources are the brilliant ones, which have a lot of space available for the installation of insertion devices. This report will concentrate on the high-brilliance intermediate- and high-energy light sources. It will be reported upon the machines which are in operation and in the commissioning, construction or design phase. It will also describe new projects, which are in the design phase, and ideas about how to upgrade already existing light sources. The next or 4th generation light sources are divided in 3 categories: the traditional light sources (LS), the energy-recovery linacs (ERL) and the free electron laser projects (FEL). The last two categories will not be considered within this review.

Key words: Synchrotron Light Source, Lattices, Emittance

Vasili Tsakanov and Helmut Wiedemann (eds.), Brilliant Light in Life and Material Sciences, 3–20.
© 2007 *Springer.*

1.   INTRODUCTION

Synchrotron radiation is emitted from circular accelerated charged particles (in most of the cases electrons) moving at relativistic velocities. To get radiation from in-vacuum undulators within an energy range of up to 20 keV, the energy of the electrons should be around 3 GeV and the circulating current within the storage ring should be within 200 and 500 mA. In order to reach these specifications the main components of a synchrotron light source are: linac, booster synchrotron, storage ring, transfer lines between the accelerators, front ends, and the beam lines. A general set up of a synchrotron light source with these components is given in Fig. 1, which shows the layout of CANDLE, the proposed light source for Armenia[1].

The linac produces an electron beam with a current of some mA (2 to 10) and energies between 100 and 200 MeV, plus a time structure required by the booster and the storage ring. Within the booster synchrotron the energy is increased up to the nominal energy of the storage ring. Each single shot from the booster increases the current in the storage ring by 2 or 3 mA. To store a current in the storage ring of 300 mA, 100 injection shots from the booster are needed. With a typical repetition rate of the injector (linac and booster) of 3 Hz, the storage ring will be filled within roughly 30 to 50 seconds.

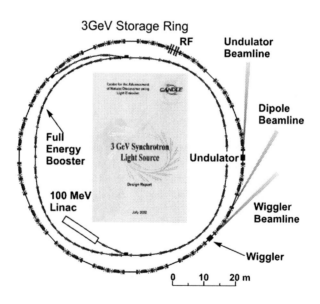

Figure 1. The layout of the synchrotron light source (LS) CANDLE.

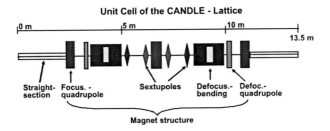

Figure 2. Layout of the unit cell of the CANDLE lattice with the corresponding arrangements of magnets.

The storage ring is built up with a number of unit cells[2,3]. In the case of CANDLE the number of unit cells is equal to 16. The layout of one of these cells is shown in Fig. 2. Each unit cell consists of a straight section and a magnetic structure. The straight sections are used for the installation of insertion devices, which provide the synchrotron radiation for the experimental beam lines. The magnetic structure consists of bending magnets, quadrupoles and sextupoles. Very often, as in the case of CANDLE, the bending magnets include a defocusing component.

The performance of a synchrotron light source is given by: 1) the emitted photon spectrum required by the users, 2) the brilliance of the emitted radiation, 3) the overall length of the straight sections, and 4) the stability of the beam. These performance factors are determined by the following parameters of the machine:

1. The emitted spectrum is proportional to the square of the energy of the electrons ($E_{ph} \approx E_{electr}^2$).
2. The brilliance of an undulator is proportional to the stored current (I), to the length of the undulator (L_{ID}), and inversely proportional to the horizontal emittance (ε_x) of the stored electron beam ($Br \approx I\, L_{ID} / \varepsilon_x^{1.5}$).
3. The overall length of the straight sections is proportional to the circumference (C) of the machine.
4. The stability of the beam is given by the rf-, vacuum-, feedback-systems, etc.

By fixing the energy, according to the required photon spectrum and assuming a stable beam, the brilliance is determined by the length of the straight sections and the emittance of the beam. For a fixed circumference the emittance is given by the magnetic structure within the unit cell and the excitation of the different magnetic elements. In general the emittance is given by the following formula[4]:

$$\varepsilon_{x0}\,[\mathrm{nm\,rad}] = 31.64\ K\ [(E/GeV)]^2\ \varphi^3 \qquad (1)$$

K is a value depending upon the magnet structure and is given by the distribution of the beta- [$\beta(s)$] and dispersion [$\eta(s)$] functions within the bending magnets. A minimum of both functions within the bending magnet results in a small K-value[4-6]. Hence a small K-value means a good design of the lattice. φ is the deflection angle per magnet. In order to understand the dependence of the emittance upon the magnet structure, it is useful to understand the machine functions: beta- and dispersion functions. On the one hand, $\beta_{x,y}$ are the betatron functions which describe the betatron oscillation of the electron beam [$x(s), y(s)$] within the machine[3]:

$$x(s) = a\sqrt{\beta_x(s)} \cos[\Phi(s) - \delta]$$
$$\text{and} \quad E(\beta_x, s) = \sqrt{A_x \beta_x(s)}; \quad \sigma_x = \sqrt{\varepsilon_x \beta_x(s)} \tag{2}$$

where a and δ are given by the starting conditions, $E(\beta_x, s)$ is the envelope of the betatron oscillations, A_x is the acceptance, and ε_x is the emittance of the machine. On the other hand, $\eta(s)$ is the dispersion function; it is given by the magnetic structure too and describes the closed orbit deviation with the energy offset ΔE.

$$x(\eta, s) = \eta(s)\,(\Delta E / E) \tag{3}$$

In order to get a small emittance, the horizontal beta- and dispersion function must run through a minimum within the bending magnet, as shown in Fig. 3. The beam size within the long straight section is given by the emittance, the beta-, and the dispersion function according to Eqs. (2) and (3). As an example, the machine functions as well as the cross section of the beam within the unit cell of CANDLE are shown in Fig. 3.

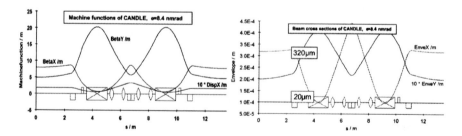

Figure 3. The machine functions (*left*) and the cross section (*right*) of the beam within one unit cell of CANDLE. Within the long straight section the cross sections (1σ) are 320 μm (horizontal) and 20 μm (vertical).

Third generation light sources use the radiation coming from insertion devices (wigglers and undulators), which are installed in the straight sections[7]. The trajectory of the electron beam within the insertion devices is a sinusoidal one with amplitude X_0 and divergence X_0'. The values of these two numbers are roughly: $X_0 \approx 20$ μm and $X_0' \approx 1.5$ mrad for wigglers and $X_0 \approx 1$μm and $X_0' \approx 0.3$ mrad for undulators, resulting in emittances of 30 nm·rad for wigglers and 0.3 nm·rad for undulators.

2.   SYNCHROTRON LIGHT SOURCES WORLD WIDE

All synchrotron- and free electron light sources have established a webpage (www.lighsource.org) from which news, information, and educational materials can be looked up. According to this webpage, there exist 49 synchrotron light sources worldwide: Australia 1, North America 9, South America 1, East Asia 19, West Asia 3, and Europe 16. There is no synchrotron light source in Middle America and Africa. According to their energy, synchrotron light sources (LS) can be divided in three categories: low-, intermediate-, and high-energy ones. Here we will concentrate in the intermediate and high-energy ranges. Within this ranges there exist 23 LS (see Table 1): 12 in operation, 3 in the commissioning phase, 4 in the construction phase, and 4 new projects.

2.1   SYNCHROTRON LIGHT SOURCES IN OPERATION

The first 3rd generation synchrotron light source which went into operation was the ERSF[8], followed by the ALS[9] and ELETTRA[10]. In the following the performance of the operating synchrotron light sources shall be described. The machine functions and the cross section of the beam within the unit cell of the ESRF are shown in the Fig. 4. The ESRF used for the magnetic arrangement a DBA-structure[11] and introduces two different types of straight sections (hybrid-structure): a high-beta and a mini-beta section. According to Eq. (2) this leads to different cross sections: the beam envelope in the mini-beta sections is reduced roughly by a factor 6 in the horizontal direction (from 360 to 61 μm) and by a factor 2 in the vertical direction (from 18.6 to 9.5 μm). The other high-energy machines, APS[12] and Spring-8[13], are using a DBA structure too, but only with high-beta straight section. The general parameters of these machines are compiled in Table 2.

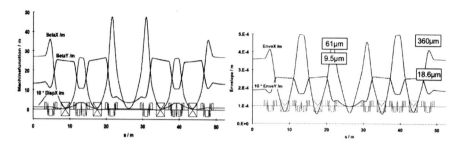

Figure 4. The machine functions and the cross section of the beam within the unit cell of the ESRF. Within the long straight section the cross sections (1σ) are 320 μm (horizontal) and 20 μm (vertical).

Table 1. Intermediate- and high-energy SL worldwide. They are divided in the following categories: operation, commissioning, construction, and new projects. The data can be found in the webpage[14].

Facility	Location	Energy	ε	C	Lifetime	Curr.	Life·Curr
		[GeV]	[nm rad]	[m]	[h]	[mA]	[A·h]
Synchrotron Light Sources in operation							
Spring-8	Himeji (Japan)	8.0	3-5.9	1432	140	100	14
APS	Argonne (USA)	7.0	3	1104	37.5	100	3.8
ERSF	Grenoble (France)	6.0	3.8	850	68	200	13.6
SPEAR-3	Stanford (USA)	3.0	18	234	16.5	500	8.3
CLS	Saskatoon (Canada)	2.9	18	171	10	170	1.7
PLS	Pohang (Korea)	2.5	12	281	30.5	174	5.3
SLS	Villigen (Switzerland)	2.4	5	288	8	400	3.2
ELETTRA	Trieste (Italy)	2.4	7	259	12	300	3.6
ALS	Berkeley (USA)	1.9	6.8	197	8	400	3.2
BESSY II	Berlin (Germany)	1.9	5.2	240	3	220	0.7
NSRRC	Hsinchu (Taiwan)	1.5	25	120	6.5	200	1.3
Synchrotron Light Sources in Commissioning							
DIAMOND	Oxfordshire (UK)	3.0	2.7	562	18	300	5.4
ASP	Melbourne (Australia)	3.0	8.6	216	18.5	200	3.7
SOLEIL	Orsay (France)	2.75	3.1	351	15	500	7.5
Synchrotron Light Sources in Construction							
PETRA III	Hamburg (Germany)	6.0	1	2300	24	100	2.4
INDUS II	Indore (India)	2.5	58	172.5	18	300	5.4
SSRF	Shangai (China)	3.5	4.8	432	10	250	2.5
ALBA	Barcelona (Spain)	3.0	3.7	266	15	400	6
SESAME	Allan (Jordan)	2.5	27	129	18	400	7.2
New projects and Upgrades							
NSLS II	Upton (USA)	3.0	1.5	620	4.5	500	2.3
MAX IV	Lund (Sweden)	3.0	1.2	285	4	500	2
TPS	Hsinchu (Taiwan)	3.0	1.7	518.4	5	400	2
CANDLE	Yerevan (Armenia)	3.0	8.4	216	19.5	350	6.8

Table 2. Parameters of the high-energy LS APS, ESRF, and Spring-8. *facility*-Distr. means that the dispersion function is distributed around the whole ring so the dispersion function within the straight sections is not zero.

Source	Lattice	E [GeV]	ε [nm·rad]	Ins. Length [m]	Angle [rad]	C [m]	Percent [%]	K	Tot. Brill.
APS-0	DBA	7	8.2	268.8	0.07854	1104	24.3	10.9	1145
APS-Distr.	DBA	7	3.7	268.8	0.07854	1104	24.3	4.9	3777
ESRF	DBA	6	4.4	201.6	0.09817	837.6	24.1	4.1	2184
Spring-8	DBA	8	4.8	416	0.0714	1440	28.9	6.5	3956
Spring-8-Distr.	DBA	8	3	416	0.0714	1440	28.9	4.1	8006

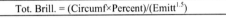

Tot. Brill. = (Circumf×Percent)/(Emitt$^{1.5}$)

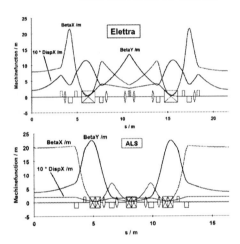

Figure 5. The machine functions (lattice) within the unit cell of the ALS and ELETTRA.

To make a comparison between different LS three parameters have been calculated: 1) the percentage of the circumference that is devoted to straight sections, 2) the *K*-value according to Eq. (1) [remember: a small *K*-value means a good lattice design], and 3) the total brilliance, which is equal to the total length of all straight sections divided by the emittance to the power of 1.5. This value gives an idea about the overall performance of a LS. A synchrotron light source has a good performance if 1) and 3) are high and 2) is small. According to Table 2, Spring-8 has the highest performance. The reason for this is that Spring-8 has a four-fold symmetry with 4 long straights of 30 m length. According to Table 2 the *K*-factor for the different LS varies from 4.1 to 10.9.

Different lattices have been used for the ALS in Berkeley and ELETTRA in Trieste. ALS has a TBA structure and ELETTRA has an expanded DBA structure (see Fig. 5 and Table 3). Both LS make use of combined bending magnets, which provide most of the vertical focusing within the lattice. For the TBA lattice, with three bending magnets per unit cell, the number of magnets is higher by a factor 1.5 compared to the DBA lattice; however, giving rise to a decrease in the emittance by a factor 3.4 according to Eq. (1). In lattices with combined bending magnets the partition function in Eq. (1) increases from 1 to around 1.4, leading to a further decrease in the emittance. The K-value for the ALS lattice is 12.9. At ELETTRA, according to Fig. 5, another focusing quadrupole has been introduced in the DBA lattice. This additional quadrupole makes the phase advance between both bending magnets roughly equal to 2π, and reduces the K-value down to 4.1 ($K = 3.0$ is the lowest K value for a DBA-structure).

The main parameters of the intermediate energy LS are compiled in Table 3 for comparison. A new step in the design of LS is the Swiss Light Source SLS[15]. The lattice of the SLS is shown in Fig. 6: it is a TBA lattice like the one at the ALS, but at the SLS magnets with different deflection angles are used. The bending in the middle magnet is larger than the one at the end magnets, thus reducing the emittance considerably.

Table 3. Parameters of the intermediate-energy Synchrotron Light Sources.

Source	Lattice	E [GeV]	ε [nm·rad]	Ins. Length [m]	Angle [rad]	C [m]	Percent [%]	K	Tot. Brill.
MAX II	DBA	1.5	9	31.4	0.3142	90	34.9	4.1	116
SRRC	TBA	1.5	25.6	36	0.1745	120	30.0	67.7	28
ALS	TBA	1.9	5.6	81	0.1745	196.8	41.2	9.2	611
BESSY II	DBA	1.9	6.4	89	0.1963	240	37.1	7.4	550
ELETTRA	DBA	2	7	74.78	0.2618	258	29.0	3.1	404
INDUS II	DBA	2.5	58	36.48	0.3927	172	21.2	4.8	8
SLS	TBA	2.4	5	63	0.1745	288	21.9	5.2	563
PLS	TBA	2.5	18.9	81.6	0.1745	280.6	29.1	18.0	99
NSLS-xray	DBA	2.5	44.5	18	0.3927	170.08	10.6	3.7	6
SESAME	TME	2.5	26.4	50.24	0.3927	128.4	39.1	2.2	37
SOLEIL	DBA*	2.75	3.72	159.6	0.1963	354	45.1	2.1	2224
CLS	DBA	2.9	18.2	62.4	0.2618	170.4	36.6	3.8	80
ROSY	DBA*	3	28.5	44.8	0.3927	148.11	30.2	1.7	29
SPEAR III	DBA	3	18.2	67	0.16535	234.13	28.6	18.7	86
ASP	DBA	3	6.88	76.72	0.2244	216	35.5	2.1	425
DIAMOND	DBA	3	2.74	218.2	0.1309	561.6	38.9	4.3	4811
ALBA	DBA*	3	4.29	103.44	0.1963	268.8	38.5	2.0	1164
CANDLE	DBA	3	8.4	76.8	0.1963	216	35.6	3.9	315
SSRF	DBA	3.5	3.9	152	0.1571	432	35.2	2.6	1974

Tot. Brill. = (Circumf×Percent)/(Emitt$^{1.5}$)

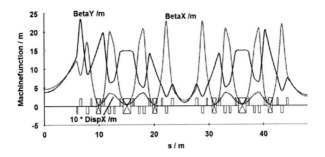

Figure 6. Lattice within half unit cell of the Swiss Light Source SLS.

On the other hand, the length of all straight sections at the ALS is much higher than that at the SLS. The total brilliance devoted to the users is roughly the same for both machines. In contrast with the other light sources, the SLS has straight sections with three different lengths: 11.8 m, 7 m, and 4 m. All straights have a mini-beta section, which helps a lot for the stability of the beam (by decreasing the impedance) and the operation with in-vacuum undulators.

The SLS introduced for the first time in a regular basis the topping-up injection[16], which from then on has been included in the design of all new LS and it is planned to be implemented in many already-operating ones. Within the topping-up injection scheme, a current of 1 mA is injected into the storage ring every 2 to 3 seconds, thus keeping the current in the storage ring constant by ±0.5 mA. Furthermore, SLS was also the first light source where both the injector and the storage ring share the same tunnel. This leads to a small emittance of the booster synchrotron, which helps to reduce the electron losses for topping-up injection.

In general one can say that all synchrotron light sources reached their specifications shortly after commissioning. They are running very well and reliably, with over 90% of the operation time devoted to the users, and there is a continuous process for upgrading each source to get better performances.

2.2   SYNCHROTRON LIGHT SOURCES IN COMMISSIONING

At present (July 2006) there are three SL in the commissioning phase: ASP[17], DIAMOND[18], and SOLEIL[19]. The lattice of ASP is roughly the same as for CLS and for CANDLE (see Fig. 3): it is a pretty good optimized lattice with a small *K*-value and a high percentage of the circumference devoted to straight sections, but it does not have any mini-beta section.

The lattice of SOLEIL is shown in Fig. 7: it is an expanded DBA lattice with a four-fold symmetry and three different straight sections (12 m, 7 m, and 3.8 m). In all 7 m straights there is a mini-beta section. According to Table 3, Soleil has almost the smallest *K*-value and the highest percentage of the circumference for the installation of insertion devices (45.1%). It is a pretty optimized design.

The lattice of Diamond is shown in Fig. 8: it is a typical DBA lattice with a 6-fold symmetry and two different lengths of the straight sections (11.2 m and 8.2 m). In all 8.2 m straights there is a mini-beta section too. 38% of the circumference is devoted to the installation of insertion devices. The *K*-factor is 4.3, higher than the one at SOLEIL by a factor of two. It is a pretty optimized design as well.

All three LS reached the design specifications pretty soon within the commissioning time.

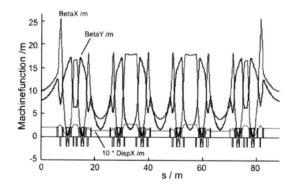

Figure 7. The lattice of the Synchrotron Light Source SOLEIL.

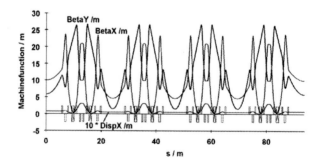

Figure 8. The lattice of the Synchrotron Light Source DIAMOND.

2.3 SYNCHROTRON LIGHT SOURCES IN CONSTRUCTION

There are four projects within the construction phase: ALBA at CELLS in Barcelona (Spain)[20], PETRA III at DESY in Hamburg (Germany)[21], SSRF in Shanghai (China)[22], and SESAME in Allan (Jordan).

After the design of ALBA was finished (see Fig. 9), it was realized that the resulting lattice is comparable with the one at SOLEIL (expanded DBA): it has straight sections with three different lengths (8m, 4.3m, and 2.3 m) and everywhere in the medium straights (4.3 m) there is a mini-beta section. However, in order to make the lattice more compact, the vertical focusing was introduced in the bending magnets. According to Table 3, ALBA has a small K-value and a 38% of the circumference is available for straight sections. By adding 30 m to the circumference the percentage would increase up to 49.6%.

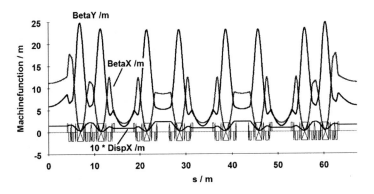

Figure 9. The lattice of the Synchrotron Light Source ALBA.

The lattice for the SSRF (see Fig. 10) in Shanghai is roughly the same as the one for DIAMOND, but the number of cells has been reduced from 24 to 20. Besides, DIAMOND has a smaller β_y-function in the mini-beta section, which leads to a reduction of the vertical beam size by roughly 30 %.

In contrast, the design of SESAME[23] is quite different from the previous ones (see Fig. 11): the unit cell consists basically of one defocussing bending magnet with a focussing quadrupole on each side. With this approach one gets the highest percentage of the circumference devoted to straight sections and a pretty small lattice factor.

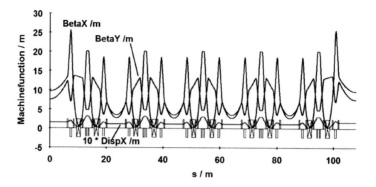

Figure 10. lattice of the Synchrotron Light Source SSRF.

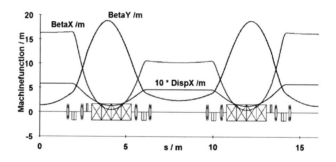

Figure 11. The lattice of the Synchrotron Light Source SESAME.

With the closing of the high energy program at DESY, the injector PETRA will be converted in a synchrotron light source, PETRA III. For that source the FODO structure with 27 magnets in one octant will be converted in a DBA structure with 18 bending magnets (see Fig. 12). The emittance of this machine will be 1 nm·rad and 14 straight sections for the installation of insertion devices will be available. At an energy of 3 GeV, the emittance will be 0.24 nm·rad, which would be the smallest value for a synchrotron light source.

2.4   SYNCHROTRON LIGHT SOURCES IN THE DESIGN PHASE

There are five new projects within the design and upgrading phase: APS-II in Chicago (USA)[24], MAX-IV at MAXLAB in Lund (Sweden)[25],

NSLS-II at BNL in Brookhaven (USA), TPS at NSRRC in Taipeh (Taiwan)[26], and the upgrading of Spring-8 in Japan. The specifications of these LS are summarized in Table 4.

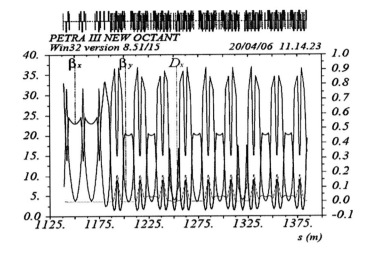

Figure 12. The lattice of the Synchrotron Light Source PETRA III.

Table 4. Parameters of the new projects of Synchrotron Light Sources

Source	Lattice	E [GeV]	ε [nm·rad]	Ins. Length [m]	Angle [rad]	C [m]	Percent [%]	K	Tot. Brill.
APS II-unit	TBA	7	0.54	392.15	0.05236	1104	35.5	2.4	98824
APS II	TBA	7	0.75	392.15	0.05236	1104	35.5	3.4	60375
Spring-8-unit	MBA	8	0.15	157.58	0.02618	1440	10.9	4.1	271247
Spring-8-upgr.	MBA	8	0.168	157.58	0.02618	1440	10.9	4.6	228843
MAX IV	MBA	3	1.28	55.2	0.0748	288	19.2	10.7	3812
MAX IV-Wiggl	MBA	3	0.86	55.2	0.0748	288	19.2	7.2	6921
NSLS-II	DBA	3	2.24	189.3	0.10472	780.3	24.3	6.8	5646
TPS	DBA	3	1.7	196.32	0.1309	518.4	37.9	2.7	8857

Tot. Brill. = $(\text{Circumf} \times \text{Percent})/(\text{Emitt}^{1.5})$

The next generation of synchrotron light sources should have a much higher brilliance than the existing one. According to Eq. (1), this can only be achieved by increasing, the number of magnets. The upgrading proposal for APS consists in changing the DBA to a TBA-structure, which decreases the emittance by a factor 3.3. Going furthermore from the classical DBA structure to a TME-structure the emittance is reduced by another factor 3, which overall makes roughly a factor 10. The proposed lattice for the upgrading of the APS is shown in Fig 13 (unit cell). With this upgrading proposal the emittance will be reduced down to 0.54 nm·rad, a factor 15 smaller than the present one (see Tables 2 and 3). For the real lattice some modifications are introduced in order to have high- and low-beta sections (see Fig. 14). This will reduce the cross section of the beam by a factor 2, leading to values similar to the ones at Diamond, Soleil or ALBA. According to Tables 2 and 3, the percentage of the circumference available for straight sections will increase from 24.3 to 35.5%.

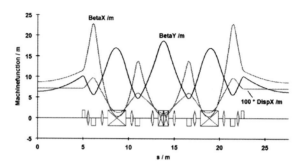

Figure 13. Lattice for the unit cell of the proposed upgrading of the APS.

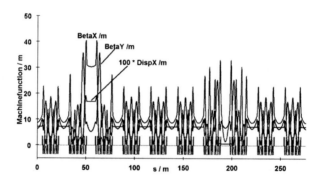

Figure 14. Lattice for the proposed upgrading of the APS.

The lattice of the proposed light source TPS in Taiwan is shown in Fig. 15, and it is more or less the same as for Diamond (see Fig. 8). Because of the higher dispersion function in the straight section the emittance is lower by a factor 1.6.

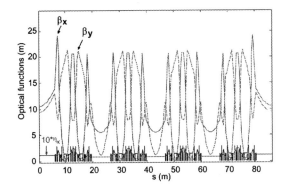

Figure 15. Lattice for the proposed light source TPS at the NSRRC at Taiwan.

The lattice of the proposed light source NSLS-II is shown in Fig. 16, and it is similar to the one at the ESRF. Taking into account the energy dependency of the emittance according to Eq. (1), the emittance for NSLS-II should be roughly 1 nm·rad. This is also reflected in the K-factor, which is pretty high for NSLS-II (see Table 3) in comparison to the ESRF. Therefore there is a lot of room for the optimization of the lattice for NSLS-II. The cross sections of the beam within the mini-beta sections are: $\sigma_x = 60$ μm and $\sigma_y = 2.9$ μm. These are pretty small numbers, smaller than the typical ones for other LS by a factor of two. The percentage of the circumference devoted to straight sections is 24.3 %, which is rather small in comparison to other LS.

Completely different is the lattice for the proposed Light Source MAX IV. It is built up with a MBA structure (see Fig. 17), consisting of five unit cells and two matching sections[4,5]. According to Table 3 the emittance of MAX IV will be 1.28 nm·rad, and by introducing some damping wigglers it will be further reduced down to 0.86 nm·rad. With such a small emittance, the cross sections of the beam will be reduced to $\sigma_x = 97$ μm and $\sigma_y = 4.8$ μm In order to make the lattice as compact as possible to save space, MAX IV is using combined function magnets with a gradient of approximately 10 T/m, which is considerably high. Because of the restricted circumference of MAX IV, only a 19.2 % of the circumference is available for straight sections.

The MBA-structure has also been proposed for a possible upgrade of Spring-8. The lattice for a cell is shown in Fig. 18. With this modifications of the lattice the emittance will be reduced down to 0.17 nm·rad. The cross sections of the beam will be decreased from $\sigma_x = 340$ μm and $\sigma_y = 14$ μm to $\sigma_x = 61$ μm and $\sigma_y = 4.2$ μm.

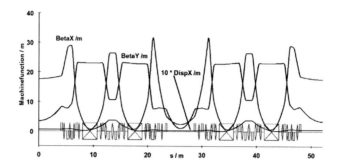

Figure 16. Lattice for the proposed Light source NSLS-II at BNL in Brookhaven.

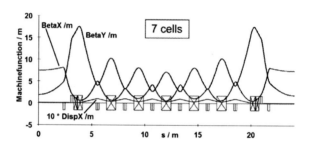

Figure 17. Lattice for the proposed Light source MAX-IV at MAXLAB in Lund, Sweden.

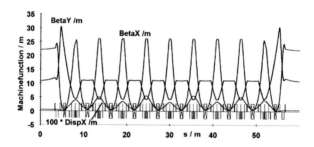

Figure 18. Lattice for the proposed upgrade of Spring-8.

3. ELECTRON BEAM CROSS SECTIONS

Table 5. Electron beam cross-sections of intermediate- and high-energy SL worldwide.

Source	Lattice	Energy [GeV]	ε [nm rad]	σ_x [μm]	σ_y [μm]	σ_x [μm]	σ_y [μm]	σ_x [μm]	σ_y [μm]
				Long straight		Straight 2		Straight 3	
MAX II	DBA	1.5	9	353	19.6				
NSRRC	TBA	1.5	25.6	515	27.1				
ALS	TBA	1.9	5.6	250	13.2				
BESSY II	DBA	1.9	6.4	329	14.8	84.3	8.7		
ELETTRA	DBA	2	7	235	12				
SLS	TBA	2.4	5	150	13.6	83.9	6.7	108.0	11.6
NSLS-xray	DBA	2.5	44.5	258	14.7				
PLS	TBA	2.5	18.9	432	27.3				
CLS	DBA	2.9	18.2	460	30				
SPEAR III	DBA	3	18.2	400	30				
ESRF	DBA	6	4.4	360	18.6	61	9.5		
APS-Distr	DBA	7	3.7	240	18				
Spring-8-Distr	DBA	8	4.8	278	11				
SOLEIL	DBA*	2.72	3.72	285	17.1	187	8	378.0	8.1
ASP	DBA	3	6.88	317	19				
DIAMOND	DBA	3	2.74	178	12.6	123	6.5		
SESAME	TME	2.5	26.4	900	18.9				
INDUS II	DBA*	2.5	68	840	62				
SSRF	DBA*	3	3.9	203	13.9	131	8.3		
ALBA	DBA*	3	4.29	270	16.2	131	7.6	315.0	15.1
PETRA III	DBA*	6	1	141	4.9	53	6	126.0	10.0
CANDLE	DBA	3	8.4	320	20				
MAX IV	MBA	3	1.28	97.4	4.8				
MAX IV-Wiggl	MBA	3	0.86						
NSLS-II	DBA	3	2.24	160	5.3	60	2.9		
TPS	DBA	3	1.7	136	12.6	97	4.6		
APS-Upgr	TBA	7	0.75	116	7	245	6.3	35.0	7.1
Spring-8-Upgr	MBA	8	3.9	61	4.2				

ACKNOWLEDGEMENTS

The Author wants to acknowledge the help of different colleagues: Klaus Balewski (DESY), Michael Borland (APS), Jean-Marc Filhol (SOLEIL), Stephen Kramer (NSLS), Boris Podobedov (NSLS), Christoph Steier (ALS), Rachel Taylor (ASP), Koji Tsumaki (SPring-8), and Richard Walker (Diamond) for their contribution. The author is indebted to Jordi Marcos (CELLS) for his excellent assistance in preparing this manuscript.

REFERENCES

1 V. M. Tsakanov, CANDLE Project Overview, Proceedings of PAC 2005, p. 629.
2 H. Wiedemann, Particle Accelerator Physics II, Springer Verlag, 1995 ISBN 3-540-57564-2.2.
3 Klaus Wille, The Physics of Particle Accelerators, Oxford University Press, 2005, ISBN 0-19-850549-3.
4 D. Einfeld, J. Schaper and M. Plesko, Design of a Diffraction Limited Light Source, 10th ICFA Beam Dynamics Panel Workshop on 4th Generation Light Sources, Grenoble (1996).
5 D. Einfled and M. Plesko, A modified QBA Optics for low Emittance Storage Rings, NIM A 335 (1993), pp. 402-416.
6 D. Einfeld, J. Schaper and M. Plesko, Design of a Diffraction Limited Light Source, Proceedings of PAC 1995, p. 177.
7 H. Onuki and P. Elleaume, Undulators, Wigglers and their applications, Taylor & Francis, 2003 ISBN 0-415-28040-0.
8 Annick Ropert, The status of the ESRF, Proceedings of EPAC 1992, p. 35.
9 Alan Jackson, Commissioning and Performance of the Advanced Light Source, Proceedings of PAC 1993, p. 1432.
10 A. Wrulich, ELETTRA Status Report, Proceedings of EPAC 1994, p. 57.
11 A. Ropert, Lattices and Emittances, in CAS-Report, CERN 98-04, p. 91.
12 G. Decker, APS Storage Ring Commissioning and Early Operational Experience, Proceedings of PAC 1995, p. 290.
13 H. Kamitsubo, First Commissioning of Spring-8, Proceedings of PAC 1997, p. 6.
14 http://www.lightsources.org/
15 A. Streun et al., Commissioning of the Swiss Light Source, Proceedings of PAC 2002, p. 224.
16 A. Lüdeke and M. Muñoz, Top-up operation experience at the Swiss Light Source, Proceedings of EPAC 2002, p. 721.
17 D. Morris, When less is more - Construction of the Australian Synchrotron, Proceedings of EPAC 2006, p. 3266.
18 R.P. Walker, Overview of the status of the Diamond project, Proceedings of EPAC 2006, p. 2718.
19 J.M. Filhol et al.,Overview of the status of the SOLEIL project, Proceedings of EPAC 2006, p. 2723.
20 J. Bordas, these proceedings.
21 K. Balewski, PETRA III: A new high brilliance Synchrotron Radiation source at DESY, Proceedings of EPAC 2004, p. 2302.
22 Z.T. Zhao and H.J. Xu, SSRF: A 3.5 GeV Synchrotron Light Source for China, Proceedings of EPAC 2006, p. 2368.
23 D. Einfeld et al.: SESAME, a third generation synchrotron light source for the Middle East Region, Radiation Physics and Chemistry 71 (2004) 693-700.
24 Michael Borland, private communication; and APS upgrade website http://www.aps.anl.gov/News/Conferences/2006/APS_Upgrade/index.html
25 M. Eriksson et al., Status of the MAX IV Light Source project, Proceedings of EPAC 2006, p. 3418.
26 C.C. Kuo et al., Design of Taiwan future Synchrotron Light Source, Proceedings of EPAC 2006, p. 3445.
27 D. Einfeld et al., The synchrotron light source ROSY, NIM B 89 (1994) 74 - 78.

THE EUROPEAN XFEL PROJECT

W. Decking
DESY

Abstract: The European XFEL will be a free electron laser based on self amplified
              spontaneous emission in the X-ray regime. The FEL is driven by a
              superconducting 17.5 GeV linear accelerator, followed by 5 separate
              undulators both for SASE FEL radiation and incoherent radiation. Start of
              operation is foreseen for 2013. This paper presents the layout of the European
              XFEL, with an emphasis on beam dynamics issues.

Key words: FEL, brilliance, coherence, linear accelerator, undulator

1.   INTRODUCTION

Free Electron Lasers (FEL) based on self amplified spontaneous emission (SASE) are viewed as one of the possibilities to go beyond the capabilities of present day 3rd generation synchrotron sources. The European XFEL project[1], presently in its preparation phase at DESY in Hamburg, Germany, targets at a radiation wavelength of 0.1 nm with a peak brilliance of up to 10^{33} photons/(s mm^2 mrad^2). The peak brilliance of this source is 100 millions times higher than that of 3rd generation synchrotron sources (see Figure 1), and even the average brilliance is increased by a factor of 10.000. Properties of the SASE radiation are short pulse length (100 fs) a high degree of transverse coherence (80 %), full polarization, and a narrow bandwidth (0.1%). The pulse intensity can be up to 10^{18} W/cm^2.

Vasili Tsakanov and Helmut Wiedemann (eds.), Brilliant Light in Life and Material Sciences, 21–30.

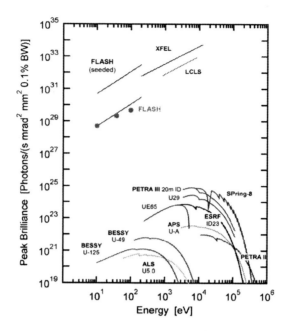

Figure 1. Peak brilliance comparison of the European XFEL with some 3rd generation synchrotron light sources.

The SASE FEL process is based on the interaction of an electron bunch with its own field in an undulator. This leads to micro-bunching along the undulator and finally the electrons within the micro-bunch will radiate coherently. For short radiation wavelengths, extremely high electron beam densities are required, in the case of the XFEL a peak current of 5 kA, a bunch length of 25 μm and normalized transverse emittance of 1.4 mm mrad is required. The SASE process starts from noise and requires long undulators (~ 100 m) to reach saturation.

2. LAYOUT OF THE EUROPEAN XFEL

The principal layout of the XFEL is shown in Figure 3. The overall site length is 3300 m. Most of the facility will be constructed underground in a 5.2 m diameter tunnel and several access-shafts. An experimental hall with dimensions of 90 m × 30 m provides space for up to 15 experimental stations.

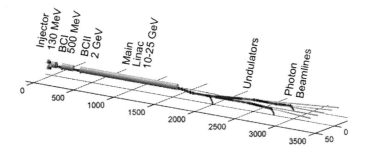

Figure 2. Layout of the XFEL

Electron bunches with a transverse slice emittance of 1 mm mrad and 1 nC charge are produced in a 1½-cell normal conducting 1.3 GHz RF gun. The design of the gun follows the well-tested design which is used in FLASH and at the Photo Injector Test Stand in Zeuthen (PITZ)[2]. Projected emittance of about 1.5 mm mrad has been already measured and is in agreement with simulations of the present setup[3]. According to these simulations the design emittance will be reached with a gradient of 60 MV/m (presently 40 MV/m) in the gun[4].

After the gun, the electrons are accelerated to 130 MeV and pass a diagnostic section before being accelerated up to 500 MeV in the first section of the linac. This linac is operated off-crest and imprints a correlated energy spread of 1.8 % on the bunch. The longitudinal phase space is linearized with a 3rd harmonic RF system (3.9 GHz, total 200 MV accelerating voltage) before entering the first bunch compressor where the bunch length is reduced from 2 mm to 100 μm. The energy is raised to about 2 GeV in the second linac section and a further compression down to 20 μm is obtained in the 2nd bunch compressor chicane[5]. The splitting into two bunch compressor sections with different energies reduces space charge effects and the distribution of the longitudinal dispersion (R_{56}) decreases coherent synchrotron radiation effects in the chicanes.

Downstream of both bunch compression chicanes, slice emittance and other parameters which vary along the longitudinal position in the bunch can be measured with vertically deflecting RF systems and wire scanner sections allow control of projected emittance and optics. Electro-optical measurement stations complement this devices and allow on-line measurements of the longitudinal beam profile[6].

Further acceleration occurs in the about 1200 m long main linac up to a maximum energy of 25 GeV (the nominal beam energy to reach 0.1 nm is 17.5 GeV). The linac consists of 100 TESLA accelerating modules, each containing eight 9-cell superconducting cavities to be operated at a gradient of 23.6 MV/m for a final beam energy of 20 GeV.

The main linac is followed by the collimation and beam distribution system. The XFEL linac can produce 10 RF pulses per second, each of 600 μs duration, and each pulse can accelerate a train of up to 3,000 electron bunches, i.e. with a minimum spacing of 200 ns between successive bunches. The users have in principle a wide variety of possibilities as the filling pattern can vary from a single or few bunches per train to full trains of 3,000 bunches. Since the facility is meant for simultaneous use of many experimental stations by different groups of users, who may have contradictory requirements, the maximum flexibility corresponds to a system of fast kickers, able to direct individual bunches to one or the other of the two electron beamlines and therefore through different sets of undulators.

Table 1. Parameters of the XFEL

Parameter	Value	Unit
Target wavelength	0.1	nm
Electron energy	17.5	GeV
# of installed 12 m long accelerator modules	116	
Acc. Gradient at 20 GeV (104 active modules)	23.6	MV/m
Beam pulse length	0.65	ms
Minimum bunch spacing within pulse	200	ns
Max. number of bunches in pulse	3250	
Average beam power	600	kW
Repetition rate	10	Hz
Average beam power	600	kW
Bunch charge	1	nC
Peak Current	5	kA
Slice emittance (normalized) at undulator entrance	1.4	mm mrad
Slice energy spread at undulator entrance	1	MeV

The two electron beam lines host the SASE undulators. The undulator parameters are tailored for different radiation wavelength and experimental requirements. Individual wavelength scanning is possible by adjusting the undulator gaps. Different energies for different bunches within one bunch train are possible within a bandwidth of ± 1.5 %, thus allowing for fast wavelength scans. The site length of the undulator and photon beam line section is about 1 km. The total installed undulator length (excluding the intersections between undulator segments) is about 580 m.

Electron bunches channelled down the electron beam line 1 pass through the undulators SASE1 and SASE3, producing hard X-ray photons with 0.1 nm wavelength in SASE1 and softer X-ray photons with 0.4 - 1.6 nm wavelength in SASE3. Electron bunches channelled through the electron beam line 2 are led through the undulator SASE2, where hard X-ray photons with wavelengths 0.1 - 0.4 nm are produced; and then through the undulators U1 and U2, where X-ray photons with wavelengths down to 0.025 and 0.009 nm are generated by spontaneous emission.

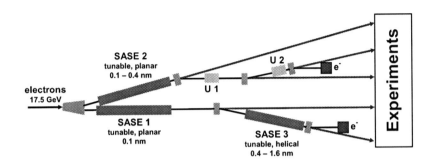

Figure 3. Schematic overview of the undulator beam lines.

The electron beam is separated from the SASE photon beam and dumped. The photon beam is transported through photon beam lines of up to 800 m length to the experimental hall. The principal layout of a photon beam line consists of a monochromator (which also serves as a bremsstrahlung absorber), switching mirrors to select up to three different experiments per beam line, photon beam collimation and diagnostics. Due to the statistical nature of the SASE radiation, each photon pulse has to be characterized individually. An example of measured radiation spectra is shown in figure 4 as an illustrative example.

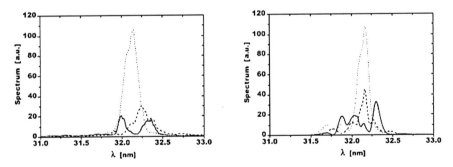

Figure 4. Three different measured (left) and simulated single-shot spectra of FLASH[7].

3.   ELECTRON BEAM DYNAMICS

3.1   Bunch compression

The bunch compression system compresses the initially 2 mm (rms) long bunch in two magnetic chicanes by factors of 20 and five, respectively, to achieve a peak current of 5 kA. The required energy chirp injected by running off-crest in the RF upstream of the first chicane is about 10 MeV, roughly compensating the energy contribution by the longitudinal wake fields of all main linac RF structures. A third harmonic RF system is used to optimize the final longitudinal charge distribution.

3.1.1   Space charge effects

In low-emittance, high-current electron beams, space charge forces can cause growth of slice emittance and mismatch of slice Twiss parameters with respect to the design (zero current) optics. Slice emittance growth directly degrades the performance of the SASE-FEL. The generated mismatch complicates accelerator operations because of projected emittance increase and the dependence of transverse dynamics on beam current.

With the second compression stage at 2 GeV beam energy and optimized optics in the beam line sections downstream of the compressor chicane, the slice emittance growth (at design parameters) due to transverse space charge is less than one per cent. The optical mismatch downstream of the first chicane corresponds to a beating of the β-function of about 5% if the bunch charge would change by 10%, and less than 1% downstream of the second chicane.

3.1.2   Coherent synchrotron radiation

The impact of Coherent Synchrotron Radiation (CSR) fields on beam emittance in the compressor chicanes was calculated with a specifically developed computer code CSRtrack[8]. The simulations show that the slice emittance growth only slightly even at peak currents of 15 kA and that very short bunch length of 5 μm can be achieved with reduced total bunch charges. Operating the XFEL like this has two (potential) rewards: The FEL pulse gets shorter and, with even smaller transverse emittance, lasing at shorter wave lengths might be achievable.

3.1.3 Instabilities

An initial bunch current ripple can be amplified by the following mechanism: a slight modulation of the initial bunch density profile produces an energy modulation due to longitudinal impedance, caused by CSR and space charge fields. In a bunch compression chicane, the energy modulation creates more density modulation. In a multi-stage bunch compression system, the gain of this amplification can be very high. A limiting factor for this mechanism is the uncorrelated energy spread. The calculated gain at the expected small values for uncorrelated energy spread from the gun (rms < 2 keV) would be sufficient to start amplification from shot noise. With a 'laser-heater'[9], an about two meter long magnet chicane with an undulator magnet which the electron beam traverses together with a laser beam, we can adjust the uncorrelated energy spread between its initial value and up to 40 keV. The initial uncorrelated energy spread must be above 10 keV to limit the modulation of the bunch after the last compressor chicane to less than 1 %.

3.2 Single and multi-bunch beam dynamics in the linear accelerator

3.2.1 Single bunch effects

The single bunch emittance dilution in linacs is determined by chromatic and transverse wakefield effects. Error sources (and magnitudes) are assumed to be the transverse injection jitter (1 σ of the incoming beam size), random cavity tilts (0.25 mrad), quadrupole misalignments (0.5 mm), and cavity and module random misalignments (0.5 mm).

The transverse-wakefield caused correlated emittance dilution along the linac for cavity and module misalignments are negligibly small due to weak transverse wake fields in the accelerating sections. In the case of cavity random tilts the particles experience the transverse Lorenz force of the accelerating RF field and the beam performs coherent oscillations.

The strongest impact is observed for a disturbed central trajectory caused by quadrupole misalignments (see figure 5). The steering of the central trajectory is supposed to use one-to-one correction algorithm: the beam trajectory is corrected in each focusing quadrupole to its geometrical axis based on the beam position monitors (BPM) reading by correction dipole coils incorporated in the previous quadrupole.

The total contributions to the single bunch emittance dilution are small and amount to below 5% after this simple steering correction[10].

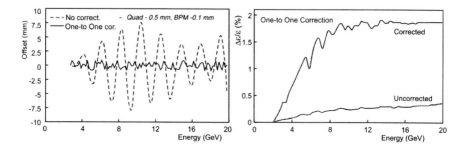

Figure 5. Coherernt betatron oscillation of the beam in the main linac with misaligned quadrupoles and steered trajectory (left). Correlated and uncorrelated chromatic emittance emittance dilution along the linac after trajectory steering (right).

3.2.2   Multi bunch effects

The effect of the long-range transverse wakefield on multiple bunches has been investigated for cases with a random cavity misalignment of 500 μm rms[11]. Typically, the bunch offset at the end of the linac reaches a steady state after the first 200 bunches of a train. The initial oscillation is largely repetitive. The average emittance growth along the linac is only 0.02 % relative to the design slice emittance. The major contribution again comes from the first part of the bunch train. Consequently the average emittance dilution gets larger for shorter trains, where the steady state is not reached. The emittance growth and the initial oscillation can be corrected by a fast intra-train feed-forward system.

The longitudinal higher order modes lead to an energy spread along the bunch train. As in the transverse case this is strongest for the first part of the bunch train (5.15 MeV rms for a 20 μs bunch train) while the rms energy spread over the whole train is 0.88 MeV. This variation again is repetitive and should be compensated by the low-level RF system.

4.   ACCELERATOR TECHNOLOGY

The XFEL accelerating cavity is a 9-cell standing wave structure of about 1m length whose fundamental TM mode has a frequency of 1300 MHz. It is identical to the so-called TESLA cavity[12], made from solid niobium, and is bath-cooled by superfluid helium at 2 K. Each cavity is equipped with a helium tank, a tuning system driven by a stepping motor, a coaxial RF power coupler, a pickup probe and two higher-order mode (HOM) couplers. The superconducting resonators are fabricated from bulk niobium by electron-beam welding of deep-drawn half cells. The tubes for the beam pipes and the coupler ports are made by back extrusion and are joined to the cavity also by electron-beam welds.

Figure 6. A 9-cell Niobium cavity before its installation in the He-vessel and cryostat.

Each cavity undergoes an extensive cleaning and preparation procedure to ensure the surface quality that is required to reach the accelerating gradient of 23.6 MV/m. In a final acceptance test the excitation curve (the quality factor Q versus the accelerating gradient) is measured. Figure 7 shows that the quality of the cavities is sufficient for the XFEL.

Eight of these 9-cell cavities are combined to one cavity string, and mounted together with a superconducting quadrupole in a 12 m long cryostat, the accelerating module. One klystron will supply rf power over a waveguide distribution system to four modules.

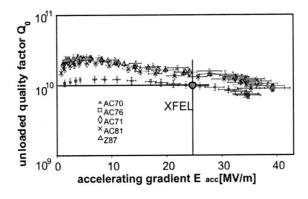

Figure 7. Excitation curve for the latest production of TESLA cavities.

5. SUMMARY

The European XFEL will provide the synchrotron radiation community with unique opportunities for research with hard X-rays. Its construction becomes possible due to the availability of low-emittance electron sources, the development of superconducting cavities by the TESLA collaboration and advances in the understanding and theory of the SASE process. Experimental proof of the viability of the SASE concept is provided by the operation of several facilities worldwide. The smallest photon wavelength of 13 nm has been reached in FLASH, a VUV user facility based on SASE radiation. The XFEL will start its operation in 2013, after a 7 year construction and installation phase.

REFERENCES

1. M. Altarelli et al. (eds.), The European X-Ray Free-Electron Laser Technical Design Report (DESY 2006-xxx, Hamburg, 2006).
2. S. Schreiber et al., The Injector of the VUV-FEL at DESY, Proceedings of FEL 2005, Palo Alto, 2005.
3. K. Abrahamyan et al., Experimental characterization and numerical simulations of the electron source at PITZ, Nucl. Instr. and Meth. A, Volume 558, Issue 1, 1 March 2006, Pages 249-252.
4. P. Piot et al., Conceptual design for the XFEL Photoinjector, DESY TESLA-FEL 01-03, 2001.
5. V. Balandin et al.: Optimized Bunch Compression System for the European XFEL. PAC 2005, Knoxville, 2005.
6. Ch. Gerth, M.Röhrs, H. Schlarb, Layout of the diagnostic section for the European XFEL, Proceedings of PAC 2005, Knoxville, 2005.
7. V. Ayvazyan, et al., First operation of a Free-Electron Laser generating GW power radiation at 32 nm wavelength, Eur. Phys. J. D37 (2006) 297.
8. M. Dohlus, T. Limberg, CSRtrack: Faster Calculation of 3D CSR effects, FEL 2004, Trieste, 2004.
9. E. Saldin, E. Schneidmiller, and M. Yurkov, Nucl. Instrum. Meth. A 528 (2004) 355.
10. G. Amatuni, V. Tsakanov, W. Decking, R Brinkmann, Single Bunch Emittance Preservation in the XFEl Linac, Proceedings of 37th ICFA Advanced Beam Dynamics Workshop on Future Light Sources, Hamburg, 2006.
11. N. Baboi, Multi-Bunch Beam Dynamics Studies for the European XFEL, Proceedings of LINAC 2004, Lübeck, 2004.
12. R. Brinkmann et al. (eds.), TESLA Technical Design Report – Part II: The Accelerator (DESY 2001-011, Hamburg, 200), http://tesla.desy.de.

SPEAR3 SYNCHROTRON LIGHT SOURCE
Accelerator Update and Plans

James Safranek for the SSRL Accelerator Physics Group
Stanford Linear Accelerator Center, 2575 Sand Hill Road, Menlo Park, CA 94025 USA

Abstract: The SSRL SPEAR3 3 GeV synchrotron light source storage ring started operation in Spring, 2004. We will briefly discuss the accelerator commissioning and the accelerator performance during the first two years of operation. An overview will be given of ongoing and planned accelerator improvements, including fast orbit feedback, double waist chicane optics, 500 mA tests, top-off injection, and short bunch length generation.

Key words:

1.   INTRODUCTION

The SPEAR storage ring was built in 1972 as an electron positron collider at the Stanford Linear Accelerator Center. Its potential as a synchrotron radiation source was recognized early, so the dipole vacuum chambers were built with ports to extract photons. Much pioneering synchrotron radiation work was subsequently performed at SPEAR. In 1989, the highly-successful high energy physics research at SPEAR was completed, and SPEAR became a dedicated light source.

By 1999, SPEAR2 was delivering light to seven insertion device beamlines and four dipole beamlines. In order to increase the performance and reliability of these beamlines, it was decided to remove the existing storage ring, and install a new storage ring, SPEAR3[1], designed to deliver higher brightness photon beams. Table 1 compares SPEAR2 to SPEAR3. In addition to the higher performance associated with lower emittance and higher current, SPEAR3 was built with a low-impedance copper vacuum chamber, mode-damped cavities, and mechanically stable girders and BPMs, designed to deliver more stable and reliable beam.

Vasili Tsakanov and Helmut Wiedemann (eds.), Brilliant Light in Life and Material Sciences, 31–43.
© 2007 *Springer.*

Table 1. SPEAR3 upgrade parameters.

	SPEAR2	SPEAR3
Energy	3 GeV	3 GeV
Emittance	160 nm*rad	18 nm*rad
Current	100 mA	500 mA
Lifetime	40 hours	16 hours
Critical energy	4.8 kV	7.6 kV
Circumference	234.14 m	234.14 m
RF frequency	358.5 MHz	476.3 MHz
Injection energy	2.3 GeV	3 GeV

Removal of the old SPEAR storage ring began in April, 2003. By December, the new SPEAR3 storage ring was installed and ready for commissioning.

2.   SPEAR3 COMMISSIONING

SPEAR3 commissioning proceeded rapidly, with the first user run starting in March, 2004, only three months after commissioning began.[2] The rapid and successful commissioning was due in part to the experience and expertise of the commissioning team. This team included series of visitors were invited typically for a week or two. The visitors came from other synchrotron light sources throughout the U.S. as well as Europe and Asia.

Anther key to rapid success was the MATLAB interface to the EPICs control system. This MATLAB interface was originally developed at ALS[3], and ported to SSRL[4], where it was used for the first time for commissioning an accelerator. Many GUI tools are included in the MATLAB software, including the GUIs for orbit control and the LOCO[5] optics correction program.

In addition to the pre-written GUIs, the MATLAB environment made it very easy to write code on the fly for making relatively complicated accelerator measurements. Figure 1 shows an example in which the horizontal dynamic aperture was measured as a function of horizontal and vertical tunes. The QD and QF quadrupole strengths were varied to set the tunes to a grid with 176 points. For each combination of tunes, the injector beam was turned on automatically with MATLAB, 1 mA was stored, QD and QF were varied to give the desired tunes, and a single injection kicker strength was increased until the beam was kicked out. The kicker strength at which the beam was lost is plotted vs. tunes. The reduction in dynamic aperture at the difference coupling resonance and at two higher-order resonances is visible. The reduced dynamic aperture is offset from the actual resonance lines due to tune shift with betatron oscillation amplitude.

The script for making this measurement was written and tested within a couple hours, and the data was collected automatically over the course of four hours on an owl shift. The ease of scripting in MATLAB greatly increases accelerator shift productivity.

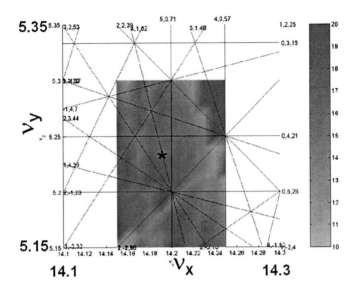

Figure 1. Measured horizontal dynamic aperture vs. tunes.

3. FAST ORBIT FEEDBACK[6]

During SPEAR3 commissioning and the first two years of running, a MATLAB-based slow orbit feedback was used to correct the orbit every six seconds. This year, this system has been replaced with a fast orbit feedback with a bandwidth of about 100 Hz. Figures 2 and 3 show the measured orbit motion when running with the slow orbit feedback.

Figure 2 shows the horizontal and vertical rms orbit variations over the course of 24 hours. Each data point is the rms difference between the measured orbit and the initial orbit on the 54 BPMs in the orbit feedback. The orbit is measured with Bergoz BPM electronics[7] digitized at 4 kHz. In Fig. 2, each orbit is the average of 2000 consecutive BPM readings (1/2 second), so only slow orbit motion is shown. For the most part the orbit is stable to within a small fraction of a micron, with occasional larger glitches due to insertion device gap changes and vehicular traffic in the vicinity of the storage ring.

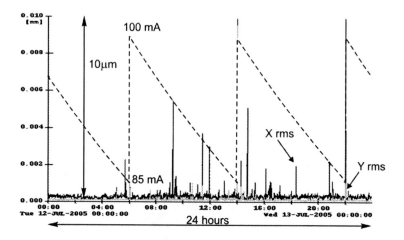

Figure 2. Orbit motion over 24 hours with slow orbit feedback; 1/2 second averaged orbit readings.

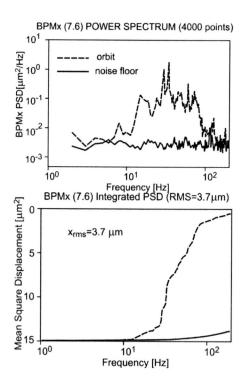

Figure 3. Orbit motion up to 200 Hz, no fast orbit feedback.

Figure 3 shows the power spectral density and PSD integral for horizontal orbit motion between 1 and 200 Hz without fast feedback. The beam is inherently stable, with only 3.7 microns rms motion, which can be attributed mostly to magnet girder resonances in the vicinity of 30 Hz. The vertical PSD is similar, with 2.2 microns rms motion integrated from 1 to 200 Hz.

Figures 4 and 5 show orbit data with fast orbit feedback for comparison to the data without fast orbit feedback shown in Figures 2 and 3. Figure 4 shows ½ second averaged orbit data over two hours. At first slow orbit feedback (SOFB) is running, and occasional orbit glitches are visible, similar to Fig. 2. At 9:12, fast orbit feedback is turned on, and the rms orbit drops to about 0.1 micron with all orbit glitches suppressed.

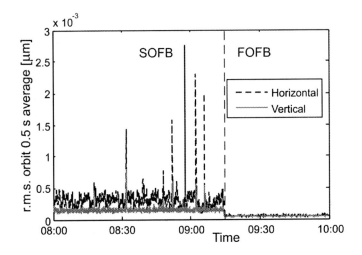

Figure 4. Slow orbit motion with slow, then fast orbit feedback.

Figure 5 shows the BPM power spectrum without fast orbit feedback and for varying fast orbit feedback gain. The fast orbit feedback successfully damps the magnet girder vibration frequencies between 20 and 80 Hz.

According to Figs. 4 and 5, the SPEAR3 photon beams should be extremely stable. Unfortunately, these figures do not tell the whole story. The BPM data in the figures comes from the same BPMs that are used in the orbit feedback. When monitoring photon BPMs or electron BPMs not in the feedback, we see orbit drift of tens of microns over the course of many minutes or hours. This is a result of drifts in the readings of the feedback BPMs.

Investigation of this drift showed that the dominant part came from temperature dependence of the BPM electronics. Figure 6 shows an example of the correlation between the temperature in the electron BPM electronics rack and the measured position at a photon BPM over 24 hours. Since this data was taken, temperature controlled rooms have been built around the electron BPM electronics, greatly improving the photon beam stability.

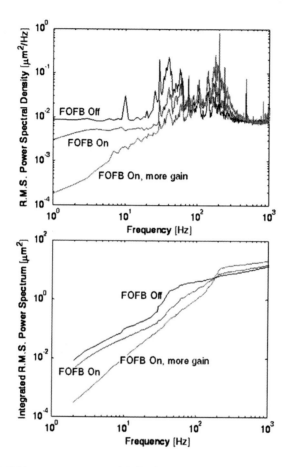

Figure 5. Orbit power spectrum with the fast orbit feedback with varying gain.

Now that the drift from BPM electronics temperature dependence has been mitigated, we are investigating other smaller sources of error in the electron BPM readings. We are building an invar stand to measure mechanical movement of the BPMs with respect to the floor. We are also working with a hydrostatic leveling system to measure variations in the height of the floor within the storage ring tunnel and at the photon beamlines. Figure 7 shows some preliminary results[8] from the hydrostatic leveling system showing relative variations in the floor height over the course of several months. We see a maximum of about 300 microns in differential floor motion between two locations in the accelerator tunnel separated by 24 meters.

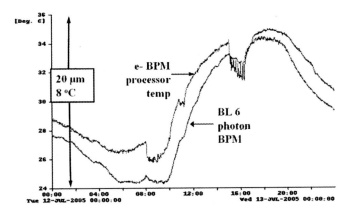

Figure 6. Correlation between electron BPM electronics temperature and photon beam motion over 24 hours.

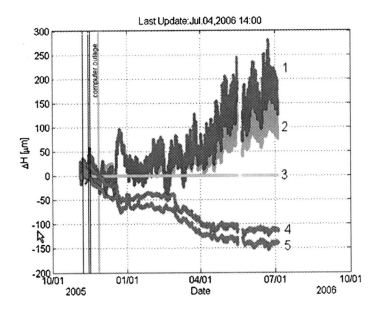

Figure 7. Measured variations in floor height over the course of several months.

4.   DOUBLE-WAIST CHICANE OPTICS

This year (2006), we are implementing an optics upgrade to reduce the vertical beam size to allow smaller gap insertion devices in up to six straight sections.[9,10] Figure 8 shows the layout of the optics upgrade. The seven existing insertions devices (IDs) in SPEAR3 are located in the 3-meter long standard straights in the arcs. In addition the 3-meter straights, SPEAR3 has two 7.6-meter straights, where the detectors for colliding beam experiments used to sit. On either side of the 7.6-meter straights are 4.8-meter straights.

The double waist chicane optics upgrade consists of adding a quadrupole triplet in the center of one of the two 7.6-meter straights. This triplet focuses the beam to a small vertical waist (β_y = 1.6 m) at each of the two new 2.2-meter straight sections. An in-vacuum undulator with a minimum gap of 5.5 mm is being installed in the second of these straights this summer, and will be commissioned for operations during the 2006-2007 run.

In addition to the new quadrupole triplet, four new chicane dipoles are being installed in the 7.6-meter straight. These chicane dipoles will separate the two 2.2-meter straights by 10 mrad, so that a future beamline can be installed in the upstream 2.2-meter straight.

The new double-waist optics will also reduce β_y to 2.5 meters (from 4.8 meters) in the four 4.8-meter straight sections. An elliptically polarized undulator will be installed in one of these straight sections during summer 2007.

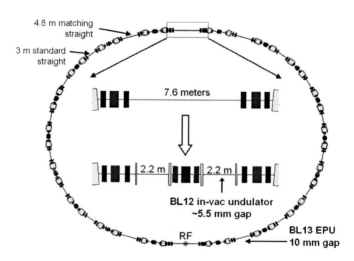

Figure 8. Layout of the new double-waist chicane optics.

5. 500 mA TESTS

The SPEAR3 vacuum chamber was designed for 500 mA. The existing photon beamlines, however, were built for the SPEAR2 operating current of 100 mA. In order to run at 500 mA, the photon beamline optics are being upgraded to handle higher power loads, and the beamline radiation shielding is being upgraded to handle the higher radiation levels.

In the mean time, accelerator physics studies have proceeded with the photon beamlines closed to show that SPEAR3 runs well at 500 mA. We have found that the SPEAR3 beam is inherently stable at 500 mA without the need for multi-bunch feedbacks, so long as the non-normalized chromaticities are set to +2 in both planes. This confirms predictions for the copper vacuum chamber with mode-damped RF cavities. Below +2 in chromaticity, we see evidence of resistive wall instability.

The measured lifetime at 500 mA is 14 hours.

We anticipate some limited operations at 500 mA with beamlines open during the 2006-2007 run.

6. TOP-OFF INJECTION

The increased power load on the photon optics associated with 500 mA running is driving our push toward top-off injection. To date, SPEAR3 has been running with three fills per day. At each fill, the photon beamlines must be closed for about 2½ minutes while the current is filled back to 100 mA. During this 2½ minutes, the beamline optics cool, causing thermal transients when the photon shutters are re-opened. The transients will be much worse for 500 mA running.

Top-off injection consists of injecting with the photon beamline shutters open, and injecting frequently to keep the stored beam current close to constant. In order to inject with photon shutters open, radiation safety requires proof that it is impossible to send the injected electrons down a photon beamline, because this would lead to very high radiation levels on the floor. As was done at APS[11], we will implement a stored-current interlock, so injection with photon shutters open can only occur when there is stored current in SPEAR3. Tracking simulations are ongoing to prove that it is impossible for an injected electron pulse to go down a photon beamline when there is stored current in the ring.

Much of the radiation shielding at SPEAR3 was built in 1972, so it does not meet the present standards for a 500 mA ring running in top-off mode. Therefore, in addition to the stored current interlock, we have had to add additional lead and steel shielding in the SPEAR tunnel. We have also installed four coaxial-cable long ion chambers (LIONs) spanning the full circumference of the storage ring. The LIONs will trip off the injector when excessive radiation levels are detected.

Top-off injection will also require significant improvements in the stability and performance of the SSRL injector. Eventually we would like to inject beam into SPEAR3 every 30 seconds to keep the stored current fluctuations below 0.1%. To minimize the number of transients, we'd like to fire the injection kickers only once each 30 second interval. We will need to improve the charge per pulse from our injector by about a factor of ten to get enough charge for single pulse injection each 30 seconds.

There is much on-going work with the injector to achieve this goal, including doubling the energy of our linac, realigning the booster, rebuilding our booster to SPEAR transport line vacuum, and adding and improving our diagnostics and controls throughout the injector.

Another important preparation for top-off is minimizing the transient seen by the users on the stored beam when the injection kickers are fired. Turn-by-turn BPM measurements were made after firing the kicker bump. The peak amplitude of these oscillations is plotted vs. kicker bump strength in Fig. 9. For this plot, a kicker bump strength of 1 corresponds to the kicker bump at its full operational strength. The horizontal kicker bump transient peaks at about 0.7 in kicker bump amplitude, because the bump spans eight sextupole magnets.

The vertical kicker bump transient is large for the kicker bump fully on. It has been determined that this vertical transient is driven by horizontal stray fields in the injection septum. We are working to design a correction magnet to cancel these fields.

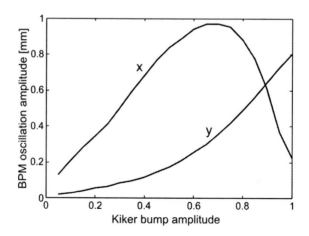

Figure 9. Measured kicker bump transient vs. kicker bump strength.

We plan to start injection with photon shutters open toward the end of the 2006-2007 run. We hope to achieve full top-off injection with a ~30 second injection interval during the 2007-2008 run.

7. SHORT BUNCHES

With the SPPS, LCLS and Ultrafast Science Center, the photon user community interested in short photon pulses at SLAC is growing. For this reason we've begun experiments to generate short pulses in SPEAR3. We have implemented low-alpha optics similar to those used to create short bunches at BESSY II.[12] Figure 10 compares streak camera measurements in low alpha optics and nominal optics. The measurements were made with the SPEAR3 visible/UV diagnostics beamline.[13]

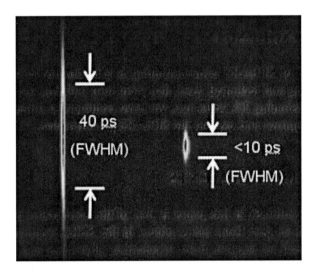

Figure 10. Streak camera bunch length measurements for nominal and low-alpha optics.

Figure 11 shows the results from streak camera bunch length measurements for varying alpha and single-bunch current. In the figure legend, alpha0 refers to alpha in the nominal optics, alpha0=1.2e-3. The solid black line is the theoretical threshold for the micro-bunching instability.[14]

Discussions are underway with the photon users to develop operation with short bunches.

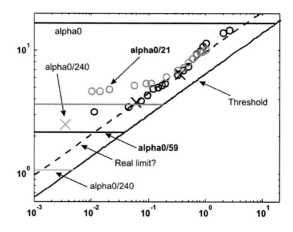

Figure 11. Bunch length vs. current and alpha in SPEAR3.

8. CONCLUSION

The speed of SPEAR3 commissioning is a testimony to the quality of the engineering and technical work that went into building the storage ring. Since commissioning 2½ years ago, we have continued a robust accelerator program to improve and expand the SPEAR3 performance.

ACKNOWLEDEGEMENTS

The work outlined in this paper is that of the accelerator physics group at SSRL with support from many others in the SSRL accelerator systems division as well as support from accelerator physicists and engineers at SLAC. Work supported in part by the US Department of Energy Contract DE-AC03-76SF00515 and Office of Basic Energy Sciences, Division of Chemical Sciences.

REFERENCES

1. R. Hettel et al., SPEAR 3 upgrade project: the final year, Proceedings of the 2003 Particle Accelerator Conference.
2. J. Safranek et al., SPEAR3 commissioning, Proceedings of the 2004 Asian Particle Accelerator Conference.
3. G. Portmann, Slow orbit feedback at the ALS using MATLAB, Proceedings of the 1999 Particle Accelerator Conference.

4. W.J. Corbett, G. Portmann, J. Safranek and A. Terebilo, SPEAR3 commissioning software, Proceedings of the 2004 European Particle Accelerator Conference.
5. J. Safranek, G. Portmann, A. Terebilo and C. Steier, MATLAB-based LOCO, Proceedings of the 2002 European Particle Accelerator Conference.
6. A. Terebilo, T. Straumann, Fast global orbit feedback system in SPEAR3, Proceedings of the 2006 European Particle Accelerator Conference.
7. http://www.bergoz.com/
8. G. Gassner, private communication.
9. J. Corbett et al., Implementation of double-waist chicance optics in SPEAR3, Proceedings of the 2006 European Particle Accelerator Conference.
10. J. Safranek, A. Terebilo, X. Huang, Nonlinear dynamics in the SPEAR3 double-waist chicane, Proceedings of the 2006 European Particle Accelerator Conference.
11. M. Borland and L. Emery, Tracking studies of top-up safety for the Advanced Photon Source, Proceedings of the 1999 Particle Accelerator Conference.
12. J. Feikes, K. Holldack, P. Kuske and G. Wustefeld, Sub-picosecond electron bunches in the BESSY storage ring, Proceedings of the 2004 European Particle Accelerator Conference.
13. J. Corbett, C. Limborg-Deprey, W. Mok and A. Ringwall, Commissioning the SPEAR3 diagnostic beamlines, Proceedings of the 2006 European Particle Accelerator Conference.
14. G. Stupakov and S. Heifets, Beam instability and microbunching due to coherent synchrotron radiation, Phys. Rev. ST Accel. Beams **5**, 054402 (2002).

STATUS OF THE ALBA PROJECT

Joan Bordas
CELLS, PO Box 68, 08193 Bellaterra

Key words: accelerator, synchrotron radiation, brightness, insertion device

1. INTRODUCTION

ALBA is a 3GeV third generation Synchrotron Radiation (SR) Facility currently under construction in Cerdanyola del Vallés, Barcelona, Spain. The Storage Ring is to deliver as high as possible photon flux densities on the samples. To this end the Storage Ring has been designed to have an electron beam emittance of ca. 4.5 nm.rad and, in spite of a relatively small perimeter, a significant number of straight sections in which to house Insertion Devices (IDs). The injector complex consists of a 100 MeV LINAC and a small emittance (ca. 9 nm.rad), full energy, 3 GeV, Booster. It is intended to operate the accelerator complex in a top-up mode from the very beginning of its operation. The project started in earnest in 2004 and its completion with seven beam-lines (BLs) in its initial phase, is expected by 2009, so that routine user operation can start in 2010.

The responsibility for the construction, commissioning and subsequent exploitation of ALBA was awarded in 2003 to the Consortium: ¨Consorcio para la Construcción, Equipamiento y Explotación del Laboratorio de Luz Sincrotrón¨, or Consortium CELLS in short. CELLS is owned and jointly supported with an equal share by the Spanish Ministry of Education and Science (MEC) and the Department of Education and Universities (DEU) of the Catalan Autonomous Government.

Vasili Tsakanov and Helmut Wiedemann (eds.), Brilliant Light in Life and Material Sciences, 45–55.
© 2007 *Springer.*

2.   BACKGROUND

For years there has been among the Spanish SR community a desire to count on a SR facility in Spain. In the early 1990s there emerged a number of proposals to construct a SR facility. In the specific case of ALBA one can trace its origin to an initiative of the Catalan Autonomous Government, or "Generalitat de Catalunya", who in 1992 commissioned a feasibility study. In 1993 the Generalitat created a Project Promoting Commission and a program of personnel training was initiated. After consolidation of these first tentative steps, an agreement was reached between the Spanish State and the Catalan Autonomous governments to carry out a first detailed design study. The Consortium LLS (Laboratorio Luz Sincrotrón), belonging to the Autonomous University of Barcelona and the Generalitat, was charged with this study that was completed in 1998. After a significant period of reflection, in March 2002 an agreement between the Spanish State and the Catalan Autonomous Government to jointly fund a SR facility in Spain was announced. During the following year the Consortium CELLS was legally constituted and in June 2003 the Governing Council of CELLS had its first meeting where the Chairman of CELLS Executive Commission, effectively a supervisory body of CELLS's operation, and the Director of CELLS were appointed. In October 2003 the process of staff recruitment was initiated and CELLS, de facto, started its technical and scientific work in January 2004.

3.   THE COMPLEX OF ACCELERATORS

The complex of accelerators[1,2] is arranged so that the Booster and the Storage Ring share the same tunnel, whilst the LINAC is placed in its own shielded area tucked against the inner wall of the shielding tunnel. The outer and inner perimeters of the Storage Ring and Booster are ca. 268.8m and ca. 249.6m, respectively, so that the mean separation between the booster and storage ring orbits is ca. 3m. This allows for a sufficiently small span between the inner and the outer wall of the shielding tunnel to close the roof with removable concrete slabs. Fig. 1 shows the layout of the accelerator's complex, whilst Fig. 2 illustrates the distribution of Storage Ring components on one of the girders.

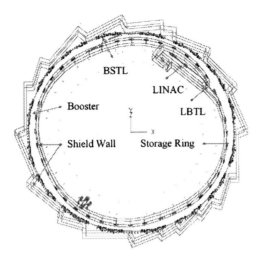

Figure 1. Lay-out of the accelerator's complex showing the distribution of Linac, Booster, Storage Ring, Linac to Booster Transfer Line (LBTL) and Booster to Storage Ring Transfer Line (BSTL). The Booster and the Storage Ring share the same shielding tunnel, whilst the safety enclosure of the Linac is tucked away against the inner shield wall. This frees significant space for services in the central area. The mean distance from the front of Shield Wall to the circular passage around the Main Hall is ca. 20 m.

4.   THE STORAGE RING

The Storage Ring has undergone a number of design iterations with the aim of: achieving the highest possible flux density on the samples; providing stable photon beams; maximizing the number of straight sections where to house insertion devices, and; not exceeding a perimeter of 270m. The final design is an expanded DBA lattice, with finite dispersion in the straight sections, and with 4 super-periods. This results in 4 straight sections of ca. 8 m long each. Within each super-period, there are 3 straight sections of ca. 4.2m length and 2 straight sections of ca. 2.6m length respectively (see Fig. 1). Therefore, the total number of straight sections is: 8 of ca. 2.6m, 12 of ca. 4.2m, and 4 of ca. 8m length, of which there are 3 of ca. 8m, 12 of ca. 4.2m and 2 of ca. 2.6m available for the eventual installation of IDs. The remainders are used for injection, installation of RF plants, accelerator diagnosis and other components.

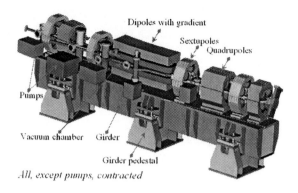

Figure 2. Illustration of the arrangement of the main Storage Ring components on one of the girders.

For the design current of 400 mA, the RF system has to provide an accelerating voltage of 3.6 MV and 520 kW of beam power. To this end, the RF system consists of six independent RF units (installed in the 2.6 m long straight sections), where in each one there is a Higher Order Mode (HOM) damped cavity, two 80 kW Inductive Output Tubes (IOTs), whose power is combined in a Cavity Combiner[3] and applied through a transmission line to the cavity.

The Storage Ring has a total of 32 bending magnets with a dipolar component of 1.42T and a gradient of 5.65 T/m each, 112 quadrupoles and 120 sextupoles[4].

The chosen lattice frees a significant amount of space for IDs, whilst it reduces the beam sizes (e.g. the σ of the electron beam sizes at the center of the 4.2 m sections has dimensions of ca. 130 and 8 μm, horizontally and vertically, respectively, assuming the usual 1% coupling), and delivers an emittance of ca. 4.5 nm.rad. Fig. 3 and Table 1, summarize the electron beam sizes that the lattice delivers around the perimeter of the ring and the main parameters, respectively.

Table 1. Summary of parameters of the storage ring lattice of interest to users.

E = 3.0 GeV C = 268.8 m	ε ca. 4.5 nm.rad
4 straight sections of 8 m length	3 useful for Beam-lines
12 straight sections of 4.3 m length	12 useful for Beam-lines
8 straight sections of 2.58 m length	2 useful for Beam-lines
32 Bending Magnets	16 useful for Beam-lines
i.e. 33 total of 33 front-ends available for beam-line.	

Table 2.

Source point	β_x (m)	β_y (m)	σ_x (μm)	σ'_x (μrad)	σ_y (μm)	σ'_y (μrad)
Long S.	10.20	5.22	263	21	15	3
Med-1 S.	1.98	1.153	133	47	7	6
Med-2-S.	1.98	1.21	133	47	7	6
Short S.	8.63	5.79	310	23	16	3
Bending M.	0.36	23.93	44	116	32	2

To keep the perimeter of the ring within bounds, a number of compromises have been made, for example: the use of a relatively high gradient in the bending magnets that does most of the beam vertical focusing, and thus reduces the number of quadrupoles required; to allow some dispersion to occur in the straight sections; to place only doublets of quadrupoles in most straight sections, and; to integrate the corrector magnets in the sextupoles. Nonetheless, the lattice delivers low enough chromaticity so that with the use of nine families of sextupoles it provides a large dynamic aperture; good energy acceptance even after considering coupling errors and realistic physical apertures, and; more than 40 hours Touschek lifetime. Moreover: there is sufficient flexibility in the configuration to allow change in the working point if/when needed; negative effects due to multipolar components are within bounds, and; the corrector strengths needed to achieve close orbit are well within acceptable tolerances[5] and allow to reach sub-μm stability.

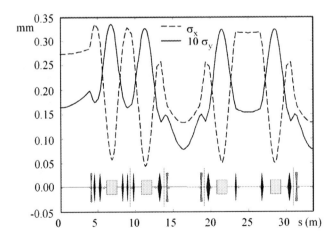

Figure 3. Electron beam dimensions around half a quadrant of the Storage Ring. The positions of the dipoles are indicated with the yellow boxes. Note that at this position is where the source has the roundest dimensions (see Table 1). The smallest source dimensions are at the center of the ca. 4.3m long straight sections.

The lattice design is now finalized, many of the major components (e.g. sextupoles, quadrupoles, bending magnets, RF systems, vacuum vessels and girders) are ordered and under construction.

5.   THE INJECTOR COMPLEX

The injector complex consists of a nominally 100 MeV LINAC and a full energy Booster[6]. The design of the injector system is now completed and tendering exercises have been/will soon be launched for the procurement of the main components.

The LINAC is a turn-key device to operate both in single and multibunch mode, that is already under construction and whose delivery is expected by early autumn of 2007. The LINAC is made up of a 90 kV thermo-ionic gun, a 500 MHz sub-harmonic pre-buncher, a 3 GHz pre-buncher, a 3 GHz/22-cells standing wave buncher and two traveling wave accelerating sections with constant gradient. Two pulsed klystrons feed the accelerating sections and the 3 GHz buncher, whilst the sub-harmonic pre-buncher and the buncher are fed from an independent RF amplifier. Beam focusing is ensured by solenoids up to the bunching section, and by a triplet of quadrupoles in between the two accelerating sections. The device should deliver 125 MeV output energies with 80% transmission efficiency from gun to exit.

The Booster is a modified FODO lattice with 4-fold symmetry. Each quadrant has 10 cells, of which 8 are regular FODO cells and 2 act as matching cells. The four quadrants connect via four 2.46 m long straight sections. These will be used for injection, for installation of the RF system and for diagnostic components. All together the Booster has 40 bending magnets with combined function (these also provide the vertical focusing), 60 quadrupoles (horizontal focusing), 16 sextupoles and 72 steering magnets. The RF system is based on a 5-cell Petra type cavity, fed with 43 kW to deliver 1 MV at 500 MHz and 5 mA (i.e. the maximum that the LINAC can supply) current. The 43 kW are provided by an 80 kW IOT, that is identical to those in the Storage Ring, thus ensuring ample spare power as well as a facility wide standard. The Booster lattice delivers an emittance of ca. 9 nm.rad, i.e, sufficiently small to ensure high enough injection efficiency for top-up operation.

6. INITIAL PORTFOLIO OF BEAMLINES

In early 2004 the Spanish SR user community was invited to submit through its user association (AUSE) bids for the initial set of beam-lines at ALBA. Although capable of housing at least 33 beam-lines, the initial capital costs for ALBA only contemplated an initial set of five BLs. The SR community submitted proposals for 13 BLs that were evaluated by ALBA's Science Advisory Committee (SAC). The SAC recommended that 7 BLs, rather than 5 as planned, should be built in the first phase. Subsequently, this recommendation was approved by the Council of ALBA.

The initial set of BLs address a range of scientific objectives primarily in the areas of Materials Science, Physics, Chemistry and Biology. The seven BLs are: a Soft X-ray BL drawing photons from an Apple II type undulator and equipped with two end stations for XMCD and resonant scattering experiments; another soft X-ray beam line, also with an APPLE II type undulator and two end stations, for PEEM and photo-emission spectroscopy from "real" surfaces, i.e. at relatively high pressures; an X-ray Absorption Spectroscopy (XAS) BL, using a conventional wiggler, and capable of coping with "soft"edges, e.g. sulphur, as well as harder ones; a macromolecular crystallography BL using an in-vacuum undulator; a BL, also using an in-vacuum undulator, for time resolved X-scattering/ diffraction experiments on non-crystalline systems; a high resolution powder diffraction BL, equipped with a low deflection parameter, low K, superconducting wiggler, and; a soft X-ray microscopy BL, primarily for imaging biological material at the wavelengths of the water window.

The state of development of these BLs is varied. Largely their progress has been dictated by the speed of incorporation of experienced staff. As of today, three of these BLs, XMCD, macromolecular crystallography and non-crystalline diffraction, are fairly well specified and calls for tender for their detailed design and subsequent construction have been issued (for more details, the interested reader is addressed to www.cells.es). The conception of a fourth one, i.e. the High Resolution Powder Diffraction BL, is advanced and detailed conceptual design leading to a call for tender is expected sometime soon. The conceptual design for the remainder of the BLs is in progress.

Note that this initial set of BLs will require 6 IDs, i.e. two APPLE types, two in-vacuum undulators, 1 superconducting multipole wiggler, and 1 normal conducting multipole wiggler. Detailed design of these IDs is currently in progress[7].

7.   THE SITE AND THE BUILDING COMPLEX

ALBA is sited in the municipal term of Cerdanyola del Vallés, a medium size town at about 25Km from the city center of Barcelona (Spain) and that has in its territory one of the largest Universities in Spain; The Autonomous University of Barcelona (UAB). The UAB is about 1 Km from the site of ALBA.

The site of ALBA has undergone extensive geological studies since the middle of 2004, including detailed identification of sub-soil composition, long-term stability, vibration levels, etc. From the data available so far there is now a reasonable level of confidence about the suitability of the site in terms of long term soil stability and vibrations levels sufficient to ensure the necessary mechanical stability of the critical floor area, i.e. the area on which the complex of accelerators and the beam lines are placed. The solution adopted for the base of the critical floor area consists of a 1m thick concrete slab, constructed from 20 segments. The segments will be produced one at a time and subsequently joined by shuttering boards and with longitudinal re-enforcing bars going through the shuttering. The area below the slab will be previously treated with a 1.5 m thick refill of selected gravel, homogenously and suitably compacted for additional stability, and sandwiched between two layers of poor concrete for protection. Ref. 8 provides a comprehensive review, and Table 2 lists a summary of the requirements on mechanical stability.

Table 3. Dimensions and stability requirements of the critical floor area in the Main Hall.

Inner diameter:	ca. 60m
Outer diameter:	ca. 120m
Slow relative displacements:	<0.25 mm/10m/year
	<0.05 mm/10m/month
	<0.01 mm/10m/day
	<0.001mm/10m/hour
Perimeter differential displacement:	<2.5 mm/year
Vertical vibrations amplitude:	<0.004 mm from 0.05-1 Hz
	<0.001 mm from 1-100 Hz
Horizontal vibrations amplitude:	<0.002 mm

The building complex consists of three main areas: a technical building, the main Hall placed over the slab but with decoupled foundations, and the office/personnel wing. The main Hall and the office/personnel wing share a common metallic roof that allows the indirect entrance of natural light, but avoids temperature variations inside the building. The combination of the roof design and the internal air conditioning and temperature regulation equipment ensures that below a height of 4m the ambient temperature will be maintained within 0.5 degrees.

Figure 4. Rendering of the Buildings' Complex. The building in the foreground, seen to spiral out of the doughnut shaped Main-Hall, is the Office/Personnel Building. Ancillary, preparatory, laboratories are disposed along the periphery of the Main-Hall. Both buildings share a metallic, thermally isolating roof that allows the entrance of natural light via scattering through the different elevations of its segments. For reasons of integration into the local environment as well as functional ones the Service Building is landscaped against the natural terrain in the background.

The mechanical installations, comprising air conditioning, cooling, treatment and distribution of water arriving from the mains network, and fluids (i.e. natural gas, diesel, compressed air and technical fluids) are included as part of the Buildings' Project.

Cold and hot energy production is carried out centrally and respectively obtained from water condensation in a cooling tower and by means of a condensation tank and a vapor tank. These plants are placed in the Technical Building. Distribution of hot and cold water is achieved via pumps also installed in the Technical Building. Distribution of cold water is carried out via 5 circuits to the: accelerator tunnel and Linac; Beam-lines; Main Hall service and experimental areas; ancillary laboratories (placed in the periphery of the Main Hall), office area and others, and; exchanger circuit for acclimatization and cooling. Water is treated with ion exchange and reverse osmosis units. The various gases and fluids are stored and/or delivered from source (e.g. natural gas) at the Technical Building and distributed to the rest of the facility thereafter.

Regarding electrical installations, one of the design criteria has been to ensure power redundancy and quality. Redundancy is achieved via two externally supplied, commutable active lines of 25 Kv/12 MW each. These are derived from transformers taking power from a 220 kV external line. Earth connection (<0.2 ohm) is achieved via a 1mx1m reticule made of naked, buried copper wire of 50 mm2 cross section. The reticule is re-enforced with copper-steel spokes and joined to an equipotential net of galvanised steel that is imbedded in the floor of the Hall. This net is also joined to a perimeter ring of naked copper, again with a 50 mm2 cross-section. All earth networks are joined together into a single equipotential net.

Two emergency diesel generators (720kW each) are installed in the Technical Building to back up static un-interruptible Power Supply units, UPS, in case of failure of the external supplies. Dynamic UPS, i.e. fly wheels, are available as filters for short lived dips in the mains with autonomy of 12 seconds. This gives enough time to commute the external 25kV supplies to the self-generated emergency power.

An illustration of the Buildings Complex is shown in Fig. 4. In the foreground offices and personnel building can be seen to spiral out of the main Hall. Note that the architectural intention behind this lay-out is to allow future expansions, in the office buildings, if/when needed, without disturbing the activities in the main Hall and still remain integrated under a common structure. The dominant main Hall is placed underneath the segmented metallic roof, whilst the service building is hidden behind the main Hall and landscaped into the undisturbed natural terrain. Access roads circle the Main Hall and Office Buildings and pass in between the latter and the Service Building.

A Fast Track approach has been chosen to implement the construction of the Buildings' Complex. Project management and works direction were contracted at the end of 2005. All the necessary work licenses are secured and the construction was initiated in May 2006.

8.   TIME SCALE

Final reception of the Buildings Complex, including on site urbanization and gardening, is expected by mid-year 2008. However, the construction program contemplates that the Main Hall should be sufficiently advanced to allow Linac installation to start in November 2007 and begin the assembly of the Booster, Storage Ring and BLs infrastructure (e.g. hutches and services therein) in April 2008. Start of BLs assembly is planned for the last quarter of 2008. It is hoped that by mid 2009 full commissioning of the accelerators´ complex will begin and soon thereafter that of the first set of beam-lines. Early 2010 will be the initiation of user operations.

REFERENCES

1. Bordas J., Campmany J., Einfeld D., Ferrer S., Muñoz M., Perez F. and Pont M. "A concept for the Spanish Light Source: ALBA", NIM in Phys. Res., vol 543, Issue 1, 28-34, 2005.
2. Einfeld D., "Status of the ALBA Project", Proceedings of the 2006 EPAC meeting, in press, and references therein.
3. Perez F., Baricevic B., Hassazadegan H., Salom A., Sanchez F., and Einfeld D., "New Developments for the RF system of the ALBA Storage Ring", Proceedings of the 2006 EPAC meeting, in press, and references therein.
4. Pont M., Boter E., and De Lima Lopes M., "Magnets for the Storage Ring of ALBA", Proceedings of the 2006 EPAC meeting, in press, and references therein.
5. Muñoz M., Einfeld D. and Güntzel T.F., "Closed Orbit Corrections and Beam Dynamics issues at ALBA", Proceedings of the 2006 EPAC meeting, (in press), and references therein.
6. Pont M., Benedetti G., Einfeld D., Falone A., Al-Dmour E, Perez F and Joho W., "Injector Design for ALBA", Proceedings of the 2006 EPAC meeting, (in press), and references therein.
7. Campmany J., Bertwistle D., Marcos J., Martí Z., Becheri F., and Einfeld D., "A general view of IDs to be installed at ALBA", Proceedings of the 2006 EPAC meeting, (in press), and references therein.
8. Carles D, and Miralles L., "Site geo-technical and vibrational characterization for the design of the foundations of the ALBA Project" Proceedings of the MEDSI 2006 Conference, Himije (Japan) In press.

STATUS AND HIGHLIGHTS OF SIBERIAN SYNCHROTRON AND TERAHERTZ RADIATION CENTER

G.N. Kulipanov[a], A.I. Ancharov[b], E.I. Antokhin[a], V.B. Baryshev[a], E.L. Goldberg[c], B.G. Goldenberg[a], B.A. Knyazev[a], E.I. Kolobanov[a], V.V. Kriventsov[d], V.V. Kubarev[a], A.N. Matveenko[a], N.A. Mezentsev[a], S.I. Mishnev[a], A.D. Nikolenko[a], V.E. Panchenko[a], S.E. Peltek[e], A.K. Petrov[f], V.F. Pindyurin[a], V.M. Popik[a], M.R. Sharafutdinov[b], M.A. Scheglov[a], M.A. Sheromov[a], O.A. Shevchenko[a], A.N. Shmakov[d], A.N. Skrinsky[a], B.P. Tolochko[b], N.A. Vinokurov[a], P.D. Vobly[a] and K.V. Zolotarev[a]

[a]*Budker Institute of Nuclear Physics of SB RAS, Novosibirsk, Russia,* [b]*Institute of Solid State Chemistry and Mechanochemistry of SB RAS, Novosibirsk, Russia,* [c]*Limnological Institute of SB RAS, Irkutsk, Russia,* [d]*Boreskov Institute of Catalysis of SB RAS, Novosibirsk, Russia,* [e]*Institute of Cytology and Genetics of SB RAS, Novosibirsk, Russia,* [f]*Institute of Chemical Kinetics and Combustion of SB RAS, Novosibirsk, Russia.*

Abstract: Main directions of Siberian Synchrotron and Terahertz Radiation Center activity and novel experimental techniques are described. Development of free electron lasers and insertion devices - new types of wigglers and undulators, for generation of synchrotron radiation beams with prescribed properties is presented. The article includes description of three novel projects of SR sources and FEL developed in Siberian Center.

Key words: synchrotron radiation, terahertz radiation, free electron laser, research center, wiggler, undulator

1.   INTRODUCTION

SR experiments at the Budker Institute of Nuclear Physics had been started in 1973, and from 1981 the Siberian Synchrotron Radiation Center (SSRC) has an official status as Research Center of the Russian Academy

Vasili Tsakanov and Helmut Wiedemann (eds.), Brilliant Light in Life and Material Sciences, 57–68.
© 2007 *Springer.*

of Sciences (from 2005 - Siberian Synchrotron and Terahertz Radiation Center)[1]. Our research center is open and free of tax for the research teams from Russia and from abroad.

The main directions of Siberian Synchrotron and Terahertz Radiation Center activity are:

- research and technological application of synchrotron radiation (user's facility)
- research and technological application of the powerful terahertz radiation beams(user's facility);
- development of the new SR sources;
- development of powerful FELs of terahertz and IR regions based on accelerators-recuperators;
- development and producing of insertion devices (new types of wigglers and undulators).

2.   SYNCHROTRON RADIATION RESEARCH

In the last years, the main activity in the works with using of SR beams was concentrated at the VEPP-3 storage ring. Near 3000 hours per year of the VEPP-3 operation time are allocated for SR experiments. In the experiments at VEPP-3 there were used 11 stations on 7 SR beamlines. We touch only on the novel directions of research program.

Reconstruction of the environment state in the past. Experimental station for scanning X-ray fluorescent analysis of the lake bottom sediments, tree rings and other data mediums for the reconstruction of the environment state and climatic conditions in the past within the intervals from hundreds to million years operates at VEPP-3 from 1998.

Climate of the last 1.5 billion years is characterized by periodical changes of the glacial and interglacial epochs. Modern theory relates the global changes of climate with periodical variations of the Earth orbit parameters: excentricity, precession, bending angle of the Earth axis. The periods of this changes, the so-called Milankovitch cycles are known from astronomy. The most completely and clearly this cycles are seen in data of changes of oxigen isotope O^{18} in the deep ocean sediments.

In our first experiments it was measured the content of 30 tracer elements in the drilling columns from the lake Baikal. A clear periodicity was founded in geochemical records of various tracers. All basic orbit frequencies, expressed in oceanic records, modes were founded in geochemical records of Lake Baikal battom sediments, for example, peaks with periods 41, 72 and 96 thousand of years.

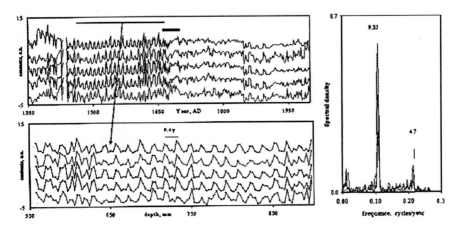

Figure 1. Left: The data on five chemical element distribution along a 100 cm bottom sediments column of Teletskoe lake. Right: Fourier analysis of the data obtained shows existence of 9.35 years main cycle of warm/cold periods.

Last years we performed a cycle of studies on climatic changes for the last century with a resolution of 1 year on the base of Lake Teletskoe bottom sediments[2]. We found out that a number of elements as K, Ca, Ti, Fe, V demonstrated a distinctly seen cycle with an average period of 9.35 years (Fig. 1). Among global processes a similar period has a cycle of the tide wave oscillation amplitudes. The principal source of the Earth's tidal force are the Sun and Moon, which continually change location to each other and this cause with nutation in Earth axis with period of 18.6 years.

Diffractometry of detonation and explosive processes. Experimental station "Extreme stations of matter"[3] is intended for contrast radiography and small X-ray scattering at study of detonation and shock-wave processes (Fig. 2). There is only one such station in the world. The station allows one to carry out experiments with as much as 30 grams of explosives. A particular feature of the station is the usage of explosion chamber with the berillium windows to pass the SR beam and 500 channels superfast one-coordinate detector DIMEX. The detector has 32 frames accumulated in such fast regime. Experiments have demonstrated the capability of the developed installation: the time resolution for measurements of density is of the order ~ 15 ns, density measurement accuracy is of 5%, time resolution for measurements of the small-angle scattering and diffraction is 125 ns at exposure of ~ 1 ns. The spatial resolution for density measurements is 0.1 mm. A cycle of experiments on a study of detonation in various explosive substances as well as investigation of formation of the diamond particles under shock wave action[4] and study of the effect of shock waves on various materials have been carried out.

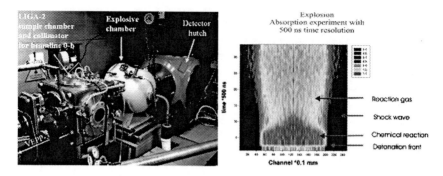

Figure 2. General view of experimental station for studies of the detonation processes and a sample of the data obtained: two dimentional map of density of explosive and explosion products in the process of detonation.

Studies and development of nanomaterials and nanotechnologies. Main directions of activity are:

- Study and optimization of the technological processes of nanopowders and nanomaterials production;
- Development of deep X-ray lithography and LIGA-technology (see example on Fig. 3, left);
- Development of the new research methods for investigation of nanoparticles, nanostructures and nanomaterials properties (Fig. 3, right);
- Development of nanobiological techniques (both with THz and synchrotron radiation).

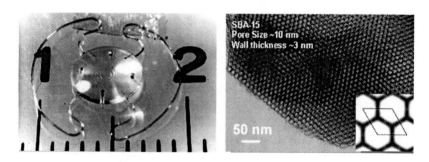

Figure 3. Left: artificial bifocal human crystalline lens produced with deep X-ray litography at SR beam. Right: mesostructured material for catalysis: electron microscopy image and structural electron density map reconstructed with synchrotron radiation EXAFS and X-ray diffractometry data.

"In situ" structural studies. "In situ" investigation of the reactions of selfpropagating high temperature synthesis (SHS) in heterogeneous systems (mixtures of powders and multilayer structures) with record time resolution up to 1 millisecond was performed using rapid time resolved X-ray diffractometry with two-coordinate X-ray detector "DED-5" developed in Budker INP. Other example - a study of clathrates (gas hydrates). This ice-like substances are a promising energy source in future. However, there was no information on the structures and properties of the high pressure (>3 Kbar) gas hydrate phases. This data were obtained at the station "High pressure diffractometry" with the use of developed "cylinder-piston" device at a pressure of up to 6.5 Kbar and diamond anvils for the range of higher pressures.

3.   DEVELOPMENT OF FREE ELECTRON LASERS AND TERAHERTZ RADIATION RESEARCH

The first stage of the Novosibirsk high power free electron laser (Fig. 4) had been commissioned in 2003. It is CW FEL based on non-superconducting, low radiofrequency (180 MHz) one pass accelerator-recuperator with the parameters: $E=12$ MeV, $Q_{bunch} = 180$ nC, bunch repetition rate $f_{rep} = (1.4, 2.8, 5.6, 11.2)$ MHz, average current $I_{av} = 2\text{-}20$ mA, peak current $I_{peak} = 20$ A, bunch duration 40-100 ps.

Figure 4. General view of first stage of the Novosibirsk high power free electron laser.

Main parameters of the first stage Novosibirsk terahertz FEL was investigated and measured. Spectral range is $\lambda = 240\text{-}120$ µm at the first harmonic, $\lambda = 117\text{-}60$ µm and $\lambda = 80\text{-}40$ µm at the second and third harmonics, maximal average power up to 0.4 kW (at $f_{rep} = 11.2$ MHz) at the first harmonic, maximal average power of second and third harmonics is

2% and 0.6% respectively to the first harmonic. Maximal peak power is 0.6 MW and repetition frequency 2.8, 5.6 and 11.2 MHz. Relative spectral width (0.25-1)%. The radiation is complete spatial coherent, the degree of radiation polarization is better than 99.6%[5].

Laser radiation is transmitted through nitrogen-filled optical beamline to the experimental hall. To provide ultrahigh vacuum in the FEL and accelerator-recuperator, their vacuum volume is separated from the beamline with the diamond window. Four user stations are now in operation (diagnostic station, photochemistry station, biological station, THz imaging station). More two stations are under construction: station for introscopy and spectroscopy, and the aerodynamics station.

High average power of the FEL enables development of imaging techniques, based on the thermal effect of radiation. Several methods for two-dimensional visualization of THz radiation with time resolution $(1-10^{-2})$ sec have been developed[6]. Instrumentation for the experimental station is developed and tested (windows, beam splitters, pyroelectric detectors, bolometers, zone Fresnel plates, kinoform lenses).

During the past year the NovoFEL was operating as a user facility. Soft ablation of biological molecules under terahertz radiation studied at NovoFEL for the last two years[7]. Precisely tuning radiation energy one can achieve the regime when "biological" molecules (DNA, proteins, etc) are "evaporated" without defragmentation (Fig. 5). These results can lead to creation of new biotechnologies.

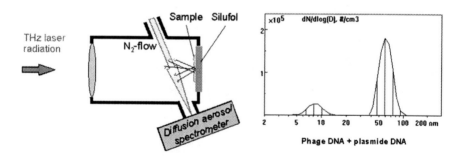

Figure 5. The scheme and a sample of ultrasoft separation by THz radiation of mixture of two types of DNA without defragmentation (right: spectrum of sizes of separated DNA molecules).

Experiments in physics, chemistry, biology, condense matter and technology at the first four user stations are now in a stage of preparing.

Next year we plan to finish construction of the full scale NovoFEL, based on the four-track 40 MeV accelerator-recuperator, using the same accelerating RF structure as the first stage. FELs in the second and fourth tracks are to generate radiation in the spectral ranges of 5-12 μm and 40-100 μm, respectively. Expected average power of each FEL is to be about 10 kW.

4. DEVELOPMENT AND PRODUCING OF INSERTION DEVICES

At the Budker INP the development and manufacture of different types of wigglers and undulators (superconducting, electromagnet, permanent magnet, hybrid) has been successfully elaborated since 1979. It was manufactured more than 50 wigglers and undulators. For example, 10.3 T superconducting wiggler for storage ring SPring-8 (Japan) is a source of especially hard SR with photon energy 1-3 MeV and intensity $\sim 10^{15}$ photon/sec (2000). New type 63-pole wiggler with a very short period of 34 mm and gap 13.5 mm with a magnetic field of 2T was constructed for protein structure beamline of the Canadian Light Source (2005).

Three electromagnetic elliptic undulators with period 256 mm was delivered and installed at SR source Soleil in 2006 (Fig. 6).

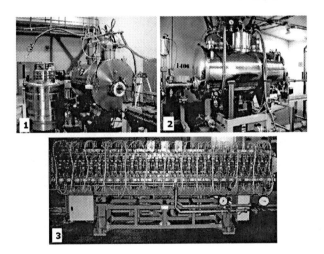

Figure 6. 1 - superconducting 3 pole 10.3 T wiggler for SR source "SPring-8" (Japan, 2000); 2 - superconducting 63 pole 2 T wiggler for Canadian Light Source (2005); 3 - elliptical electromagnet undulator HU256 for SR source "Soleil" (France, 2006).

5.   DEVELOPMENT OF NOVEL SR SOURCES AND FEL

Project of a compact storage ring - SR source. In 2002-2204, Budker INP developed and manufactured a prototype of superconducting bending magnet with 9.6 T maximal field for the storage ring BESSY-II (Berlin, Germany). The usage of such magnets allows realization of a compact scheme of a storage ring with rather low electron energy for generation of X-ray. Conceptual scheme and main parameters of 1.2 GeV compact storage ring - SR source cited in Fig. 7 and Table 1.

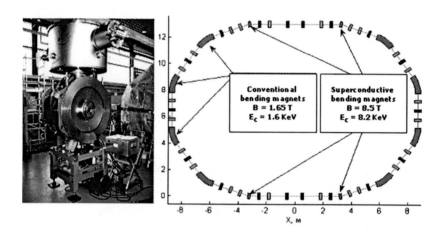

Figure 7. Left: superconducting bending magnet for BESSY-II developed and produced at Budker INP; right: conceptual scheme of 1.2 GeV compact storage ring: SR source.

Table 1. Main parameters of the compact storage ring - SR source

Electron energy	1.2 GeV
Field in bending magnets	8.5 T in superconducting magnets, 1.9 T in conventional (warm) magnets
Critical energy of SR quanta	8.6 keV for SR from superconducting magnets, 1.9 keV for beams from conventional magnets
Number of bending magnets	4 superconducting magnets, 8 conventional (warm) magnets
Bending angle in magnets	20° in superconducting magnets, 35° in conventional (warm) magnets
Beam emittance	16-20 nm-rad
Beam current	500 mA
Beam lifetime	8-10 hours
Perimeter of orbit	52 m

Project MARS. In the storage rings, the beam emittance and energy spread are determined by equilibrium between the radiation damping ($t_r \sim 10^{-2} - 10^{-1}$ sec) and two basic diffusion processes: quantum fluctuations of synchrotron radiation and intra beam scattering. The analysis shows that in practice, there is no real solutions how to reduce the beam emittance in the storage ring $\varepsilon_{x,z} < 10^{-10}$ mrad and an energy spread $\delta E/E < 10^{-3}$.

Realization of a fully spatial coherent source is possible in case of a shift from the electron storage rings to accelerators with energy recovery[8, 9].

In the accelerators-recuperators, the normalized emittance of electron beam ε_n can be kept the same during acceleration. Having a good injector with $\varepsilon_n < 10^{-7}$ mrad, due to adiabatic damping at an energy $E > 5$ GeV the beam emittance $\varepsilon_{x,z} \sim 10^{-11}$ mrad and energy spread $\delta E/E < 10^{-4}$ can be obtained. In the accelerators-recuperators, the time of acceleration is shorter compared to the time of radiation damping in the electron storage rings ($\sim 10^{-3}-10^{-4}$ time) and because of this fact, the diffusion processes cannot spoil the electron beam emittance and its energy spread. Therefore, for experimentalists the accelerator-recuperator based source is similar to the source based on the electron storage ring with the only difference that in the accelerator-recuperator every time it is used a new electron beam with small emittance ($\varepsilon_{x,z} \sim 10^{-11}$ mrad) and energy spread ($\delta E/E \sim 10^{-4}$).

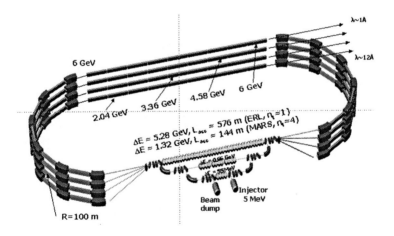

Figure 8. Conceptual scheme of multi-turn accelerator-recuperator SR source MARS and comparison of basic overall dimensions and parameters of accelerating RF system for ERL and MARS (for Emax = 6 GeV).

Conceptual scheme of MARS is shown at Fig. 8 as an example of 4-turn accelerator-recuperator at an energy of 6 GeV. This value was chosen for the task of obtaining the first harmonics of undulator radiation with $\lambda \sim 1$Å.

We will use for estimation the superconducting TESLA module ($L = 12$ m, $f_{RF} = 1.3$ GHz, $\Delta E = 110$ MeV as the accelerating gradient in cells of 15 MeV/m), the total length of a structure L_{acc} is about ~ 150 m for $n_t = 4$. Therefore, let us choose the stright section length to be $L_{sect} = 200$ m. The arcs radii should be of $R > 100$ m to prevent the beam emittance degradation by the quantum fluctuations of synchrotron radiation[8].

It is quite probable that in some cases, a multi-turn accelerator-recuperator can be used for improving the brightness and for obtaining the fully spatially coherent source at the upgrade of the available SR sources of the 2nd and 3rd generations[10]. In this case, the existing storage ring with the available generation systems, beam lines, etc can be used as the last track (Fig. 9).

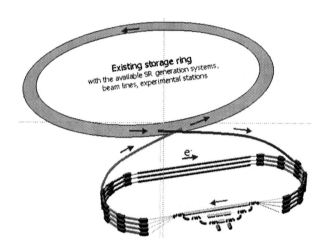

Figure 9. Multi-turn accelerator-recuperator for improving the brightness and for obtaining the fully spatially coherent radiation from existing SR sources of the 2nd and 3rd generations.

5.1   High gain ring FEL

Main conception of ring FEL – undulators, separated by isochronous bends allow to obtain master oscillator for X-ray generation and high power radiation source[11]. Problems of contemporary FELs, which may be solved by high gain ring FELs:
- poor quality of SASE FEL radiation;
- power limitation due to mirror heating in FEL-oscillators.

The scheme of high gain ring FEL, based on accelerator-recuperator ($E_e = 0.1 - 0.5$ GeV, $\lambda_{FEL} = 5$ μm $- 500$ Å) is shown at Fig. 10. For wavelength shorter then 100 Å the bend section can be made up of a number of isochronous bends on a small angle. Such a system can be thought of as a bended undulator.

Current status of project of high gain ring FEL: the feasibility study is finished; technical design may be performed, if demand appears.

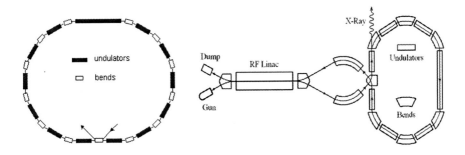

Figure 10. Left: conception of ring FEL – undulators, separated by isochronous bends; right: scheme of high gain ring FEL, based on accelerator-recuperator (Ee = 0.1 – 0.5 GeV, lFEL = 5 mm - 500 Å).

REFERENCES

1. Ancharov A.I., Baryshev V.B., Chernov V.A., et al. Status of the Siberian synchrotron radiation center // Nuclear instruments and methods in physics research. Sec. A. – 2005. – Vol. A543, No. 1. - P. 1-13.
2. Goldberg E.L., Grachev M.A., Phedorin M.A., et al. Application of synchrotron X-ray fluorescent analysis to studies of the records of paleoclimates of Eurasia stored in the sediments of Lake Baikal and Lake Teletskoe // Nuclear instruments and methods in physics research. – 2001. – Vol. A470, No 1/2. – P. 388-395.
3. Tolochko B.P., Aleshaev A.N., Fedotov M.G., Kulipanov G.N., et al. Synchrotron radiation instrumentation for "in situ" investigation of explosion with nanosecond time resolution // Nuclear instrument and methods in physics research. – 2001. – Vol. A467-A468, pt 2. – P. 990-993.
4. Titov V.M., Tolochko B.P., Ten K.A., Lukyanchikov L.A., Zubkov P.I. The formation kinetics of detonation nanodiamonds // Synthesis, properties and applications of ultrananocrystalline diamond / D.M.Gruen et al., eds. - Springer, 2005. - P. 169-180.
5. V.P. Bolotin, N.A. Vinokurov, D.A. Kayran et al., "Status of the Novosibirsk terahertz FEL", Nuclear Instruments and Methods in Physics Research, vol. A543, pp. 81-85, 2005.
6. Cherkassky V.S., Knyazev B.A., Kubarev V.V., Kulipanov G.N., et al. Imaging techniques for a high-power THz free electron laser // Nuclear instruments and methods in physics research. Sec. A. – 2005. – Vol. A543, No. 1. - P. 102-109.

7. Petrov A.K., Kozlov A.S., Taraban M.B., Goryachkovskaya T.N., Malyshkin S.B., Popik V.M., Peltek S.E. Mild ablation of biological objects under the submillimeter radiation of the free electron laser // The Joint 30th International conference on infrared and millimeter waves and 13th International conference on terahertz electronics: IRMMW-THz2005, Williamsburg, Virginia, USA, Sept. 19 – 23, 2005. – Piscataway: IEEE, 2005. – Vol. 1. – P. 303-304.

8. Kulipanov G.N., Skrinsky A.N., Vinokurov N.A. Synchrotron light sources and recent development of accelerator technology // J. of synchrotron radiation. – 1998. – V. 5, pt. 3. – P. 176-178. – (Proceedings of the SRI-97).

9. Kulipanov G., Skrinsky A., Vinokurov N. Multi-pass accelerator-recuperator (MARS) as coherent X-ray synchrotron radiation source // The Ninth International Conference on Synchrotron Radiation Instrumentation (SRI 2006) May 28 - June 3, 2006, , Daegu, Korea: Proceedings. Submitted to Review of Sci. Instruments.

10. I. Koop. Design of the ERLSYN Light Source. Proceedings of Workshop on Scientific Applications of ERL, 2002, Erlangen, Germany.

11. Vinokurov N.A., Shevchenko O.A. High-gain ring FEL as a master oscillator for X-ray generation // Nuclear instruments and methods in physics research. Sec. A. – 2004. – Vol. A528, No 1/2. – P. 491-496.

CANDLE SYNCHROTRON LIGHT SOURCE PROJECT

V.M. Tsakanov
CANDLE, Yerevan, Armenia

Abstract: CANDLE – Center for the Advancement of Natural Discoveries using Light Emission – is a 3 GeV energy synchrotron light facility project in the Republic of Armenia. The main design features of the new facility are given. The results of the beam physics study in the future facility is overviewed including the machine impedance, single and multi-bunch instabilities, ion trapping and beam lifetime. The preliminary list of first group beamlines is discussed.

Key words: synchrotron light source, brightness, beam, stability

1.   INTRODUCTION

The research highlights, based on the usage of synchrotron radiation in biology, medicine, chemistry, material and environmental sciences, the broadband application field for the results in pharmacy, electronics and nano-technology, promoted the design and construction of a number of third generation light sources worldwide at the intermediate energies 2.5-3.5 GeV[1].

Since 1967 the 6 GeV electron synchrotron in Yerevan Physics Institute (Armenia) was in operation. Number of unique results obtained on this facility includes the study of pion photoproduction, eta-meson generation, transition and channelling radiations. Nevertheless, even in 80's it was well understood that the next accelerator facility in Armenia should be a life and material sciences oriented project[2].

The new synchrotron light source project named CANDLE is a 3 GeV nominal energy electron facility, the spectrum of synchrotron radiation from

Vasili Tsakanov and Helmut Wiedemann (eds.), Brilliant Light in Life and Material Sciences, 69–80.
© 2007 *Springer.*

bends, wigglers and undulators of which covers the most essential region of photons energy 0.01- 50 keV suitable for investigations at the cell, virus, protein, molecule and atomic levels. The conceptual design of the new facility[3] has been completed in 2002 and received a favourable recommendation by the Review Panel[4].

The basic approaches that underlie the project are:

- CANDLE will be a brand new facility with the infrastructure corresponding to state-of-the-art in accelerator and experimental techniques.
- The facility will operate as an international laboratory, open for all the qualified specialists.
- The facility is designed to be a world-class machine with the competitive spectral flux and brightness, the stable and reproducible photon beam, the high rate of the control with user-friendly environment.

Resonant X-rays techniques such as absorption fine structure (XAFS), x-

2.   DESIGN OVERVIEW

The CANDLE general design is based on a 3 GeV electron energy storage ring, full energy booster synchrotron and 100 MeV S-Band injector linac (Fig. 1).

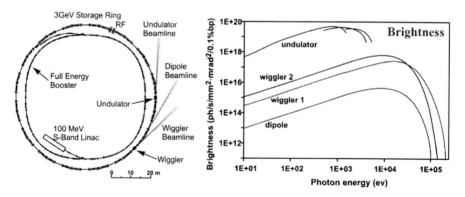

Figure 1. The general layout of CANDLE facility and the photon spectral brightness.

The full energy booster synchrotron operating with the repetition rate of 2 Hz and the nominal pulse current of 10 mA provides the storage of 350 mA current in less than 1 min. The storage ring of 216m in circumference has 16 DBA type periods. The harmonic number of the ring is h=360 for the accelerating mode frequency 499.654 MHz. The main parameters of the facility are given in Table 1.

Table 1. The facility main parameters.

Energy E (GeV)	3
Circumference (m)	216
Current I (mA)	350
Number of lattice periods	16
Betatron tunes	13.2 /5.26
Horiz. emittance (nm-rad)	8.4
Beam lifetime (hours)	18.4

In total 13 straight sections of 4.8 m in length are available for insertion devices (ID). The photon beams from the dipoles and the conventional ID's are covering the energy range of 0.01-50 keV with high spectral flux and brightness. Fig. 1 presents the CANDLE spectral brightness for the dipole (1.35 T), undulator (0.3 T) and wiggler (1.3 T, 2 T) sources.

The design of the machine is based on conventional technology operating at normal conducting conditions. The 6 ELETTRA type cavities provide the total gap RF voltage of 3.3 MV thus ensuring the Touschek lifetime of about 40 hours. With 1 nTorr vacuum pressure the total beam lifetime is at the level of 18.4 hours. The vacuum chamber of the ring is based on the stainless steel antechamber geometry design. The facility design implies the operation in the single, multi-bunch and top-up modes thus providing the broadband application of the experimental techniques.

3. OPTICS AND DYNAMIC APERTURE

The CANDLE design has 16 identical Chasman-Green type cells with non-zero horizontal dispersion in the middle of the long straight section $\eta_x = 0.18m$ that provides 8.4 nm-rad horizontal emittance of the beam.

The spectral brightness of the photon beam from undulators is one of the main figures of merits that define the advance of the facility design to utilize the whole capacity of the insertion devices. The optimization of the optical parameters of the lattice to obtain a high spectral brightness of the photon beams from insertion devices keeping large the dynamical aperture of the ring was an important issue of the R&D study[4].

Fig. 2 (left) shows the dependence of the normalized brightness on the emitted photon energy for different horizontal beta values at the source point. Dashed line corresponds to the optimal beta values associated with each photon energy. The improvement of the brightness with small horizontal beta is visible only for the photon energies below 0.1 keV. Starting from 0.5 keV the brightness increases with larger betatron fuction, and in the energy range of higher than 5 keV the brightness actually reaches its maximum for the 8 m of beta value.

In vertical plane the beam emittance is given by the coupling of the horizontal and vertical oscillations. Fig. 2 (right) shows the normalized brightness versus of the vertical beta function for CANDLE nominal lattice and 1% coupling. The small vertical emittance of the beam shifts the characteristic regions of the brightness behaviour to harder X-ray region. The increasing of the spectral brightness with low beta is now visible in the photons energy range of 0.5-8 keV and starting from about 10 keV the brightness increases for high beta function. Taking into account the requirement to have sufficient dynamical aperture of the CANDLE storage ring, the horizontal and vertical betatron functions in the middle of the straigth sections are optimized to $\beta_x = 8.1m$, $\beta_y = 4.85m$.

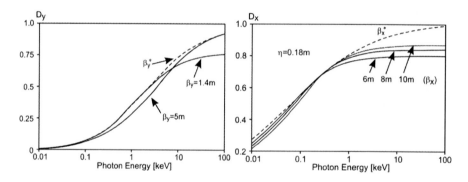

Figure 2. The normalized CANDLE brightness versus photon energy for various horizontal (left) and vertical (right) beta functions.

The comparetively high beta values in middle of straight section are significantly improving the dynamical aperture of the ring. Fig. 3 shows the CANDLE storage ring dynamic aperture with 3% energy spread which is sufficient for facility stable operation.

The reduction of the dynamics aperture due to magnets fringe field effects has been carefully analyzed[6]. The corresponding technique has been developed to adjust the quadrupole strengths around the ring.

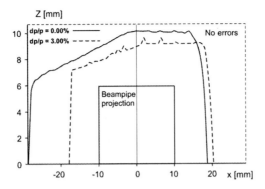

Figure 3. The storage ring dynamic aperture (ID).

4.   BROADBAND IMPEDANCE

To prevent the single bunch instabilities in the ring, the impedance caused by the walls resistivity and roughness, BPM's and transitions has been carefully analyzed. Fig. 4 presents the longitudinal and transverse impedances of the storage ring for various distributed impedance sources.

The normalized longitudinal broadband impedance for the CANDLE storage ring without ID's is at the level of 0.35 Ω. The corresponding single bunch threshold current is 8.9 mA. The transverse impedance of the ring without ID's is 12.6 kΩ/m that defines the threshold current of 113 mA for the transverse single bunch instability. The CANDLE nominal operation current of 350mA implies the single bunch current of 1.24 mA that is far below of the threshold currents.

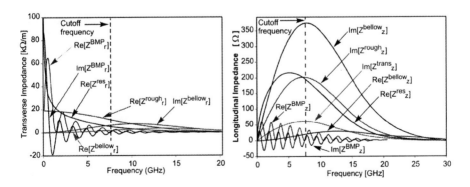

Figure 4. Longitudinal and transverse impedances of the ring caused by the walls resistivity, roughness, BPM's and transitions.

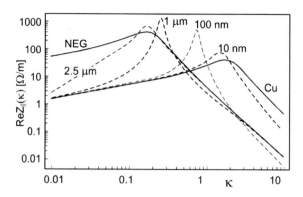

Figure 5. Real part of copper-NEG tube longitudinal impedance for various cover thickness.

An additional contribution to the ring impedance will be dominated by the small gap undulator vacuum chamber installed in the straight section of the ring.

To reduce the resistive impedance of the chamber the laminated walls are usually used, i.e the copper chamber covered by the NEG (Non-Evaporated Getter). Fig. 5 presents the real part of the point charge longitudinal impedance[7] for the cooper-NEG, 5 mm aperture vacuum chamber. The evaluation of the transverse impedance of laminated vacuum chamber is given in[8]. The results of this study will be used to evaluate the single bunch instabilities with the real broadband impedance of the ring equipped with the wigglers and undulators in the straight sections.

5.   COUPLED BUNCH INSTABILITIES

The narrow band impedance of the storage ring, basically the longitudinal and transverse High Order Modes (HOM) excited by beam in the RF cavities, determine the longitudinal and transverse multi-bunch instabilities.

The longitudinal and transverse coupled bunch instabilities for the CANDLE storage ring have been studied for the original ELETTRA cavity option. Fig. 6 presents the growing rate of longitudinal and transverse coupled bunch instabilities versus the relative mode index *n* for the 282 beam oscillation modes. The growing rates of instabilities are mostly below the synchrotron radiation damping coefficients, except for L6 longitudinal mode that excites the instability at relative oscillation mode of n=115. After the cavities RF measurements, the instability cures will be developed to ensure the stable operation of the facility.

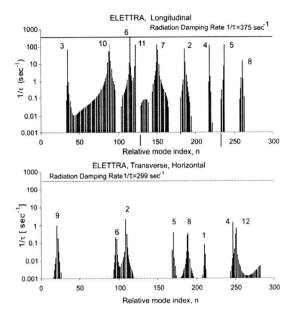

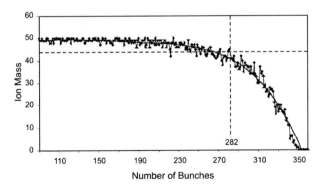

Figure 6. The longitudinal and transverse coupled bunch instabilities growing rate versus beam oscillation modes.

6. ION TRAPPING

To prevent the reduction of the beam lifetime due to ion trapping, the stability of the transverse motion of the ions has been checked along the regular lattice of the storage ring. The results are presented in Fig. 7.

Figure 7. Number of unstable ions versus number of bunches in the ring.

The effect of the ion trapping is observed when the number of bunches in the ring exceeds 90. The optimal value of the number of bunches for the CANDLE storage ring has been defined $h_b=282$ which provides an ion-cleaning gap of 78 RF buckets not filled with the electrons. With such a cleaning gap, only 6 ions number are trapped. An additional criterion for the optimization of the number of bunches in multi-bunch operation mode is an analysis and comparison of the list of trapped ions with the mass numbers of real residual gas species. Fig. 8 presents the trapped ion masses versus filled RF buckets.

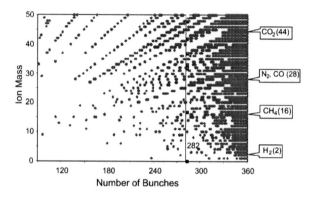

Figure 8. The trapped ion masses versus number of the filled RF buckets.

The trapped ion mass numbers in CANDLE storage ring are given in Table 2; the number of bunches is 282. The comparison with species of the ions in the residual gas that can occur in the chamber shows that for the given number of bunches 282 no components of residual gas will lead to trapped ions in the ring.

Table 2. Trapped ion masses and residual gas species.

Trapped ion masses	Residual gas species
-	2, H_2
-	16, CH_4
17	-
-	28 , N_2,CO
32	-
33	-
37	-
42	-
-	44, CO_2
48	-

The chosen number of bunches in the storage ring, 282 bunches from available 360 RF buckets, provides practically trapped ions-free operation of the machine at almost 80% filling.

7.   BEAM LIFETIME

Beam lifetime in the storage ring is dominated by three beam loss-processes: the quantum excitation, intra-beam scattering (Touschek effect), and scattering off of residual gas molecules (elastic and inelastic). The Touschek lifetime is the most critical one. Fig. 9 shows the Touschek lifetime evolution along the lattice for various energy acceptances. For a circulating current of I = 350 mA, the charge per bunch is 0.9 nC assuming 282 RF buckets of the total 360 are filled. For an energy of 3 GeV and RF energy acceptance of 2.38%, the average Touschek lifetime over the ring is then about 39 hours.

Fig. 10 shows the Touschek lifetime dependence on the coupling coefficients for different energy acceptance. In particular, for a coupling of 2% and energy acceptance of 2.4% the Touscheck lifetime is at level of 54 hours.

The summary of the electron beam lifetimes in CANDLE storage ring for 1% coupling and N_2 equivalent gas pressure of 1 nTorr is given in Table 3. An integrated beam lifetime in storage ring is at the level of 18.4 hours.

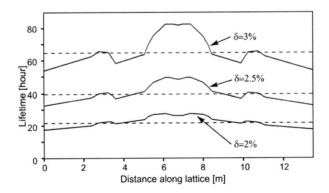

Figure 9. Touschek lifetime evolution along the lattice for various energy acceptances (solid lines).

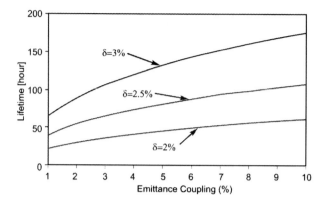

Figure 10. Averaged Touschek lifetime vs emittance coupling and energy acceptance.

Table 3. Averaged Touschek lifetime vs emittance coupling and energy acceptance.

Gap Voltage MV	3.3
Energy Acceptance %	2.376
RMS bunch length (mm)	6.5
Average beta hor./vert. (m)	5.2/10.5
Beam lifetimes	
Elastic scattering (hours)	91.4
Inelastic scattering (hours)	55.4
Touschek lifetime (hours)	39.5
Quantum lifetime (hours)	$>10^{38}$
Total lifetime (hours)	18.4

8. FIRST GROUP BEAMLINES

The increasing demand of synchrotron radiation usage worldwide drives the scenario for the first stage beamlines that are an integrated part of the facility construction. Based on the existing synchrotron radiation usage statistics, the following beamlines have been preliminary selected to build at the first phase of CANDLE construction[9]: LIGA (dipole), General Diffraction and Scattering Beamline (dipole), X-ray Absorption Spectroscopy Beamline (dipole), Soft X-ray Spectroscopy Beamline (undulator), Imaging and Small Angle X-ray Scattering Beamlines (Wiggler). The technical design of the beamlines is based on the reasonable freedom in optical elements to meet user demand.

LIGA beamlines. The LIGA beamlines will cover the experiments with three photon energy regions: 1-4 keV for fabrication of X-ray masks and thin microstructures up to 100 μm height, 3-8 keV for standard LIGA microstructures fabrication with resists height up to 500 μm and 4-35 keV for deep lithography.

General Diffraction and Scattering Beamline. The beamline is based on the dipole source and will produce a moderate flux of hard (5-30keV) focused or unfocused tuneable monochromatic X-rays sequentially serving two experimental stations: EH1 for roentgenography routine or time resolved experiments, low or high temperature studies of polycrystalline materials, for reflectivity investigations of thin films and multi-layers; EH2 for single crystal structure determination, the charge density studies and anomalous dispersion experiments.

XAS Beamline. The XAS beamline will cover a photon energy range up to about 35 keV. Using a differential pumping system this beamline can be operated with no window before the monochromator and will produce sufficient intensity in soft X-ray region of the spectrum. This region covers the K edges of elements such as Si, S, P and Cl, which are of high technical interest. Using double crystal monochromator and gold coated total reflection mirror, the beamline will be able to operate in hard X-ray region allowing users to measure EXAFS of all elements either at K or L_3 edges.

Imaging Beamline. Using the radiation from 3 T permanent wiggler this beamline will provide a high flux "white" or tuneable monochromatic coherent radiation in 6-120 keV photon energy range at about 150m from the source. The experimental program will include: phase contrast and diffraction-enhanced imaging; hard x-ray microscopy; holographic imaging and tomography; micro-focusing; X-ray topography, diffractometry; micro-fluorescence and high resolution inelastic scattering.

SAXS Beamline. The beamline will be based on the advanced Bonse-Hart camera. The scientific applications include small and ultra small angle X-ray scattering and nuclear resonance. The primary optical elements of the beamline include the high resolution and high heat load monochromators.

Soft X-ray Spectroscopy Beamline. This beamline has unique capabilities to address complex problems in materials, environmental and biological sciences. The high brightness photon beam from the undulator will support this beamline. Two types of microscopes are proposed to be build: a zone plate based scanning transmission X-ray microscope and a photoelectron emission microscope.

9. SUMMARY

The progress of the new facility project after completion of facility design report is summarized in the CANDLE laboratory activity reports. The next stage of the project development implies an extensive prototyping program, the RF, magnet and vacuum test stands establishment, number of machine and user international workshops. The international collaboration is highly appreciated and we express our deep gratitude to all colleagues for their interest, support and cooperation.

REFERENCES

1. J. Corbett, T. Rabedau, Synch. Rad. News **12**, 22, 1999.
2. A. Amatuni *et al*, Proc. of 9th All-Union Conf. On Part. Accel., Dubna, 1984.
3. A. Abashian *et al*, CANDLE Design Report, ASLS-CANDLE R-001-2002, 2002.
4. http://www.candle.am/revrep/revrep.pdf
5. M. Ivanyan,Yu. Martirosyan,V. Tsakanov, NIM (A), **531/3** 651-656, 2004.
6. Yu. Martirosyan, NIM (A), **521**, 556-564, 2004.
7. M. Ivanyan, V. Tsakanov, Phys. Rev. ST-AB, v.7, 114402, November 2004.
8. M. Ivanyan, A.V. Tsakanian, Phys. Rev. ST Accel.. Beams **9**: 034404, 2006.
9. M. Aghasyan *et al*, SRI2003, AIP Conf. Proc. v. 705, New York, p. 494, 2004.

SOCIAL AND ECONOMIC IMPACT OF THE CANDLE LIGHT SOURCE PROJECT
CANDLE project impact

M. Baghiryan
Center for the Advancement of Natural Discoveries using Light Emission
31 Acharyan St., 375040 Yerevan, Armenia

Abstract: Social and economic progress related to the realization of the CANDLE synchrotron light source creation project in Armenia is discussed. CANDLE service is multidisciplinary and long-lasting. Its impacts include significant improvement in science capacities, education quality, industrial capabilities, investment climate, country image, international relations, health level, restraining the "brain-drain", new workplaces, etc. CANDLE will serve as a universal national infrastructure assuring Armenia as a country with knowledge-based economy, a place for doing high-tech business, and be a powerful tool in achieving the country's jump forward in general.

Key words: knowledge-based economy, high-tech, CANDLE impact

Scientific knowledge and technological change are the major drivers of economic growth today. Already in 1883 Werner von Siemens - a man who turned a humble little workshop into one of the world's largest enterprises - said: "Scientific research forms always the safe foundation of technical progress and the industry of a country will never attain an internationally leading position and maintain itself, if the country does not at the same time take a leading position at the frontier of scientific progress. To strive for this, is the best way to promote industry[1]". The relation of scientific research and economic growth can be expressed in a discovery of a new phenomena leading to significant qualitative changes in technology or bring to completely new technological applications; the transformation of laboratory-scale techniques into industrial versions; concept-design-implementation objective leading to innovations which require close

Vasili Tsakanov and Helmut Wiedemann (eds.), Brilliant Light in Life and Material Sciences, 81–88.
© 2007 *Springer.*

cooperation between research institutions and industry to link together all the key elements crucial for translating scientific and technical knowledge into new products and services; formation of new technology- and knowledge-led enterprises (research spin-off companies) which are considered as the prime drivers of knowledge-based economy; promotion of international collaboration and formation of national and international networks providing free access to experts and information (knowledge)[2]; benefits to industry from skilled graduates having ability to acquire and use knowledge in new and powerful ways, bringing enthusiasm and a critical approach[3].

The driving force of the knowledge economy is the development of technologies, namely information technology and communications, biotechnology and nanotechnology. Providing advances in various research fields and offering competitive edge for various industries the synchrotron plays a crucial role as an infrastructure element for building knowledge-based economy acting as a catalyst to engender collaborations required on the way to the development of a country, region and the world in general.

When discussing the benefits of the CANDLE synchrotron light source creation in Armenia we should consider that the facility service is multidisciplinary and long-lasting. A Review Panel established by the US State Department including leading scientists and World Bank experts held two-day meeting in Washington in 2002 to discuss various aspects of the proposed CANDLE project including its economic impact. The Panel identified seven positive economic impacts: direct creation of jobs; potential demonstration of political stability for prospective investors; workforce enhancement; attraction of new high quality industry; multiplier on annual budget coming from outside the country by up to 3 times (World Bank estimate); usage of currently under-subscribed electrical generating capacity[4].

Increasing employment is undoubtedly the most socially acceptable and economically efficient means of poverty reduction. A number of workplaces will be directly created both in the phase of the CANDLE facility construction and operation, and in the framework of the CANDLE center, and as a result of long-term spin-off activities as well. Staffing for the facility is expected to grow from 120 in Year 1 to 140 in Year 4. On contract basis much scientific and technical workforce (over 1000 people) and constructors will be involved in the activities of prototyping, construction and commissioning of the synchrotron complex.

The synchrotron investment will support the overall investment climate and reputation of Armenia as a place for doing high tech business. It will indicate a sense of political stability that will bootstrap serious thinking on other investments. The advances in R&D and technical capabilities (new manufacturing techniques) presented by synchrotron will attract entrepreneurs and large private corporations worldwide in many high quality industries of the 21st century as pharmaceuticals, biotechnology, advanced materials, telecommunications, and electronics to invest in CANDLE beamlines, to center their R&D activities around CANDLE and to establish their offices and branches in Armenia.

If the CANDLE operating budget of 7 million USD can be secured outside the country a positive impact of about 20 million USD per annum to the Armenian economy might be anticipated at this stage of Armenia's economic recovery. As the economy and infrastructure improve the multiplier could improve considerably[4]. American University of Armenia performed a financial study for evaluating the feasibility of the CANDLE facility creation and operation in Armenia. The utilization of the beamlines in foreign synchrotron facilities have been studied and scenarios predicting the viability of the CANDLE project are evaluated. Table 1 shows three scenarios of the facility operation by mean of the generated income from the synchrotron radiation beam time usage. The most likely scenario, when 34% of beam time is charged, the facility income completely covers the facility annual operation cost for 10 beamlines in used.

Table 1. Income from non-public usage of light at CANDLE.

Scenario	Income5 beamlines	Income10 beamlines	Payable Users (%)
Optimistic	8.4 mln	16.7 mln	58
Pessimistic	1.32 mln	2.6 mln	9
Most likely	**4.85 mln**	**9.6 mln**	**34**

Power requirements for the facility would likely provide approximately $1 million/yr in revenues for the local electric utility. This could aid financial stability for power generation, which currently has underutilized capacity[4].

One of the distinctive features of the Armenian reality is the human capital of the nation. The maintenance and modernization of the present educational and scientific capacity is essential to preserve the high levels of literacy, higher education and scientific potential. The most substantial impact of the CANDLE facility creation is connected with long-term science benefits which are difficult or impossible to translate into monetary terms. About 20000 scientists and engineers now use synchrotron radiation at about 50 operational facilities around the world. Representing a vast complex of diverse world-level equipment simultaneously operating contemporary experimental methods, CANDLE will provide the Armenian, regional and international scientific community with a new powerful tool offering advances in scientific understanding in such essential areas of basic research as molecular and micro-biology, chemistry, solid state physics, materials research, environmental science, medicine and pharmacology, geology and other related areas, with applications in drug design, materials analysis, medical diagnostics and therapy, environmental remediation, nanotechnology, etc. The possibility to access synchrotron radiation from this state of the art facility will allow researchers to extend their investigations with greater accuracy and sensitivity. Large number of scientists and engineers from diverse fields will utilize the CANDLE facility competitive with those available in other leading industrial nations. Works to be conducted are likely to be frontier and to have a greater likelihood of leading to new discoveries which may result in major benefits to society.

Synchrotron radiation facility is arguably the most effective instrument for training graduate students in basic and applied research and technology. To hundreds of Armenian and regional graduate students and postdoctoral researchers CANLDE will serve as a brilliant place for learning new techniques and performing their studies effectively, getting high-level training and be familiar with the latest developments in fast developing fields in broad spectrum of basic and applied research and technology. Thus, CANDLE will help retain highly qualified personnel at educational, research and industrial institutions, as well as assist the recruitment of new talent. When fully operational, CANDLE will produce several tens to one hundred PhD's per year[4]. With the growing demand for technology-related products and services more and more technology-based start-ups are formed focusing mainly on bio-technology, information technology, advanced materials, high-tech services, designs, bio-medicine, drugs and pharmaceuticals, and instrumentation. As the experts of the Review Panel noted the experience in both Western Europe and the US shows that the construction and operation of synchrotron facilities result in a large number of highly skilled scientists and engineers who can contribute in a major way in developing high technology industries contributing to local economies. This is an effective way in which CANDLE will also contribute to the technology transfer in Armenia with its very high percentage of literacy[4].

The realization of the CANDLE project, in both construction and operation stages, will have a significant impact on improving the industrial capacities and building in-country capability in high-tech manufacturing. The synchrotron would generate in-country purchases of highly specialized equipment for the accelerator complex and beamline configuration. Approximately 60% of the equipment in the initial capital budget is expected to be produced in-country. For the participating Armenian manufacturers that will enable an upgrade in general manufacturing capabilities of the republic. The first dipole magnet has been successfully produced by "ArmElectroMash" company in 2003.

Providing new possibilities in terms of size, shapes, materials, production process, the information generated through synchrotron research is of immense importance to various industries, namely pharmaceutical, petrochemical, cosmetics, advanced materials, micromachining, nanoelectronics and communications, mining and minerals, metallurgy, construction, food products, plastics, papermaking, etc. CANDLE synchrotron light source will enhance the existing methods in a wide range of industrial applications offering full service in their usage and analysis to industrial customer which will not need to set aside its own experienced manpower. The reliable and dedicated apparatus at CANDLE will facilitate the usage of the offered analytical results by industries that make the core of Armenian economy (namely machinery, non-ferrous metallurgy, chemical and petrochemical industry, logging, wood-working, pulp and paper industry, building materials industry, light industry) providing a key link in emerging product and process developments. That will be especially beneficial for medium size enterprises to improve their production and become or remain competitive.

The greatest power of today's economic success is due to free exchange of scientific ideas and technological advances that occurred since the last century. Economic growth depends more and more on the continuous supply and use of technologies, wherever they are produced. The established network of R&D institutions with qualified professionals will facilitate the dissemination and use of technologies making the transfer of industrial technology from technologically advanced countries feasible in a shorter period of time. Science in general is a brilliant area for international collaboration. Thousands of researchers of multidisciplinary fields from different countries are working in synchrotron centers, forming international collaborations and networks. CANDLE will be the first of today's third-generation synchrotrons to appear within a radius of 2000 kilometers from the selected site. There is no equivalent facility in the wide region including FSU, East European countries, and the Middle East. Therefore, the creation of CANDLE synchrotron light source in Armenia might be viewed as having important repercussions over a wider area. The establishment of a scientific center like CANDLE will have a crucial influence on the development of not only Armenian science, technology and industry, but also those in the neighboring region and even for countries outside the region. CANDLE will provide the regional users an easier access to the facilities, since now have to travel to Europe or the US and compete for synchrotron beam time. Taking into account that depending on the scientific field the demand on the synchrotron light worldwide is 3-10 as greater as the existing capacities, some foreign scientists from far-flung countries will also come to conduct independent investigations, as these researches will in addition cost them much cheaper. On the way of becoming an international laboratory CANDLE will promote the establishment of strong national, regional and international linkages, networks and collaborations in the fields of common interest leading to more efficient use of resources and to improved dissemination of results. Synchrotron is a brilliant venue that will bring together university, R&D institutions and industry to initiate joint programs. International collaboration can take many different forms, ranging from the creation of joint institutes to joint venture, sharing of results and data, and setting up informal networks.

The early nineties and the last decade were a critical period for the Armenian scientific community leading to internal and external "brain drains" in and from Armenia. Highly qualified scientists, engineers and other professionals with a wealth of experience have been lost by emigration or simple lace of productive employment opportunities. It resulted in the loss of the nation's best talent which was concentrated to a significant degree in scientific and technical areas. The establishment of CANDLE center will serve as a major step to restrain the brain drain from Armenia which is a substantial problem and is especially acute now. Thousands of highly educated and skilled young researchers, scientists and engineers continue to abandon their fields of specialization and their positions at once well regarded institutions to seek employment either in less skilled fields or outside of their native country. Having the possibility to get high-level training and to do world-class research in appropriate conditions, the scientific-technical workforce will not only stay in Armenia, but the "departed" ones will be back to work in their home-country. The presence of the CANDLE synchrotron center, with the concentration of its vast capabilities for R&D in Armenia, will attract the foreign high-tech companies to set their R&D laboratories here thus enabling the local scientists to do also contract research on demand. The latter will not necessitate the loss of the scientific-technical resources, but will also allow having it fulfilled for the benefit of the country.

Among the indirect impacts is the multiplier effect from visits by international visitors. CANDLE will attract foreign students and professionals to study and work in Armenia. As foreign scientists will make the lion's share of CANDLE users, they will greatly promote the tourism infrastructure development in Armenia. Perhaps ~1000 users and other visitors per year could be anticipated, for terms ranging from week-long conferences to year-long sabbaticals[4].

For numerous medical techniques, namely computed microtomography imaging technique particularly to discover and identify brain tumors, microbeam radiation therapy method for the treatment of brain tumors, mammography for early detection of breast cancer, imaging of tumors, etc., synchrotron light sources provide higher resolution, high beam precision, enhanced contrast, application of lower doses for getting better results than possible by conventional sources. These privileges over the presently used medical equipment have been yet confirmed in the experiments on animals and human phantoms in many leading laboratories worldwide. While the advantages of light from synchrotron are currently used mostly with the purpose of pre-clinical protocols, demonstrating the enhancement of the above-mentioned and other techniques, the clinical use for human diagnostics and also treatment is practiced only for angiography. The usage of other techniques for diagnosis and treatment of people will make up the progress of the 21st century. Actually all modern synchrotron centers are accompanied by a satellite medical center. The services and possibilities provided by CANDLE medical beamline and the biomedical satellite established in the future will be significant resource to keep and improve the health of the overall population of Armenia and the region. They will definitely be an excellent laboratory for scientists to do medical research getting enhanced experimental materials. The modern conditions offered for the training of high-level physicians is of related benefits in the field of medicine.

CONCLUSION

CANDLE will be the locomotive in building science-intensive industries and a knowledge-based economy in Armenia, since it represents the essential infrastructure element in creating the necessary prerequisites: a strong research base; targeted strategic research investments; highly qualified human resources to manage the knowledge economy; national research facilities and infrastructure; national, regional and international networks and interconnections between them; appropriate support mechanisms for innovative technology-based firms.

The presented possible progress induced by CANDLE proves this project as a rare critical and real chance to maintain and enhance the scientific-technical potential of the nation having it fulfilled for the country's socio-economic growth and to maintain high growth rates in the long perspective.

M. Baghiryan

REFERENCES

1. Schopper H., "Physics in the Next Century" (Talk at the International Workshop on the Future of Physics and Society, Debrecen, 4-6 March, 1999).
2. Liewellyn-Smith C., "International Collaboration in Science: Lessons from CERN", In: Science for the Twenty-First Century: A New Commitment, pp. 132-135.
3. Salter A. J. and Martin B. R., "The Economic Benefits of Publicly Funded Basic Research: A Critical Review".
4. Report to the US Department of State Relative to the CANDLE Project Proposed for Armenia, August 20, 2002.

KHARKOV X-RAY GENERATOR NESTOR

V. Androsov[a], A. Agafonov[b], E. Bulyak[a], J.I.M. Botman[c] V. Grevtsev[a],
A. Gvozd[a], P. Gladkikh[a], Yu. Grigor'ev[a], A. Dovbnya[a], I. Drebot[a],
V. Ivashchenko[a], I. Karnaukhov[a], N. Kovalyova[a], V. Kozin[a], A. Lebedev[b],
V. Markov[a], N. Mocheshnikov[a], A. Mytsykov[a], I. Nekludov[a], F. Peev[a],
A. Ryezayev[a], V. Skirda[a], A. Shcherbakov[a], R. Tatchyn[d], Yu. Telegin[a],
V. Trotsenko[a], O. Zvonareva[a], A. Zelinsky[a]

[a]National Sciece Center "Kharkov Institute of Physics and Technology" 1 Akademicheskaya
Str., Kharkov, Ukraine; [b]P.N. Lebedev Physical Institute, Russian Academy of Sciences,
117924, Moscow, Leninsky Prosp. 53, Russia; [c]Eindhoven University of Technology (TU/e),
P.O. Box 513, NL-5600 MB Eindhoven, The Netherlands; [d]Stanford Synchrotron Radiation
Laboratory, Stanford, CA 94305, USA

Abstract: NSC KIPT proposed to construct a new X-ray and soft ultraviolet source
               NESTOR with a 40 - 225 MeV electron storage ring and Nd:YAG laser in
               Kharkov. NESTOR is a new type radiation source on the base of Compton
               scattering. In the paper the progress in development and construction of
               Kharkov X-ray generator NESTOR is presented. The current status of the
               main facility systems design and development are described. The facility is
               going to be in operation in the middle of 2007 and generated X-rays flux is
               expected to be of about 10^{13} phot/s.

Key words: Compton scattering, X-rays generator, magnets, vacuum system

1.   INTRODUCTION

Nowadays the sources of the X-rays based on a storage ring with low
beam energy and Compton scattering of intense laser beam are under
development in several laboratories. The paper is report of the supported by
SfP NATO Grant #977982 project of NESTOR (New-generation Electron
STOrage Ring) X-ray generator based on Compton scattering of an intense
laser beam on electron beam with low energy circulating in a storage

Vasili Tsakanov and Helmut Wiedemann (eds.), Brilliant Light in Life and Material Sciences, 89–93.
© 2007 *Springer.*

ring[1-5]. The layout of the facility is presented and the current status of the main facility systems design and development are described.

2. X-RAY PARAMETERS OF NESTOR FACILITY

Luminosity and brightness are the main characteristics of any light source. Table 1 shows luminosity and spectral brightness of NESTOR source in the main operation modes.

Table 1. NESTOR X-ray parameters

	Angiography	Biology	Hard X-ray
X-ray energy, keV	33	5-16	900
X-ray luminosity, phot/(mm^{2}s)	10^{15}	4×10^{13}	4×10^{14}-4×10^{15}
Spectral brightness, phot/(s mm^2 mrad^2 0.1%BW)	5×10^{12}	2×10^{11}	5×10^{12}-5×10^{13}

3. MAIN NESTOR FACILITY PARAMETERS

Radiation intensity and brightness are the main characteristics of any light source. Based on the strategy of maximum X-ray intensity with feasible parameters of technological systems the main NESTOR facility parameters were worked out (Table 2).

Table 2. The main NESTOR facility parameters.

Parameter	Value
Storage ring circumference, m	15.418
Electron beam energy range, MeV	40-225
Betatron tunes Q_x, Q_z	3.155; 2.082
Amplitude functions β_x, β_z at IP, m	0.14; 0.12
Linear momentum compaction factor α_1	0.01-0.078
RF acceptance, %	> 5
RF frequency, MHz	700
RF voltage, MV	0.3
Harmonics number	36
Number of circulating electron bunches	2; 3; 4; 6; 9; 12; 18; 36
Electron bunch current, mA	10
Laser flash energy into optical cavity, mJ	1
Collision angle, degrees	10; 150
Scattered photon energy (Nd laser, ε_{las} = 1.16 eV), keV	6-900
Spectral brightness, phot/(s mm^2 mrad^2 0.1%BW)	5×10^{12}-5×10^{13}

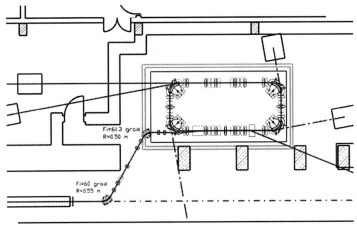

Figure 1. NESTOR facility layout.

4.   MAGNETIC LATTICE

The detail description of the magnetic lattice of the NESTOR storage ring is presented in[6,7]. The lattice corresponds to racetrack. Long straight section with IP is dispersion free while dispersion on opposite long straight section is non-zero. The designed lattice includes 4 dipole magnets with combined focusing functions, 20 quadrupole magnets and 19 sextupoles (9 with horizontal-vertical correction, 4 with octupole component of magnetic field). Such huge number of magnetic elements as for compact storage ring with low electron beam energy is dictated by very specific and contradictory requirements to RF acceptance value, beam size at the IP and chromatic effects strength. It is supposed the reference orbit correction system will be used 12 pick-up monitors and 9 correctors. The system will provide global reference orbit correction along circumference with maximal RMS reference orbit displacement 240 μ.

5.   MAGNETIC ELEMENTS OF NESTOR

NESTOR bending magnets have been manufactured. The bases for the new magnets were N-100 dipole magnets.

Quadrupole magnets of three different size-types but with similar design are under manufacturing. The main quadrupole coils are winded with cupper rectangular pipe (10×10 mm) with inner diameter of 7 mm. All qaudrupoles are connected in series with bending magnets. To meet parameters of quadrupoles and bending magnets additional low current coils are provided. NESTOR magnet parameters are listed in[5].

6.   INJECTION SYSTEM

It is supposed that on the first stage NESTOR will use the 60 MeV accelerator section with the main parameters are listed in[5].

Injection channel is "nearly parallel displacement" scheme with bending angles in line magnets $\varphi_{B1} = 60^0$ and $\varphi_{B2} = -61^0$. The final parallel displacement of the beam is fulfilled with inflector. As a base for the channel the classic 5 lenses scheme of parallel displacement was chosen for the reason of its flexible focusing properties. Such achromatic scheme can provide focusing with linear and angle enlargement factor values are equal near 1, so called "I" optics.

We have chosen D-F-D-F-D focusing variant for such scheme provides better beam focusing in the vertical plane. Since final part of the injection channel after parallel displacement (bending magnet fringe field, quadrupole and inflector) is a dispersion system the parallel displacement part of the injection channel has to compensate the dispersion produced with the final part. Single turn horizontal injection will be performed by means of three fast pulsed electromagnetic inflectors. Fast switches will regulate the voltage on the inflectors. The output of the switches will be matched to the complex load of the inflectors[4]. Two blocks of the inflectors will be based on existed two sections of N-100 inflector.

7.   VACUUM SYSTEM

The vacuum system of the storage ring NESTOR is intended for production and maintaining of average pressure $\leq 5 \times 10^{-9}$ *Torr* in all operation modes with storage current up to 1 A[8, 9].

To provide mentioned above pressure of residual gas both concentrated, and distributed pumping equipment where considered under design of NESTOR bending magnet vacuum chamber.

As the material for manufacturing of cameras leaf of a stainless steel 316L (according to classification of the USA) with parameters: thickness is 20 mm, contents of carbon of about 0.03 %, factor of a magnetic permeability $\mu_0 < 1.005$ (for inductions 0.1 T) was selected. It is supposed that each camera will be manufactured from two trough-shaped parts and welded with argon-arc welding.

8. SUMMARY

X-ray source NESTOR designing at NSC KIPT may be used for medical and biological studies, science of materials etc. X-rays over energy range $6\,keV \le \varepsilon_\gamma \le 900\,keV$ with long-term stable intensity up to 10^{13} phot /s can be generated under realizable parameters of the injector, storage ring and laser system. Maximum allowed Compton beam intensity limited by energy acceptance of the storage ring is approximately 10^{15} phot /s over all energy range.

REFERENCES

1. E. Bulyak et al, "Compton scattering in the 100 MeV Kharkov storage ring," PROC. PAC-99, June, 1999, New York, USA, v.2, pp. 3122-3124.
2. P. Gladkikh et al, "Lattice design for the compact X-ray source based on Compton scattering," Proc. Of EPAC-2000, Vienna, Austria, pp. 696-698.
3. E. Bulyak et al, "Compact X-ray Source Based on Compton Backscattering," NIM A, 2002, # 487, pp. 241-248.
4. A. Agafonof et al, "Spectral Characteristics of an Advanced X-ray Generator at the KIPT Based on Compton Back-Scattering", Proc. SPIE48th annual meeting, August, 2003, San Diego, USA.
5. V. Androsov et al, "Status of Kharkov X-ray Generator based on Compton Scattering NESTOR", Proc. of EPAC 2004, pp.2412-2414.
6. P. Gladkikh et al, "Lattice of NSC KIPT Compact Intense X-ray Generator NESTOR," EPAC-2004, Lucerne, Switzerland, pp. 1440-1442.
7. P. Gladkikh, "Lattice and beam parameters of compact intense x-ray sources based on Compton scattering", Physical Review Special Topics - Accelerators And Beams 8, 050702 (2005), 10p.
8. V. Anashin, "Vacuum systems of charged particle storage rings". BINP USSR, 5 Transactions of All-Union meeting on charged particle accelerators(boosters), Vol. I, "Science", Moscow, 1977, pp.273-276.
9. V.G. Grevtsev, A.Yu. Zelinsky, I.I. Karnaukhov, N.I. Mocheshnikov, The analysis and choice of the system for attaining vacuum in a 300 MeV electron storage ring. Problems of atomic science and technology. Series " Nuclear Physics Investigations " (41), N2, 2003, pp. 126-130.

PART 2

Research in Biophysics and Biochemistry

X-RAY ABSORPTION SPECTROSCOPY IN BIOLOGY AND CHEMISTRY

Graham N. George and Ingrid J. Pickering
Department of Geological Sciences, University of Saskatachewan, 114 Science Place, Saskatoon, SK S7N 5E2, Canada

Abstract: X-ray absorption spectroscopy has widespread applications to biology and chemistry. Here we review the principles of the technique and discuss some chemical and biological applications.

Key words: X-ray absorption spectroscopy, Chemistry, Biology, Metalloenzymes, *in-situ* probes, XAS imaging, X-ray microprobe.

1. INTRODUCTION

X-ray absorption spectroscopy (XAS) provides a powerful probe of both physical and electronic structure. As spectroscopic methods go it is a relatively new technique, originating in the early 1970's, and in recent years it has become increasingly applied to a very wide variety of fields. X-ray absorption spectra arise from core-level excitation by absorption of X-rays, and are thus associated with an absorption edge (*e.g.* 1s excitation for a K-edge – see Figures 1 and 2). They are usually separated into two different regions: the Extended X-ray Absorption Fine Structure or EXAFS, which occurs at energies higher than the absorption edge; and the near-edge region which consists of features before the major inflection, and any after the inflection which are not part of the analyzable EXAFS. XAS is element specific and can be used to investigate solids, liquids (including solutions), gaseous materials, and any mixtures thereof. It probes all of an element within a sample with moderate sensitivity, and is applicable to a very wide range of elements. XAS theory is well developed, and the methods of analysis of the EXAFS part of the spectrum to give a local radial structure

97

Vasili Tsakanov and Helmut Wiedemann (eds.), Brilliant Light in Life and Material Sciences, 97–119.

for the absorbing atom, are also well established. We will discuss the different information available from these different regions below. This review is not intended to be a comprehensive study of the literature, but instead seeks to illustrate the capabilities and scope of the technique in the hard X-ray energy regime with examples taken predominantly from the authors' own research.

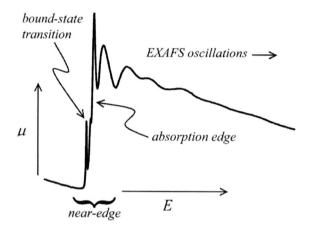

Figure 1. A typical X-ray absorption spectrum.

1.1   Experimental aspects

In an XAS experiment the absorption of an X-ray photon creates a core hole, which is then filled by decay of a higher level electron with concomitant emission of an Auger electron or a fluorescent X-ray photon (Figure 2). Thus, X-ray absorption can be monitored either directly by measuring the transmission of X-rays through the sample, or by measuring the X-ray fluorescence, or the electron yield. In the hard X-ray energy range (i.e. ≥ 5 keV) X-ray transmission is most useful for concentrated samples, while X-ray fluorescence has a much greater sensitivity and is useful for dilute compounds. XAS requires an intense, tunable source of X-rays and realistically it is only feasible using a synchrotron radiation source. Two decades ago this was a major limitation of the technique; however the current availability of modern synchrotron radiation facilities has gone a long way towards alleviating this problem. Figure 3 shows a typical XAS experimental setup, with a liquid helium cryostat to cool the sample and an X-ray fluorescence detector; typically this might be a solid state

Germanium array detector or a fluorescent ion chamber detector, which are used for different concentration ranges. Keeping the sample at a low temperature is important for two main reasons – firstly samples can often suffer damage from exposure to intense X-ray beams, and low temperatures help to minimize this. Secondly, the EXAFS part of the spectrum, which will be discussed below, can be much more intense and easier to detect and interpret, due to freezing out of thermal vibrations.

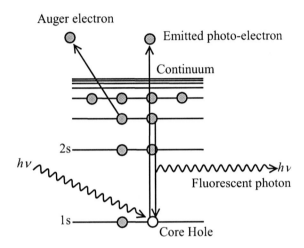

Figure 2. Diagram of the various processes which are associated with X-ray photo-absorption.

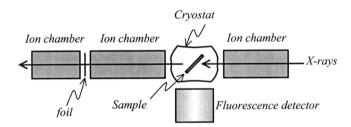

Figure 3. Schematic (plan view) of a typical XAS experiment.

1.2   Near-edge spectra

We noted above that XAS provides a probe of both electronic and physical structure. The features observed in near-edge spectra arise from predominantly dipole-allowed (i.e. $\Delta l = \pm 1$) transitions to bound state molecular orbitals, and analysis of near-edge spectra can thus yield information on electronic structure (i.e. oxidation state, and in some cases geometry).

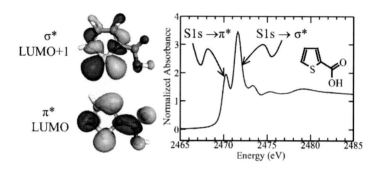

Figure 4. Transitions to bound states.

Figure 4 shows an example of a sulfur K near-edge spectrum of an organic sulfur compound. The near-edge spectrum shows two intense absorptions, which can be assigned as 1s→(S—C)π* and 1s→(S—C)σ* transitions, respectively. It is important to realize that the relative orbital energies measured by separation of features in the near-edge spectra correspond to excited states. The so-called $Z + 1$ approximation states that, because of the presence of a core-hole, the various transition energies correspond more closely to those of the element in the next group (i.e. atomic number increased by one). For the K-edges of first transition metal ions the smallest feature in the near-edge spectrum, the formally dipole-forbidden, quadrupole-allowed 1s→3d feature can be the most informative. This feature gains dipole intensity from admixture of metal *p*-orbitals in non-centrosymmetric environments and from a small but significant quadrupole-allowed (*i.e.* $\Delta l = 2$) cross-section. For L_{III} and L_{II} near-edge spectra, which arise from 2p excitation, the dipole selection rules mean that transitions to the d-manifold are intense. This has been applied to high-valent molybdenum complexes to yield estimates of the ligand field splittings of the 4d-manifold. For the first transition ions the L-edges are complicated by significant exchange interactions between the core and valence levels. Nevertheless, with appropriate analysis, important information upon electronic structure can be obtained from first transition element L-edge spectroscopy. Investigations of L-edge X-ray magnetic circular dichroism suggest that this related technique will be an additional probe of electronic structure.

Near-edge spectra are often highly distinctive, and in many cases can be used to fingerprint an unknown chemical species by a simple comparison with spectra of known compounds (usually referred to as model compounds). Complex mixtures can be quantitatively speciated by fitting the near-edge spectrum to the sum of the spectra of model compound spectra. A major limitation with this method is that one needs to be in possession of XAS data for models of all the species that are likely to be present. Nevertheless, it has been successfully applied in many cases.

Principal component analysis is becoming increasingly useful. Here a set of spectra from samples which have related compositions are analyzed to provide information on the minimum number of components that the set of spectra can represented by. We start with a set of m spectra each consisting of n data points, arranged in an array $\mathbf{A}$. These can be written using a single-value decomposition in which $\mathbf{A}$ is written as the product of an $m \times n$ matrix $\mathbf{U}$, an $m \times m$ diagonal matrix $\mathbf{V}$, and the transpose of an $m \times m$ orthogonal matrix $\mathbf{W}$. The matrix $\mathbf{U}$ contains the Eigenvectors or principal components, while $\mathbf{V}$ contains the Eigenvalues:

$$
\begin{pmatrix} a_{11} & a_{12} & a_{1m} \\ a_{21} & a_{22} & \\ a_{n1} & a_{n2} & a_{nm} \end{pmatrix} = \begin{pmatrix} u_{11} & u_{12} & u_{1m} \\ u_{21} & u_{22} & \\ u_{n1} & u_{n2} & u_{nm} \end{pmatrix} \bullet \begin{pmatrix} v_{11} & & \\ & v_{22} & \\ & & v_{mm} \end{pmatrix} \bullet \begin{pmatrix} w_{11} & w_{21} & w_{m1} \\ w_{12} & w_{22} & \\ w_{1m} & & w_{mm} \end{pmatrix} \quad (1)
$$

or

$$\mathbf{A} = \mathbf{U} \bullet \mathbf{V} \bullet \mathbf{W}^t \qquad (2)$$

It is important to remember that the principal components thus obtained are not the component spectra but instead are mathematical constructs. The significance of each individual principal component is then tested by setting the respective Eigenvalue v_{ii} to zero, reconstructing the original data set $\mathbf{A}$, and comparing it with the original. Figure 5 shows an example of principal component analysis where five spectra are subjected to single value decomposition.

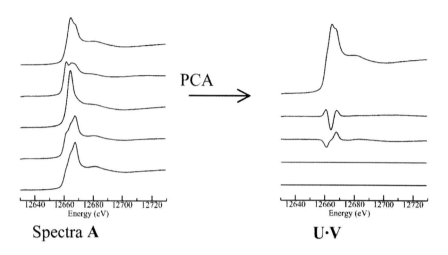

Spectra A   U·V

Figure 5. Principal component analysis (PCA) of a set of five spectra **A** (left panel). The analysis, shown by the scaled Eigenvectors **U·V** (right panel), indicates that only three components are needed to reproduce the entire set of five spectra.

Once the number of principal components has been determined a technique called Target transformation can be used to test spectra of candidate models. The Target transform **T*** of a candidate spectrum **T** is computed from the Eigenvector matrix obtained from principal component analysis as follows:

$$
\begin{pmatrix} t_1^* \\ t_2^* \\ \vdots \\ t_n^* \end{pmatrix} = \begin{pmatrix} u_{11} & u_{12} & & u_{1m} \\ u_{21} & u_{22} & & \\ & & & \\ u_{n1} & u_{n2} & & u_{nm} \end{pmatrix} \bullet \begin{pmatrix} u_{11} & u_{21} & & u_{n1} \\ u_{12} & u_{22} & & \\ & & & \\ u_{1m} & u_{2m} & & u_{nm} \end{pmatrix} \bullet \begin{pmatrix} t_1 \\ t_2 \\ \vdots \\ t_n \end{pmatrix} \tag{3}
$$

Comparison of the **T*** with **T** tests whether the candidate spectrum can be reproduced from the matrix **U** obtained from principal component analysis. Figure 6 shows two alternative target transforms, using **U** obtained from figure 5, one (A) where the spectrum tested is clearly present (**T-T*** being essentially zero) and the other (B) where it is clearly not present (**T-T*** showing a significant miss-match).

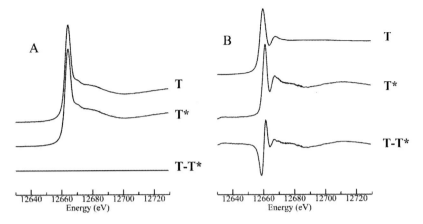

Figure 6. Target transforms; in A T is part of the set, and in B T is not part of the set.

Quantitative analysis of near-edge spectra in structural terms has been approached by two different methods. The first is the use of molecular orbital theory (e.g. figure 4) and the second is the use of full multiple scattering calculations which sum large numbers of photo-electron scattering pathways to simulate the near-edge. While success has been reported for both methods, neither is at present routinely used for determination of structural information (although it may be in the future), and for this most workers turn to the EXAFS. We will now discuss the basic physics and the analysis of EXAFS spectra.

1.3   EXAFS spectra

At X-ray energies below an X-ray absorption edge, the incident X-rays have insufficient energy to eject a core electron, and above the edge the electron leaves the atom possessing a kinetic energy which is a function of the difference in the X-ray energy E above the absorption edge threshold energy E_0. For a photoelectron of de Broglie wavelength $\lambda_e = h/m_e v$ (where h is Planck's constant, m_e the electron rest mass, and v the electron velocity), we can define the wave-vector $k = 2\pi/\lambda_e$. The photoelectron kinetic energy is $(E - E_0) = m_e v^2/2$, and k is thus given by equation (4).

$$k = \sqrt{\frac{2m_e}{\hbar}(E - E_0)} \qquad (4)$$

We now consider the emitted photoelectron leaving the absorber atom as a photoelectron wave. For a K-edge (i.e. 1s excitation) this will have p-symmetry, which will confer polarization sensitivity to the EXAFS. This photoelectron wave will be scattered by any neighboring atoms. Quantum mechanical analysis tells us that the X-ray absorption coefficient will be at a maximum when the backscattered wave is in phase with the outgoing wave at the nucleus of the absorber, and at a minimum when it is out of phase. This can be considered as being due to constructive and destructive interference, respectively. Thus, as we increase the X-ray energy E above the edge, λ_e will decrease, and the absorption will go through successive minima and maxima. It is this oscillatory absorption which is known as the EXAFS. The frequency of the EXAFS depends upon the atomic separation, and the amplitude of the EXAFS, which is a reflection of the effectiveness of the backscatterer atoms, upon the size (atomic number) and the number of backscatterer atoms. Because the backscatterer atom modulates the phase of the photoelectron, the phase of the EXAFS will also depend upon the nature of the backscatterer. Thus, both phase and amplitude of the EXAFS depend on the backscatterer, and this allows identification of the backscatterer. In general it is not possible to discriminate atoms with similar atomic number (*e.g.* sulfur and chlorine give nearly identical EXAFS, while sulfur and oxygen are quite different).

The EXAFS $\chi(k)$ is described by the simple equation which is summed over all EXAFS absorber backscatterer pairs i :

$$\chi(k) = \sum_i \frac{N_i S_0^2 \Im\{\delta_i^c(k)\} f_{eff}(\pi, k, R)}{k R_i^2} e^{-2\sigma_i^2 k^2} e^{-2R_i/\lambda(k)}$$
$$\times \sin\left[2kR_i + 2\Re\{\delta_i^c(k)\} + \Phi_{eff}(k)\right] \tag{5}$$

In which N is coordination number, R is the absorber-backscatterer distance, σ^2 is the Debye-Waller factor – the mean-square deviation in R, δ^c the central-atom phase shift, Φ_{eff} the backscatterer phase-shift, f_{eff} the effective curved-wave backscattering amplitude, $\lambda(k)$ is the photoelectron mean free path, and S_0^2 is a many-body amplitude reduction factor. This simple formalism is applicable to multiple scattering by simply considering i as the index of a scattering path, in which case N is then the degeneracy of each multiple scattering path. We can simplify the expression further by combining variables into a total amplitude function $A(k,R)$ and a total phase function $\varphi(k)$:

$$\chi(k) = \sum_i \frac{N_i A_i(k,R)}{kR_i^2} e^{-2\sigma_i^2 k^2} e^{-2R_i/\lambda(k)} \sin[2kR_i + \varphi_i(k)] \tag{6}$$

The EXAFS spectra are extracted from the raw XAS spectra by what are now standard techniques. It is common practice to weight the EXAFS by k^3 (or less often by k^1 or k^2) in order to counteract the damping of the spectra caused by thermal vibrations, and from the fall-off in scattering amplitude. In addition, the theory used to analyze the spectra is more accurate at higher k (increasing approximately as k^1). EXAFS spectra are usually Fourier transformed in order to visualize the frequency contributions to the EXAFS:

$$\rho(R) = \frac{1}{4\pi^{1/2}} \int_{k_{min}}^{k_{max}} \chi(k) k^3 e^{i2kR} dk \tag{7}$$

The results of the Fourier transform are usually displayed as the transform magnitude, $|\rho(R)|$ verses R:

$$|\rho(R)| = \sqrt{\Re\{\rho(R)\}^2 + \Im\{\rho(R)\}^2} \tag{8}$$

The Fourier transform shows peaks with positions corresponding to the interatomic distances plus half the average phase-shift slope (usually this is about -0.3 Å). It is common practice to phase-correct Fourier transforms. This is done using an EXAFS phase function, typically a theoretical one, for the dominant absorber-backscatterer interaction in the EXAFS. The effect of phase-correcting the transform is to move the peaks to values close to the actual absorber-backscatterer distances (R_i). In the absence of any phase correction the peaks will be at positions $R+\Delta$, where Δ is approximately half the average phase-shift slope for a given interaction. In the absence of phase-correction, peak shapes are often asymmetric due to structure in the phase-shift function, and with heavy backscatterers multiple peaks can be present. The phase-corrected Fourier transform can be computed using eqn 9, and Figure 7 illustrates the effects of phase-correction on the Fourier transform.

$$\rho(R) = \frac{1}{4\pi^{1/2}} \int_{k_{min}}^{k_{max}} \chi(k) k^3 e^{i2kR + i\varphi(k)} dk \tag{9}$$

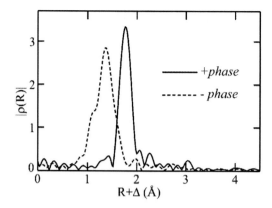

Figure 7. Effects of EXAFS Fourier transform phase-correction. The EXAFS spectrum of Na_2MoO_4 was transformed between $k=1$-18 Å^{-1} with and without Mo—O phase-correction

Strictly speaking, phase-correction is only valid in systems with just one kind of backscatterer. This is because only a single $\varphi(k)$ can be used (typically for the EXAFS with the largest backscattering amplitude), and thus the other interactions will be miss-matched. In practice the use of phase-correction in these cases still generally yields much clearer looking transforms, but the abscissa is usually labeled $R+\Delta$, to indicate the presence of a partial miss-match.

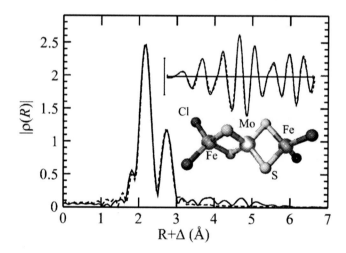

Figure 8. Mo K-edge EXAFS Fourier transform (phase-corrected for Mo—S backscattering) of [Cl2FeS2MoS2FeCl2]2- (structure shown in inset). The intense peak at ~2.2 Å arises from the first-shell of Mo—S ligands, while the smaller peak at ~2.8 Å arises from the more distant Mo·Fe EXAFS. The k3-weighted EXAFS is shown in an inset.

Figure 8 shows the phase-corrected Fourier transform of an EXAFS spectrum containing two major components in the EXAFS. In practice, the EXAFS Fourier transform is not usually interpreted directly (although some analysis packages do analyze data in Fourier space). Instead, a radial structural model is fit to the extracted experimental EXAFS oscillations, typically using *ab initio* theory to approximate the model, employing what is commonly called EXAFS curve-fitting analysis to determine values for N, R and σ^2. N and σ^2 are typically highly correlated in the refinement and because of this these can be determined with an accuracy of around $\pm 20\%$. Similarly, while ligand identity can to an extent be established from the characteristic $A(k,R)$ and $\varphi(k)$, miss-assignment of ligands is quite possible and occasionally occurs with data of limited k-range. A common value for a shift in E_0 (or ΔE_0) between computed and experimentally measured EXAFS is also often refined. With modern theory such as FEFF a common ΔE_0 value for all components can be co-refined, but the wise experimenter will be guided by values from model compounds, particularly if data with only limited k-range are available. A major strength of EXAFS is the accuracy with which it can determine R, the average inter-atomic distance for a given type; for directly coordinated ligands this is approximately ± 0.02 Å. The EXAFS resolution ΔR is the minimum distance between similar ligands that can be discerned; this is directly related to the extent of the data and is approximately given by $\Delta R \approx \pi/2k$. For typical data ranges ΔR is rather poor at around 0.15 Å, and this is a limitation of EXAFS.

Multiple scattering occurs when the photo-electron is scattered by more than a single atom before returning to the absorber atom. Appropriate analysis can in some cases give information on geometry, particularly when linear or close to linear arrangements of atoms are present. Multiple scattering paths are usually referred to by the number of legs in the multiple scattering pathway, for example with single scattering this is just two. It is important to include all relevant multiple scattering paths for a given atomic arrangement. For an arrangement of three atoms multiple scattering with up to four legs is important, while for four atoms multiple scattering with up to six legs must be used. For linear arrangements of two backscatterer atoms with similar size, the multiple scattering EXAFS from the three-leg path will be approximately 90° out of phase with the single scattering (two-leg) path, while that from the four-leg path will be approximately 180° out of phase with the single scattering path (Figure 9).

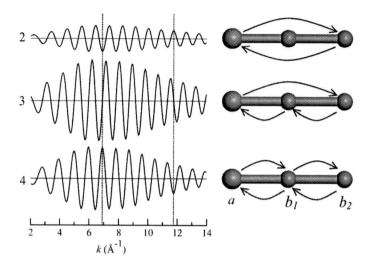

Figure 9. EXAFS Multiple Scattering. Computed multiple scattering paths are shown for a hypothetical linear arrangement of equally spaced atoms.

1.4   XAS imaging

Biological samples are, by their very nature, highly structured, and XAS imaging is providing important new insights into these materials. Here a micro-focused X-ray beam is used to illuminate the sample which is raster scanned while the X-ray fluorescence is monitored, so that an image or map of the sample is built up. Methods of quantitative analysis have been described[5].

2.   APPLICATIONS TO BIOLOGY AND CHEMISTRY

The scientific areas in which XAS has been employed are very diverse; areas of application encompass the biological, chemical, physical and many other branches of sciences. As a complete review of all applications in these fields is impossible, we will illustrate the use of the technique by example and will focus on examples from our own research. In general there is some overlap between fields, for example many biological studies have a strong chemical component, and all depend upon the physics of X-ray absorption. We will discuss chemical applications first.

2.1 Applications in the chemical sciences

Applications of XAS to the chemical sciences are numerous, and can again be divided by category. One broad category is structural studies of materials which, for one reason or another, are not amenable to X-ray crystallography. This may be because they are amorphous or liquid, or perhaps because they are kinetic transients in a chemical reaction. A good example of the first type is provided by studies of catalysts in which XAS has provided important information since the earliest days of the technique. XAS has also been extensively used in studies of the chemical nature of sulfur in coals and other heavy hydrocarbon fuels. Some of this work overlaps significantly with the environmental sciences.

2.1.1 Fuel chemistry

The chemistry of hydrocarbon fuels has been studied by XAS since the early 1980's. Of particular interest is sulfur in heavy hydrocarbons and in coal[6,7]. Early work showed that sulfur K-edge XAS could be used to understand the chemical forms in coal, but that precautions against air exposure needed to be taken for the results to make sense[8]. Studies of fuel additives such as zinc dialkyldithiophosphates (ZDDP) have yielded important information on the structures of these metal complexes *in situ* (Figure 10)[9].

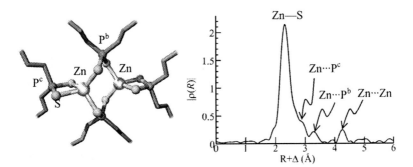

Figure 10. Structural model (left, long alkyl side-chains truncated) and EXAFS Fourier transform (right) of commercial ZDDP dissolved in "150 solvent neutral" base oil. The EXAFS clearly indicates a binuclear cluster in solution.

2.1.2   Environmental chemistry

There has been some change in emphasis in environmental chemistry in recent years, and the field is now predominantly focused towards anthropogenic environmental problems. It is in this area where XAS has been applied the most, although much important work is still done upon natural systems. Examples of environmental applications include selenium K-edge XAS studies of selenium contaminated soils and sediments from the Kesterson reservoir in central California[10], and studies of actinide-contaminated radioactive waste[11]. XAS has been used to study how heavy elements move through the food chain, from plants into animals and then into their predators. In particular selenium K-edge XAS was used to study alfalfa grown on selenate enriched media, the leaves of which were eaten by the beet army worm *Spodoptera exigua*, which in turn was preyed upon by the parasitoid wasp *Cotesia marginiventris*. This study showed that the chemical form of selenium changed (becoming increasingly methylated) as it moved up through the tropic levels of this compact model ecosystem[12]. As we will see, this work overlaps somewhat with the whole tissue XAS discussed below.

2.1.3   Physical chemistry

Here we will briefly consider some modifications of the technique of X-ray absorption spectroscopy for which new physical methods have been or are being developed.

Grazing incidence XAS (GIXAS) exploits the evanescent wave propagated in the plane of a sample when the incident X-ray beam is under conditions of total external reflection. It can yield information on the chemical nature of a surface[13], and has important applications in a number of areas, including the materials sciences and environmental sciences. X-ray Magnetic Circular Dichroism (XMCD) spectroscopy has been used in recent years in studies of materials[14]. In particular spin-polarized EXAFS (SPEXAFS)[15] shows promise as a combined probe of physical and electronic structure, in its sensitivity to magnetically coupled neighbor atoms. High-resolution fluorescence also shows much promise[16]. In XAS detected by X-ray transmittance, the line-widths of spectral features are governed by the lifetime of the core-hole. For X-ray fluorescence-detected XAS, if the X-ray fluorescence is measured with high resolution then the spectral line-widths are governed by the lifetime of the hole created by decay of the outer electron to fill the initial core-hole (the same goes for XAS measured by monitoring Auger emission). Thus, one can obtain very high resolution near-edge spectra, allowing resolution of features not normally visible[16]. Finally, Diffraction Anomalous Fine Structure is a promising technique in which the fine structure of X-ray diffraction in the vicinity of an absorption is used to effectively provide site-selective XAS information on complex samples[17].

2.1.4   Inorganic chemistry

XAS is not generally considered a useful analytical tool by most Inorganic Chemists, and thus it has been little applied in general Inorganic Chemistry. The reason for this is clearly that it less readily available than other spectroscopic tools commonly employed by Chemists. A good example of an application of XAS in Inorganic Chemistry is the investigation of the structure of a novel low-temperature intermediate of N_2 reduction. Cummins and co-workers reported a Mo(III) species that reacted 2:1 with N_2 to form a Mo(VI) nitrido (Mo≡N) species via an intense purple intermediate that was stable only at low temperatures[18]. EXAFS was used to show that this species contained a novel Mo—N=N—Mo core with a nearly linear arrangement of atoms. Multiple scattering analysis allowed measurement of the N=N distance in the intermediate, and provided vital information for understanding the nature of this chemically important intermediate[19].

2.2   Applications in the biological sciences

Applications in the biological sciences can be broadly divided into two areas. The first is studies of purified proteins, usually metalloproteins, with a view to understanding the structure-function relationships that are the cornerstone of biochemistry. The second area is studies of whole tissues or cultures, with a view to understanding gross chemical changes in the sample. We will discuss each of these applications in turn.

2.2.1   Metalloprotein XAS

Metalloprotein XAS was among the very first applications of the technique. Early workers sought to exploit the high accuracy of EXAFS-determined bond lengths to provide additional structural information on crystallographically characterized samples. In fact, the very first EXAFS study of a biological molecule, that of rubredoxin, pointed out errors in the bond-lengths determined from crystallography[20]. Later crystal structures at slightly higher resolution were in agreement with the EXAFS results[21].

While the role of XAS in providing ancillary information to crystallography is increasingly important, a large number of metalloprotein EXAFS applications are studies of systems for which we have little or no structural information. An excellent early example of this is provided the enzyme nitrogenase. This enzyme is the key to the global nitrogen cycle and catalyzes the six-electron reduction of the very stable N_2 molecule to ammonia. Very early EXAFS unambiguously indicated that the active site of this important protein was a novel molybdenum-iron-sulfur cluster[22]; a possibility which had not previously been considered, and which was at that time without chemical precedent. Later crystal structures revealed a novel elliptically shaped molybdenum-iron-sulfur cluster with the molybdenum positioned near one end[23]. Returning to comparisons with crystallography, another early success of metalloprotein XAS concerned Fe_3S_4 clusters, although at the time the XAS conclusions were widely disbelieved. Early crystal structures of the Fe_3S_4 cluster-containing *Azotobacter vinelandii* ferredoxin I produced what was at the time a quite startling structure for the three-iron active site (Fig. 11A)[24]. This was a Fe_3S_3 cluster with an open, almost flat, twisted-boat conformation six-membered ring with alternating iron and sulfide. The Fe··Fe separation of 4.2 Å and the structural type was quite unprecedented, and the structure generated considerable excitement, representing a fundamental new kind of iron-sulfur cluster. However, the crystal structure was incompatible with the results of resonance Raman spectroscopy[25], and furthermore it did not agree with the conclusions

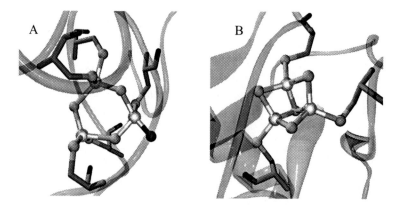

Figure 11. Crystallographic structures of the three-iron site of Azotobacter vinelandii ferredoxin I. A shows the initial structure obtained with an incorrect point group (the enantiomer of the actual point group), and B shows the correct structure obtained later.

of XAS which indicated a 2.7 Å Fe··Fe separation, typical of more conventional structural types[26]. Unfortunately, the spectroscopic data was widely disbelieved[27], and only later did a re-evaluation of the crystallography uncover errors in the original work[28,29], yielding a structure fully consistent with all spectroscopic data (Fig 11B), and that had previously been proposed from spectroscopy[27].

Our final example of XAS providing corrections to crystallography is provided by the molybdenum-containing enzyme DMSO reductase. In the interests of brevity we will focus only on the structures of the oxidized Mo(VI) enzyme. The first structural information on the active site came from EXAFS[30] which indicated a six-coordinate metal site, with four sulfurs from two pyranopterin-dithiolene cofactors, one Mo=O ligand, and one longer oxygen interpreted as a coordinated serine. Subsequent protein crystallography from three different groups yielded active site structures that were at odds with both the EXAFS and with each other[31-33]. Later re-evaluation of the EXAFS, plus critical examination of the active site structures proposed from crystallography indicated that the original EXAFS conclusions were correct, and that the active site structures proposed from crystallography were either chemically impossible or unlikely[35]. Subsequent protein crystallography at very high resolution revealed the likely problem – that the crystals contained two different active structures each with fractional occupancy in the crystal; an inactive one, and an active one[36]. The latter closely resembled the structure suggested by the original EXAFS study, and thus our understanding of the molybdenum coordination of this enzyme came almost full circle, returning close to the point at which the original EXAFS work left it.

Any comparison of this nature would not be complete without a mention of cases where the analysis of the EXAFS was misleading, and crystallography provided the definitive information on the active site structure. Ni-Fe hydrogenases are bacterial enzymes which have a short Ni··Fe inter-atomic distance, but despite extensive XAS studies[37] this key feature of the active was not detected until it was observed crystallographically[38,39]. Another example is given by the Cu$_A$ site of cytochrome oxidase. Here again Cu$_A$ is a binuclear cluster with a short metal-metal inter-atomic distance (~2.45 Å). In fact Cu K-edge EXAFS detected this interaction, and while a short Cu··metal interaction was considered, it was dismissed as unlikely both on chemical grounds and because of ambiguities in the number of coppers in cytochrome oxidase. Hence it was miss-interpreted as a long Cu—S interaction[40]. In this case also, the discovery of a novel short Cu···Cu inter-atomic distance came as a surprise when the crystal structure was reported[41]. In both of these cases the available *k*-range of the EXAFS was truncated by a nearby absorption edge (in the Ni case by Cu, and in the Cu case by Zn) and this was a contributing factor in the miss-assignment of EXAFS that fooled the researchers. A major point here is that EXAFS analysis is driven by the structural models being considered – unless one has the correct model, it is often very hard to arrive at the correct answer.

Finally, there are some systems for which structural information on the metal binding site is not available from crystallography. An interesting example of this comes from the copper metallochaparones. Metallo-chaperones are proteins that can be thought of as molecular taxi-cabs; they carry metal ions with great specificity within the cell to a particular destination molecule. They must bind the metal ion with great tenacity, but must also readily give it up upon arrival at the destination. Copper metallochaparones have been extensively studied by X-ray crystallography, but despite considerable effort, crystals containing the native metal ion are almost impossible to make. This is generally because the process of crystallization triggers the chaparones to give up their bound metal ions. Thus, techniques such as XAS are the primary tool for understanding the metal ion coordination in these cases[42]. Particularly elegant has been the combination of XAS and NMR spectroscopy to build up a detailed picture of the solution structure of copper metallochaparones[43].

2.2.2   XAS of intact tissues

The application of XAS to the study of whole tissues or cultures is a relatively new development. Examples of such studies include uptake and transformation of heavy metals in plants, and the biochemistry of sulfur in blood. These studies typically exploit the sensitivity of the near-edge spectrum to chemical type. As mentioned above, chemical quantification is to some extent possible by fitting experimental data to a linear combination of model compound spectra. The study of whole tissues by XAS has recently been revolutionized by improvements in sensitivity of the technique due to advances in detector technology and beamline design. Thus, until fairly recently the accepted low concentration limit for most samples was about 1 mM, but concentrations of 1 µM are now possible. This allows many metals and metalloids to be studied at physiological levels. An example is a recent investigation of the chemical form of mercury in marine fish[44]. Examination of Hg L_{III} near-edge spectra of different species of fish otherwise intended for human consumption indicated that they contained methylmercury coordinated to an aliphatic thiol. This was not a surprising result because of mercury's well-known affinity for thiols, nevertheless an *in-situ* probe is the only way that this information can be recovered, as conventional analysis destroys the chemical form present *in vivo*.

XAS imaging studies have been performed in both biological and environmental studies. As we have discussed above, the sample is raster-scanned with a small X-ray beam (*ca.* 2-100 μm in size) so that different spatial regions are interrogated, while the X-ray fluorescence is monitored. Regions of particular interest can then be investigated further by recording an XAS spectrum, which is often referred to as micro-XAS. A recent example is shown in figure 12 where As K-edge XAS imaging was used to monitor details of how the hyperaccumulating fern *Pteris vittata* transports arsenate and transforms it to arsenite[45].

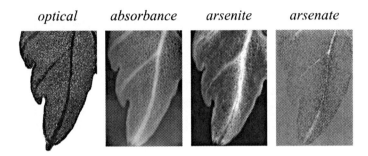

optical   absorbance   arsenite   arsenate

Figure 12. XAS imaging of a leaf-tip of *Pteris vittata* showing arsenate in the transport vessels and arsenite in the leaf tissue. The horizontal dimension is approximately 1mm.

Finally, we relate an example where XAS was used to study an obscure phenomenon in chemical toxicology resulting in a discovery which may dramatically improve the lives of tens of millions of people. The two most toxic environmentally common oxy-anions of arsenic and selenium are arsenite and selenite, respectively. When administered in isolation these have approximately equal toxicities, but when given together their toxic effects cancel. The molecular basis of this surprising phenomenon was uncovered by As and Se K-edge XAS. A novel arsenic-selenium compound – the seleno-bis (S-glutathionyl) arsinium ion – is formed in erythrocytes, and subsequently excreted in the bile[46, 47].

In Bangladesh low-levels of arsenic contamination of drinking water have caused health problems on a massive scale. In what has been called the world's worst mass poisoning, some 35-80 millions are affected. There are several curious things about the Bangladeshi poisoning. Firstly, there are other locations in the world that have higher arsenic in drinking water but where there are no apparent associated health problems, and secondly in Bangladesh the poisoning is curiously selective, affecting some and sparing others in a community where all are consuming the same contaminated water. Now, selenium is an essential element, being important in antioxidant biochemistry (for example at the active site of glutathione peroxidases). Selenium deficiency can result in symptoms that are very similar to those experienced by Bangladeshi sufferers of chronic arsenic poisoning. Excretion of a molecular entity containing both arsenic and selenium would inevitably deplete the body's reserves of essential selenium and in cases where the dietary intake is close to inadequate this might cause a selenium deficiency. Because dietary selenium intake varies between individuals then this would explain why only some individuals are stricken. The regions of Bangladesh afflicted with arsenic poisoning in the water are indeed very low in dietary selenium, and those parts of the world with high arsenic water but no arsenicosis have ample dietary selenium. Thus, the pathology experienced by those suffering from chronic low-level arsenic poisoning in Bangladesh may in fact be a selenium deficiency[46], and administering selenium supplements might provide an effective treatment. If indeed this turns out to be the case occurring, then this is an excellent example of how fundamental research with synchrotron radiation can make a major impact upon a human population.

REFERENCES

1. I. Iwasaka, (Ed.) X-ray Absorption Fine Structure for Catalysts and Surfaces, World Scientific Press, Singapore (1996).
2. D. C. Koningsberger and R. Prins (Eds.) X-ray Absorption. Principals, Applications, Techniques of EXAFS, SEXAFS, and XANES, John Wiley and Sons (1988).
3. E. I. Stiefel and G. N. George, in Ferredoxins, Hydrogenases and Nitrogenases : Metal-Sulfide Proteins in Bioinorganic Chemistry (Eds. I. Bertini, H. B. Gray, S. J. Lippard, and J. Valentine) University Science Books (1994), Mill Valley, California pp 365-453.
4. J. Stöhr, NEXAFS Spectroscopy, Springer-Verlag (1992).
5. I. J. Pickering, R. C. Prince, D. E. Salt, and G. N. George, Quantitative, chemically specific imaging of selenium transformation in plants. Proc. Natl. Acad. Sci. USA. 97(20), 10717-10722 (2000).
6. C. L. Spiro, J. Wong, F. W. Lytle, R. B. Greegor, D. H. Maylotte, and S. H. Lamson, X-ray absorption spectroscopic investigation of sulfur sites in coal: organic sulfur identification. Science, 226(4670), 48-50 (1984).
7. G. N. George and M. L. Gorbaty, Sulfur K-edge X-ray absorption spectroscopy of petroleum asphaltenes and model compounds. J. Am. Chem. Soc. 111(9), 3182-3186 (1989).
8. G. N. George, M. L. Gorbaty, S. R. Kelemen and M. Sansone. Direct determination and quantification of sulfur forms in coals from the Argonne premium sample program. Energy & Fuels 5(1), 93-97 (1991).
9. G. N. George and M. Sansone, unpublished observations.

10. I. J. Pickering, G. E. Brown Jr., and T. K. Tokunaga. Quantitative speciation of selenium in soils using X-ray absorption spectroscopy. Environ. Sci. Technol., 29(9), 2456-2459 (1995).

11. J. G. Catalano, J. P. McKinley, J. M. Zachara, S. M. Heald, S. C. Smith, and G. E. Brown Jr. Changes in uranium speciation through a depth sequence of contaminated Hanford sediments. Environ. Sci. Technol. 40(8), 2517-2524 (2006).

12. V. B. Vickerman, J. T. Trumble, G. N. George, I. J. Pickering, and H. Nichol, Selenium biotransformations in an insect ecosystem: effects of insects on phytoremediation. Environ. Sci. Technol. 38(13), 3581-3586 (2004).

13. H.-M. Christen, K. S. Harshavardhan, M. F. Chisholm, E. D. Specht, J. D. Budai, D. P. Norton, L. A. Boatner and I. J. Pickering, The effect of size, strain, and long-range interactions on ferroelectric phase transitions in KTaO3/KNbO3 superlattices studied by X-ray, EXAFS, and dielectric measurements. Electroceramics, 4(2/3), 279-287 (2000).

14. T. Funk, A. Deb, S. J. George, H. Wang and S. P. Cramer, X-ray magnetic circular dichroism - a high energy probe of magnetic properties. Coord. Chem. Rev. 249(1-2), 3-30 (2005).

15. G. Schütz, R. Frahm, P. Mautner, R. Wienke, W. Wagner, W. Wilhelm, and P. Kienle, Spin-dependent extended x-ray-absorption fine structure: Probing magnetic short-range order. Phys. Rev. Lett. 62(22-29), 2620-2623 (1989).

16. P. Glatzel and U. Bergmann, High resolution 1s core hole X-ray spectroscopy in 3d transition metal complexes - electronic and structural information. Coord. Chem. Rev. 249(1-2), 65-95 (2005).

17. I. J. Pickering, M. Sansone, J. Marsch and G. N. George, Diffraction anomalous fine structure: a new technique for probing local atomic environment. J. Am. Chem. Soc. 115, 6302-6311 (1993).

18. C. E. Laplaza, and C. C. Cummins. Dinitrogen cleavage by a three-coordinate molybdenum(III) complex. Science, 268(5212), 861-863 (1995).

19. C. E. Laplaza, M. J. A. Johnson, J. C. Peters, A. L. Odom, E. Kim, C. C. Cummins, G. N. George, and I. J. Pickering, Dinitrogen cleavage by three-coordinate molybdenum(III) complexes: mechanistic and structural data. J. Am. Chem. Soc., 118(36), 8623-8638 (1996).

20. R. G. Shulman, P. Eisenberger W. E. Blumberg, and N. A. Stombaugh. Determination of the iron-sulfur distances in rubredoxin by x-ray absorption spectroscopy. Proc. Natl. Acad. Sci. USA, 72(10), 4003-4007 (1975).

21. K. D. Watenpaugh, L. C. Sieker, and L. H. Jensen, Crystallographic refinement of rubredoxin at 1.2 Å resolution. J. Mol. Biol., 138(3), 615-633 (1980).

22. S. P. Cramer, K. O. Hodgson, W. O. Gillum, and L. E. Mortenson, The molybdenum site of nitrogenase. Preliminary structural evidence from X-ray absorption spectroscopy. J. Am. Chem. Soc., 100(11), 3398-3407 (1978).

23. O. Einsle, F. A. Tezcan, S. L. A. Andrade, B. Schmid, M. Yoshida, J. B. Howard, and D. C. Rees, Nitrogenase MoFe-protein at 1.16 Å resolution: A central ligand in the FeMo-cofactor. Science, 297(5587), 1696-1700 (2002).

24. D. Ghosh, S. O'Donnell, W. Furey Jr., A. H. Robbins, and C. D. Stout. Iron-sulfur clusters and protein structure of Azotobacter ferredoxin at 2.0 Å resolution. J. Mol. Biol., 158(1), 73-109, (1982).

25. M. K. Johnson, R. S. Czernuszewicz, T. G. Spiro, J. A. Fee, and W. V. Sweeney Resonance Raman spectroscopic evidence for a common [3-iron-4-sulfur] structure

among proteins containing three-iron centers J. Am. Chem. Soc. 105(22), 6671-6678 (1983).

26. H. Beinert, M. H. Emptage, J. L. Dreyer, R. A. Scott, J. E. Hahn, K. O. Hodgson and A. J. Thomson. Iron-sulfur stoichiometry and structure of iron-sulfur clusters in three-iron proteins: evidence for [3Fe-4S] clusters. Proc. Natl. Acad. Sci. USA., 80(2), 393-396 (1983).

27. G. N. George and S. J. George, X-ray crystallography and the spectroscopic imperative: the story of the [3Fe-4S] clusters. Trends. Biochem. Sci. 13(10), 369-370 (1988).

28. G. H. Stout, S. Turley, L. Sieker, and L. H. Jensen, Structure of ferredoxin I from Azotobacter vinelandii. Proc. Natl. Acad. Sci. USA., 85(4), 1020-1022 (1988).

29. C. D. Stout, 7-Iron ferredoxin revisited. J. Biol. Chem., 263(19), 9256-9260 (1988).

30. G. N. George, J. Hilton, and K. V. Rajagopalan, K.V. X-ray Absorption spectroscopy of dimethylsulfoxide reductase from Rhodobacter sphaeroides. J. Am. Chem. Soc. 118(5), 1113-1117 (1996).

31. H. Schindelin, C. Kisker, J. Hilton, K. V. Rajagopalan, and D. C. Rees. Crystal structure of DMSO reductase: redox-linked changes in molybdopterin coordination. Science, 272 (5268), 1615-1621 (1996).

32. F. Schneider, J. Loewe, R. Huber, H. Schindelin, C. Kisker, J. Knaeblein. Crystal structure of dimethyl sulfoxide reductase from Rhodobacter capsulatus at 1.88 Å resolution. J. Mol. Biol., 263(1), 53-69 (1996).

33. A. S. McAlpine, A. G. McEwan, A. L. Shaw, and S. Bailey. Molybdenum active center of DMSO reductase from Rhodobacter capsulatus: crystal structure of the oxidized enzyme at 1.82 Å resolution and the dithionite-reduced enzyme at 2.8 Å resolution. J. Biol. Inorg. Chem., 2(6), 690-701 (1997).

34. A. S. McAlpine, A. G. McEwan, and S. Bailey, The high resolution crystal structure of DMSO reductase in complex with DMSO. J. Mol. Biol., 275(4), 613-623 (1998).

35. G. N. George, J. Hilton, C. Temple, R. C. Prince, and K. V. Rajagopalan. The Structure of the Molybdenum Site of Dimethylsulfoxide Reductase. J. Am. Chem. Soc. 121(6), 1256-1266 (1999).

36. H-K. Li, C. Temple, K. V. Rajagopalan, and H. Schindelin, The 1.3 Å crystal structure of Rhodobacter sphaeroides dimethyl sulfoxide reductase reveals two distinct molybdenum coordination environments. J. Am. Chem. Soc., 122(32), 7673-7680 (2000).

37. C. Bagyinka, J. P. Whitehead, and M. J. Maroney, An x-ray absorption spectroscopic study of nickel redox chemistry in hydrogenase. J. Am. Chem. Soc. 115(9), 3576-3585, (1993).

38. A. Volbeda, M-H., Charon, C. Piras, E. C. Hatchikian, M. Frey, and J. C. Fontecilla-Camps, Crystal structure of the nickel-iron hydrogenase from Desulfovibrio gigas. Nature, 373(6515), 580-587 (1995).

39. H. Ogata, S. Hirota, A. Nakahara, H. Komori, N. Shibata, T. Kato, K. Kano, and Y. Higuchi, Activation process of [NiFe] hydrogenase elucidated by high-resolution X-ray analyses: conversion of the ready to the unready state. Structure 13(11), 1635-1642, (2005).

40. G. N. George, S. P. Cramer, T. G. Frey, and R. C. Prince, X-ray absorption spectroscopy of oriented cytochrome oxidase. Biochim. Biophys. Acta 1142, 240-252, (1993).

41. T. Tsukihara, H. Aoyama, E. Yamashita, T. Tomizaki, H. Yamaguchi, K. Shinzawa-Itoh, R. Nakashima, R. Yaono, and S. Yoshikawa. Structures of metal sites of oxidized bovine heart cytochrome c oxidase at 2.8 Å. Science 269(5227), 1069-1074 (1995).

42. H. S. Carr, G. N. George and D. R. Winge, Yeast Cox11, a protein essential for cytochrome c oxidase assembly, is a Cu(I)-binding protein. J. Biol. Chem. 277(34), 31237-31242 (2002).

43. F. Arnesano, L. Banci, I. Bertini, S. Mangani, and A. R. Thompsett, A redox switch in CopC: An intriguing copper trafficking protein that binds copper(I) and copper(II) at different sites. Proc. Natl. Acad. Sci. USA 100(7), 3814-3819 (2003).

44. H. H. Harris, I. J. Pickering and G. N. George. The chemical Form of mercury in fish. Science, 301(5637), 1203 (2003).

45. I. J. Pickering, L. Gumaelius, H. H. Harris, R. C. Prince, G. Hirsch, J. A. Banks, D. E. Salt, and G. N. George, Localizing the biochemical transformations of arsenate in a hyperaccumulating fern. Environ. Sci. Technol. (in the press, 2006).

46. J. Gailer, G. N. George, I. J. Pickering, R. C. Prince, S. C. Ringwald, J. E. Pemberton, R. S. Glass, H. Younis, D. W. DeYoung, and H. V. Aposhian, A metabolic link between arsenite and selenite: the seleno-bis(S-glutathionyl) arsinium ion. J. Am. Chem. Soc., 122(19), 4637-4639 (2000).

47. S. A. Manley, G. N. George, I. J. Pickering, R. S. Glass, E. J. Prenner, R. Yamdagni, Q. Wu, and J. Gailer, The seleno bis(S-glutathionyl) arsinium ion is assembled in erythrocyte lysate. Chem. Res. Toxicol. 19(4); 601-607 (2006).

BIOLOGICAL APPLICATION OF SYNCHROTRON RADIATION: FROM ARTEM ALIKHANYAN TO NOWADAYS

A.A. Vazina
Institute of Theoretical and Experimental Biophysics of RAS, 142290 Pushchino, Moscow region, Russia

Abstract: Considerable progress in the development of structural biology me-thods has been achieved recently due to the use of synchrotron radiation. The purpose of this leaflet is to display a dual aspect of the field of biological and medical application of synchrotron radiation: first, some basic programs of structural biology, and second, programs aimed at medical applications.

Key words: time-resolved technique; muscle contraction; titin; protein crystals; EXAFS-spectroscopy; phase contrast; microbeam diffraction; biological tissues; mucus; proteoglycan; extracellular matrix.

1.   INTRODUCTION

Advances in the modern molecular biology enabled us to elucidate many mechanisms of functioning of living systems at the molecular level. However, a great number of functions in the organism that are realized at higher levels: the cellular, supracellular, tissue, and organ levels, have been little investigated. The problem of biological motility occupies one of the central places in modern biology. Muscle contraction is one of those biological phenomena that we can all appreciate in our everyday life. The way in which muscles produce force has exercised the minds of poets, philosophers, scientists, engineers and physicians. We know a very great deal about muscle structure, physiology and biochemistry, but we still do not know exactly what is the molecular process that produces movement.

Vasili Tsakanov and Helmut Wiedemann (eds.), Brilliant Light in Life and Material Sciences, 121–131.
© 2007 *Springer.*

Figure 1. Collage of photos demonstrates the portraits of the Director of the Yerevan Physical Institute Prof. A.I. Alikhanyan and the Director of the Institute of Biological Physics of USSR Academy of Sciences Prof. G.M. Frank.

An inaccessible dream of scientists studying the muscle contraction was to carry out X-ray diffraction experiments on a living muscle during biological functioning. Gleb Frank, the director of the Institute of Biological Physics of the USSR Academy of Sciences, the scientist who has devoted all his scientific life to studying the structural bases of biological mobility, perceived that such experiments could be realized only in a tandem with powerful experimental techniques of nuclear physics. In summer 1971, together with G. Frank, we arrived to Armenia to the director of the Yerevan Physics Institute Artem Alikhanyan to discuss the reality of such experiment. Physicists have met this proposal with great enthusiasm. The use of synchrotron radiation and sensitive detectors made it possible to put forward and solve the problems of dynamics of structural transformations during biological functioning, in the first place in muscle contraction.

The application of synchrotron radiation to studies of the structure of biological objects started in 1971 at DESY (Hamburg)[1], and in the USSR the Government Scientific Program of SR research also appeared in 1971. Two powerful accelerators were available at that time as SR sources in the X-ray range: the synchrotron ARUS[2] in the Yerevan Physics Institute and the storage ring VEPP-3[3] in the Budker Institute of Nuclear Physics, Novosibirsk; the latter remains up to now the major site for SR research in Russia.

Over a period of 35 years Pushchino's institutes: Institute of Theoretical and Experimental Biophysics of RAS, and Institute of Cell Biology of RAS, have been working with synchrotron radiation using it in the field of biology and medicine in collaboration with some institutes of the Siberian Department of RAS (Budker Institute of Nuclear Physics, Institute of Inorganic Chemistry, Institute of Solid State Chemistry), Kurchatov Institute of Atomic Energy, Moscow State University, Timiryazev Agricultural Academy, Karayev Institute of Physiology of Azerbaijan AS, Blokhin Cancer Research Center of RAMS and some institutes of the Ministry of Public Health of the Russian Federation (Pulmonology Research Institute, Central Dermal Venereological Institute). We use SR from SRS VEPP-3 and VEPP-4, Novosibirsk; CLRC, Daresbury Laboratory and DESY-DORIS, Hamburg; ESRF, Grenoble. At present, X-ray diffraction studies of biological objects are systematically carried out on the small-angle station DICSI of the "SIBERIA-2" storage ring (RRC "Kurchatov Institute", Moscow). Figure demonstrates a group of first designers and the station DICSI, 'diffraction movie', which has been constructed by us in cooperation with Budker INP SD RAS with the participation of designers and scientists from St. Petersburg, Moscow, and Armenia.

This article summarizes experimental approaches to the application of SR in the solution of problems of structural biology.

2.   GENERAL RESULTS OF USING THE SYNCHROTRON RADIATION FROM DIFFERENT SOURCES

2.1   Time-resolved X-ray diffraction technique

The central problem of cell biology is the study of the structural aspects of the functioning of various cellular organelles, including the cytoskeleton motility, contractility, membrane transport, excitation conductivity, cell division and differentiation, age-related transformations, and pathological processes. These functions are maintained by unique cellular structures, which, however, have one property in common: all of them represent regular macromolecular ensembles with nanoscale structural periodicities between 1 and 100 nm. This suggests the occurrence of small-angle diffraction and/or small-angle diffuse scattering. The difference between functionally important structures is displayed in the character and the degree of their regularity, as well as in the time parameters of their structural dynamics. Relevant information can be obtained by the small-angle time-resolved X-ray diffraction technique, also called "diffraction movie".

For the first time the time-resolved X-ray diffraction technique was created by us at the Siberian Synchrotron Radiation Center in 1973. We have used an original X-ray optical scheme, which involves a monochromator-mirror, a one-dimensional proportional detector, and a computer. In the following years, several versions of this technique with 4, 8, 32, 64 time-frames and a time resolution of up to 1 ms were developed. For about 30 years, the "diffraction movie" technique has been substantially improved. Each version of the "diffraction movie" technique was provided with data treatment programs and appropriate software packages serving to control the experimental device, to collect and to process the data obtained[4-6]. The new "diffraction movie" station DICSI-1 equipped with the OD-3 detector has been set in operation at the RRC "Kurchatov Institute"[7].

2.2   X-ray time-resolved study of muscle contraction

Devices to keep the muscle in a living state and to measure simultaneously its mechanical parameters and diffraction data with the same time resolution have been designed. They enable us to realize various types of contraction: isometric, isotonic, quick release or quick stretch, and double stimulation with different interpulse intervals. The ability of striated muscle to potentiate its twitch response after preliminary stimulation at short-term stimulus history and/or fatigue has been studied. Model calculations of diffraction patterns for various configurations of structure of cross-bridges were carried out. The conception proposed by us postulates the key role of the supramolecular structural level in the contraction cycle of a muscle whose contractile apparatus is built up of two types of helical structures with different symmetries and incommensurable periods. The principle of the dynamic coupling of symmetries in such systems may define the mechanism of biological motility in its different manifestations. Our general results of the research of structural dynamics of a muscle are presented only briefly and are described in detail in some articles[8-12].

2.3   Structural principles of multidomain organization of giant polypeptide chain of the muscle protein titin

The elasticity of titin is a key parameter that determines the mechanical properties of the muscle. These include reversibility, i.e. the capacity of the muscle to change its length manifold and return to its original state, and the transduction of passive tension generated by the stretched muscle. The morphology and elastic properties of oriented fibres of titin molecules were studied using small- and wide-angle X-ray scattering, and mechanical techniques. We succeeded in obtaining oriented filaments of purified titin suitable for diffraction measurements[13]. Our X-ray data suggest a model of titin as a nanoscale, morphological, aperiodical array of rigid Immunoglobulin and Fibronectin3-type domains covalently connected by conformationally variable short loops. The line group symmetry of the model can be defined as SM with axial translation τ_∞. Both tension transduction and high elasticity of titin can be explained in terms of crystalline polymer physics. Titin stretching experiments show that each individual titin macromolecule can adopt a novel two-phase state within the fibre. The conversion between high elasticity and strength can be explained as a phase transition under external tension. In terms of the concept of orientational melting, the origin of the functional heterogeneity along the titin strand becomes interpretable[14,15].

2.4   Study of protein crystals

The spatial structure of the covalently bounded spin label has been determined to analyze how dynamics of its immediate surroundings in a lysozyme crystal depends on relative humidity. It was shown that the change of local conformations in the vicinity of spin label has a two-stage character upon dehydration of tetragonal lysozyme crystal from 100% to 40% of the relative humidity[16]. The effect of relative humidity on protein crystallization has been estimated[17].

The crystal structure of the nicking endonuclease N.BspD6I from Bacillus species at 0.18 nm resolution was solved[18]. The endonuclease recognizes the sequence 5'-GAGTC/5'-GACTC and selectively cleaves the GAGTC-strand. Selective nicking of double-stranded DNA is frequently employed in biological research and diagnostics. This is the first known 3D structure of site- and strand-specific nicking endonucleases. The study provides a insightful basis to investigate nicking mechanism on a molecular level.

2.5   EXAFS-spectroscopy of Ca^{2+}-binding proteins

The studies of the Ca2+-binding proteins (parvalbumin, troponin C, calmodulin and α-lactalbumin) have been carried out. An original method of double isomorphous replacement of light metals by heavier ions was used. For comparison analysis of the local structure of two metal-binding centers with different binding constants, Ca^{2+} ions were sequentially replaced in these centers by rare-earth Eu^{3+} and Tb^{3+} ions. The similarity of EXAFS spectra of the proteins demonstrated the identity of the stable configurations within the first coordination sphere and some general features of the structural arrangement of metal-binding centers. It was shown that the first coordination sphere is formed by six oxygen atoms around the calcium atom; the experimentally determined radius of the sphere, R = 0.243 + 0.002 nm, is an averaged Ca-O distance within these proteins[19,20].

2.6   Phase contrast X-ray imaging of biological objects

The synchrotron radiation of VEPP-3 at the energies of 8.9 and 33.2 keV was used for obtaining the X-ray images of biological objects by the method of refraction radiography[21]. Phase-contrast X-ray images can be produced in various ways. We demonstrated a great potential of the method of phase contrast imaging for a study of muscles and animal organism in normal and pathological states. The method was applied to image biological tissues that have the unique structural feature, the translation symmetry of 0.1 - 10 (m periodicity. The cross-striated muscle is the most interesting example of such objects. The experiment was done using high-brilliant coherent X-rays delivered by a synchrotron radiation source of the third generation (ESRF, Grenoble) and a high-resolution 2D-detector[22]. We obtained for the first time direct phase contrast images of the fine periodical structure of the matter density of live frog muscles with a resolution better than a fraction of micron. Despite the poor visibility of the images, the structures are seen clearly. On the other hand, the small value of phase shift variation allows us to make a straightforward reconstruction of the phase variation profile from the local visibility profile. The diffraction introscopy method is potentially suitable for clinical applications.

2.7   Microbeam diffraction technique for living tissue

X-ray diffraction patterns from a single hair cross section and from functionally different submicron fragments of a frog striated skeletal muscle sarcomere were obtained for the first time by using a microbeam on the small-angle ID-18 station of the ESRF (Grenoble, France). The microbeam technique is a potential tool for studying the local microstructures of various biological objects at a micron scale[23].

2.8   Structural studies of proteoglycans of mucus and extracellular matrix of biological tissues

The medical diagnostics is one of the key goals of structural biology research. Every disease is linked to changes in the cell and/or tissue biochemistry with a subsequent effect on tissue structure; therefore, the presence and type of disease can be determined by examination of the molecular structure of tissue. The science of the structural biology of tissues is at its very beginning. Tissue cells are imbedded in an extracellular matrix consisting of tissue-specific fibrillar proteins and a basic substance of proteoglycan nature, in which polysaccharide chains are covalently linked to the protein core.

In contrast to the well-studied structures of protein filaments, the structure of the proteoglycan components of the extracellular matrix and mucins has been largely out of view of the modern structural biology because of the absence of adequate physicochemical methods. The heterogeneity of polysaccharides in composition, length, electrical charge, and the extent of branching makes difficult the application of many physical methods designed to study homogenous systems: sedimentation, electrophoresis, chromatography, light scattering, etc. Also, the powerful store of genetic methods is not so much applicable to polysaccharides as to proteins and nucleic acids since there is no direct influence of genes on the synthesis of polysaccharides. The adequate methods for investigation of the structural peculiarities of intact tissues could be the X-ray structural and fluorescent analyses.

Earlier we have obtained highly concentrated samples of mucins from several mammalian species. The X-ray patterns of these samples display a large number of sharp diffraction rings at a main spacing of 4.5 nm. It has been established using the methods of preparative column chromatography that the X-ray diffraction patterns are due to the high-ordered structure of proteoglycan components of mucins[24,25]. The set of reflections is due to the periodical packing of the polysaccharide chains covalently bound to the protein core. In 1999 an analogous set of Debay rings was registered on X-ray patterns of hairs of patients with breast cancer, the 4.5 nm ring was superimposed to the normal fiber diffraction pattern of α-keratin. The X-ray pattern of "ring" or "no ring" type was suggested by the authors to screen for breast cancer[26]. Not surprisingly, this report triggered a run by several groups to reproduce these studies, however diagnostic signal "ring" or "no ring" was not confirmed. We attributed these set of reflections to the proteoglycan structures of the extracellular matrix of hair tissue[27].

It was found that the X-ray pattern transformation ("ring" – "no ring") is determined by the element contents in hairs so that the rearrangement of the extracellular matrix proteoglycan structures is a dose-dependent process. It was also shown that the structure of the extracellular matrix can be transformed by the environment cations that penetrate into the hair tissue only after the destruction of the hair protector lipid mantle by strong detergents including sanitary and hygiene reagents. Thus, X-ray diffraction and fluorescence analysis of hairs must be recommended as a criterion for the sanitary and hygiene norms during certification of new medical and cosmetic hair care reagents.

In the last years we began large-scale X-ray diffraction and fluorescent studies of various native and modified human and animal tissues[28,29]. A collection of about 500 samples of human epithelial tissues obtained during biopsies, surgeries and pathologoanatomic revisions, as well as tissues and mucus of healthy experimental animals were used. All X-ray patterns of biological tissues, both intact and transformed by tumor, may be classified under two types: "no ring" and "ring". The 4.5 nm spacing was attributed to proteoglycans; the intensity and angular width of the 4.5 nm reflection change within a wide range, whereas the spacing varies insignificantly. The type of X-ray pattern ("ring" or "no ring") for intact tissues can reversibly be transformed by salt solutions or chelating agents. The 4.5 nm reflection should be considered as a marker of solely structural modifications of proteoglycans of intact tissue under various exogenous influences. A variety of changes in X-ray patterns of tumor-modified tissues were observed, but any specific pattern of tumor modification was not revealed[30].

A comparative analysis of the X-ray patterns of mucus and various biological tissues showed that the 4.5 nm spacing is a structural nanoinvariant of proteoglycans of mucus and the extracellular matrix of biological tissues. Mobile proteoglycan structures of the extracellular matrix can play a significant role in the structural homeostasis of biological tissue. The biological tissues and mucins can be used as markers of pathological states of an organism.

Future prospects of using SR in structural biology studies are excellent. It is also evident that techniques being developed for time-resolved analysis of muscle may be applicable to other dynamic biological systems, especially when microbeams suitable for very small specimens and ultrafast high-resolution X-ray detectors can be used.

ACKNOWLEDGMENTS

The author is grateful to N.F. Lanina, V.N. Korneev, V.D. Vasiliev and A.E. Naumov for discussion and help in the preparation of the manuscript.

The study was supported by the Russian Foundation for Basic Research, grants No. 04-02-97260, 04-02-17389, 04-02-17301, 04-04-97313, 05-02-17708 and 06-02-16933.

REFERENCES

1. G. Rosenbaum, K.C. Holmes and J. Witz, Synchrotron radiation as a source for X-ray diffraction, Nature, 230, 434-437 (1971).
2. A.A. Vazina, V.S. Gerasimov, L.A. Zheleznaya, A.M. Matyushin, B.Ya. Son'kin, L.K. Srebnitskaya, V.M. Shelestov, G.M. Frank, Ts.M. Avakyan and A.I. Alikhanyan, The experience of using the synchrotron radiation for the X-ray study of biopolymers, Biofizika (Biophysics), 20(5), 801-806 (1975) (in Russian).
3. A.A. Vazina, V.S. Gerasimov, L.A. Zheleznaya, V.B. Savelyev, L.K. Srebnitskaya, G.M. Frank, G.N. Kulipanov, A.N. Skrinskii, V.B. Khlestov and M.A. Sheromov, Use of synchrotron radiation for the study of great periods in biopolymers structure, Apparatura i methody rentgenovskogo analiza (Apparatus and methods of X-ray analysis), 19, 73-81 (1975) (in Russian).

4. A.A. Vazina, A.M. Gadzhiev, P.M. Sergienko, V.S. Gerasimov and V.N. Korneev, Time-resolved small-angle X-ray diffraction from contracting muscle, Review of Scientific Instruments, 60, 2350-2353 (1989).
5. V.M. Aul'chenko, A.B. Bessergenev, O.A. Evdokov, V.S. Gerasimov, Yu.A. Gaponov, V.N. Korneev, N.A. Mezentsev, P.M. Sergienko, M.R. Sharafutdinov, M.A. Sheromov, A.A. Titov, B.P. Tolochko and A.A. Vazina, The station for time-resolved investigation in wide and small angles of diffraction, Nuclear Instruments and Methods, A405, 487-493 (1998).
6. V.V. Boldyrev, N.Z. Lyakhov, B.P. Tolochko, A.A. Vazina, N.A. Mezentsev, V.F. Pindyurin, M.A. Sheromov and A.G. Khabakhpashev, The diffractometry with using the synchrotron radiation (Nauka, Novosibirsk, 1989) (in Russian).
7. V.N. Korneev, P.M. Sergienko, A.M. Matyushin, V.A. Shlektarev, N.I. Ariskin, V.I. Shishkov, V.P. Gorin, M.A. Sheromov, V.M. Aul'chenko, A.V. Zabelin, V.G. Stankevich, L.I. Yudin and A.A. Vazina, Current status of the small-angle station at Kurchatov center of synchrotron radiation, Nuclear Instruments and Methods, A543, 368-374 (2005).
8. A. Vazina, M. Volkenshtein, A. Gadzhiev, V. Gerasimov, A. Korystova and P. Sergienko, The discovery of the dynamic structure in the process of the skeletal muscle isometric contraction, Dokl. AS USSR (Proceedings of AS USSR), 274, 941-945 (1984) (in Russian).
9. A.M. Gadzhiev, V.S. Gerasimov, N.P. Gorbunova, A.F. Korystova, V.N. Korneev, P.M. Sergienko and A.A. Vazina, Studies of the muscle structure during contraction initiated by pair wise stimulation (new results), Nuclear Instruments and Methods, A282, 502-505 (1989).
10. A. Vazina, P. Sergienko, A. Gadzhiev, G. Rapp and N. Kunst, The time-resolved investigation of muscle structure during contraction initiated by double stimulation, in: Hamburger Synchrotronstrahlungslabor, An. Report 11 (DESY, EMBL, 1996), pp. 191-192.
11. A.A. Vazina, The synchrotron radiation in structural investigations, Vestnik AS USSR (The bulletin of AS USSR), 8, 14-23 (1978) (in Russian).
12. A.A. Vazina, Application of synchrotron radiation to small-angle X-ray analysis of biological objects, Nuclear Instruments and Methods, A261, 200-208 (1987).
13. Q. Li, A. Vazina, K. Ranatunga, D. Alexeev, A. Soteriou and J. Trinick, X-ray diffraction and mechanical studies of oriented fibres of purified titin, Fibre diffract. review, 5, 49 (1996).
14. A. Vazina, N. Gorbunova, N. Lanina, I. Dolbnya, W. Bras and I. Snigireva, X-ray study of oriented gels of titin, Nuclear Instruments and Methods, A543, 148-152 (2005).
15. A.A. Vazina, N.F. Lanina, D.G. Alexeev, W. Bras, I.P. Dolbnya, The structural principles of multidomain organization of the giant polypeptide chain of the muscle titin protein: SAXS/WAXS studies of oriented titin fibres, Journal of Structural Biology (2006), in press.
16. R.I. Artyukh, G.S. Kachalova, N.F. Lanina, D.O. Nikolskii, V.P. Timofeev and H.D. Bartunik, A local dynamic structure of lysozyme in a spin-labeled tetragonal crystal at varying humidity, Biofizika (Biophysics), 47(5), 795-805 (2002).
17. T. Hudaverdyan, G. Kachalova and H. Bartunik, Estimation of effect of relative humidity on protein crystallization, Kristallografiya (Crystallography), 51(3), 554-559 (2006).

18. G.S. Kachalova, E.A. Rogulin, R.I. Artyukh, T.A. Perevyazova, L.A. Zheleznaya, N.I.Matvienko and H.D. Bartunik, A crystal structure of the site-specific DNA nickase N.BspD6I, Acta Crystallographyca, F61, 332-334 (2005)

19. A. Korystova, V. Shelestov and A. Vazina, X-ray spectral studies of Ca-binding proteins upon isomorphous replacements, Nuclear Instruments and Methods, A282, 506-509 (1989).

20. A.F. Korystova, V.M. Shelestov, A.A. Vazina and D.I. Kochubei, EXAFS studies of different centers of Ca2+-binding proteins, in: X-ray absorption fine structure, edited by S.S. Hasnain (Ellis Horwood, London, 1991), pp. 187-190.

21. V. Gerasimov, V. Korneev, G. Kulipanov, A. Manushkin, P. Sergienko, V. Somenkov, S. Shilstein and A. Vazina, Search for biological objects by refraction radiography using SR of VEPP-3 storage ring, Nuclear Instruments and Methods, A405, 525-531 (1998).

22. V. Kohn, C. Rau, P. Sergienko, I. Snigireva, A. Snigirev and A. Vazina, The live lattices become visible in coherent synchrotron X-rays, Nuclear Instruments and Methods, A543, 306-311 (2005).

23. M. Drakopoulos, P.M. Sergienko, I. Snigireva, A. Snigirev and A.A. Vazina, Application of microbeam technique for the small-angle diffraction studies of a living tissue, Nuclear Instruments and Methods, A543, 161-165 (2005).

24. E.A. Denisova, P. Lazarev, A.A. Vazina and L.A. Zheleznaya, Intestinal mucus and juice glycoproteins have liquid crystalline structure, Studia biophysica, 108(2), 117-124 (1985).

25. L.A. Zheleznaya, E.A. Denisova, P.I. Lazarev and A.A. Vazina, X-ray diffraction studies on fine structure of mucus glycoprotein, Journal of Nanobiology, 1, 107-115 (1992).

26. V. James, J. Kearsley, T. Irving, Y. Amemiya and D. Cookson, Using hair to screen for breast cancer, Nature, 398, 33-34 (1999).

27. A. Aksirov, V. Gerasimov, V. Korneev, G. Kulipanov, N. Lanina, V. Letyagin, N. Mezentsev, P. Sergienko, B. Tolochko, V. Trounova and A. Vazina, Biological and medical application of SR from the storage rings of VEPP-3 and "Siberia-2". The origin of specific changes of small-angle X-ray diffraction pattern of hair and their correlation with the elemental content, Nuclear Instruments and Methods, A470, 380-387 (2001).

28. A. Vazina, P. Sergienko, V. Gerasimov, N. Lanina, I. Snigireva, A. Snigirev, M. Drakopoulos, I. Dolbnya and W. Bras, X-ray diffraction study of human hair as a model of epithelial tissue, Fibre diffraction review, 10, 93-94 (2002).

29. A.A. Vazina, W. Bras, I.P. Dolbnya, V.N. Korneev, N.F. Lanina, A.M. Matyushin, P.M. Sergienko and A.V. Zabelin, Peculiarities of human hair structural dynamics, Nuclear Instruments and Methods, A543, 153-157 (2005).

30. A.A. Vazina, A. Budantsev, W. Bras, N.P. Deshcherevskaya, I. Dolbnya, A.M. Gadzhiev, V.N. Korneev, N.F. Lanina, V.P. Letyagin, E.I. Maevsky, A.M. Matyushin, V.A. Trunova, V.M. Vavilov and A.L. Chernyaev, X-ray diffraction and spectral studies of biological native and modified tissues, Nuclear Instruments and Methods, A543, 297-301 (2005).

RECENT ADVANCES IN AUTOMATION OF X-RAY CRYSTALLOGRAPHIC BEAMLINES AT THE EMBL HAMBURG OUTSTATION

Matthew R. Groves
EMBL Outstation Hamburg, DESY, Hamburg, Germany

Abstract: The use of synchrotron radiation has had a marked affect on the rate at which high resolution models of biomolecules are obtained. While there are a number of sources at which protein crystallographic experiments can be performed, with others under construction, there still remains a high demand for data collection beamtime. As a result efforts have been made to automated various stages of a protein crystallographic experiment in order to improve efficiency of beamtime usage. While a number of the steps in a successful protein crystallographic structure determination have been automated it is not yet possible to build a full automated beamline as a number of central technologies have yet to be developed. Recent developments in automation and robotics will be discussed and areas in which development is required in order to produce a fully integrated and automated beamline will be outlined.

Key words: protein crystallography, X-ray beamline, MAD

1. SYNOPSIS

In this paper the current status of automation at protein crystallography beamlines will be summarized, with particular focus on developments at the EMBL crystallography beamlines at DESY, Hamburg.

Vasili Tsakanov and Helmut Wiedemann (eds.), Brilliant Light in Life and Material Sciences, 133–139.
© 2007 *Springer.*

2.   INTRODUCTION

Protein crystallography remains the predominant technique for the production of high resolution models of biomolecules. While other techniques, such as nuclear magnetic resonance (NMR), electron microscopy (EM) also provide images of biomolecules, the vast majority of the 38198 structures deposited within the protein data bank (www.rcsb.org;[1]) are from protein structures solved on by X-ray crystallography (32377). The use of synchrotron radiation for protein crystallography has had a major impact in the rate at which structures are solved, with over 80% of the currently deposited X-ray structures solved with data collected on synchrotron beamlines.

The current demand for synchrotron beamtime is still high and a number of synchrotrons are being built or commissioned in order to provide further access for biological scientists. The development of multiwavelength anomalous dispersion (MAD) phasing[2] allows crystal structures to be solved from a single protein crystal, provided that a tuneable source, such as that provided by a wiggler or undulator insertion device, is available.

A number of structural genomics projects have also demonstrated that current wet-lab technology can produce a large number of proteins for structural analysis. Additionally, ligand or drug screening using X-ray crystallography also contributes to the large number of biologically relevant samples that require provision of synchrotron beamtime.

Therefore, the total number of samples awaiting analysis on X-ray beamlines shows no sign of decreasing and, although the commissioning of new sources will reduce some of the pressure on existing sources, it is obvious that the structure solution process has to be made as time efficient as possible. Ideally, such developments will be portable between existing sources and machines that are under construction.

The structure solution pipeline can be subdivided into a number of different stages: initial protein preparation, protein crystallization, crystal mounting on a synchrotron beamline, initial crystal characterization, determination of data collection strategy, structure solution and structure refinement. This paper will attempt to summarize the developments being made at the EMBL Hamburg Outstation towards the development of a fully integrated and automated structure determination platform.

3.   INITIAL PROTEIN SAMPLE PREPARATION

It is still widely accepted that the use of recombinant expression systems is essential in order to generate protein samples in sufficient quantities for structural analysis (typically 5-10mg of pure material). The use of the pETM expression series in *E. coli* (G. Stier and A. Geerlof, manuscript in preparation) enables genes of interest to be cloned into expression plasmids with a wide variety of tags. These tags may aid solubility or facilitate purification. As all pETM series vectors share an identical multiple cloning site, the system lends itself to automation. As yet we have made no steps in this direction to automate the cloning process, but structural genomics companies such as Structural GenomiX have demonstrated that automation in this step is achievable. However, delays are inherent in this stage of the structural determination pipeline (ligations, sequencing, colony picking, etc.) and there are many individual steps that can be performed efficiently by researchers themselves.

Once a DNA construct has been produced experiments are required in order to establish the optimal expression host, whether it be bacterial, insect or mammalian. Again these steps require specialist knowledge that does lend itself easily to automation. We have developed a simple assay, which may be fully automated, in which DNA constructs may be screened against a library of E. coli expression hosts for optimal production of soluble protein[3].

Expression host culturing may be performed in small volume, high cell density fermentors in which optical densities of over 100 may be achieved. Recent expression tests of the yeast phosphatase pph22p demonstrated that soluble expression is achieved with a total biomass of over 500g.

Protein purification is also automated at the EMBL Outstation, with the use of the Akta 3D-kit (GE Healthcare), in which up to 6 samples may be simultaneously purified over affinity matrix and size exclusion chromatography.

4.   AUTOMATED PROTEIN CRYSTALLIZATION

Once sufficient pure sample is available, the next stage of the process requires that a large sample of chemical space be sampled in order to identify conditions under which the sample crystallizes. We have recently installed a crystallization robot at the EMBL Outstation[4] with which small volumes (100nl) of concentrated protein sample may be pipetted against crystallization screens. Storage and imaging of the crystallization experiments is carried out in a Robodesign storage hotel, with images taken periodically over the time course of the experiment. A standard commercial screen of 96 reagent conditions requires total sample volumes as low as 35µl. Manual hanging drop experiments would typically require up to 200µl for similar screen. The crystallization plates are pipetted, sealed and observed with minimal human intervention.

5.   CRYSTAL IDENTIFICATION AND CRYSTALLIZATION PLATE SCORING

The use of crystallization robotics provides a significant increase in the number of conditions that may be screened for crystal growth. However, the increased volume of experiments also produces a significantly increased number of images that require examination for crystal growth. Development have been made in other laboratories towards the automated identification of crystals within crystallization experiments (e.g. CEEP[5]). Image recognition of crystals within the experiments is hampered by the low signal to noise ratio of these images, as the crystals are clear objects that grow within clear drops, or precipitation. As such edge-detection algorithms have been used.

We have established a novel technique, whereby the protein sample is mixed with a non-covalent florescent dye immediately prior to crystallization. The interaction between the proteins and the dye results in a florescence signal that is easily detectable and varies with the local protein concentration. Thus, the regions of the crystallization experiment with higher protein concentration are strongly contrasted against all regions[6]. Advantages of this technique, when compared with other methods of crystal visualization, are that minimal additional handling is required and that illumination of the dye is at a wavelength at which proteins alone do not strongly absorb. No influence on the crystallization rates or quality of a trial set of proteins could be discerned, indicating that it represents a general solution to the problem of protein crystal recognition.

6.   CRYSTAL SAMPLE CHANGE

In order to minimize loss of beamtime due to manual sample mounting many synchrotron sites have been interested in developing an automated sample changer for pre-flash cooled protein crystals. Such a sample changer significantly increases throughput as the majority of the time lost when screening crystals is due to the need for manual sample mounting and safety checks within the experimental hutch. We have developed a sample changer based on a six-axis industrial robot that is also compatible with the SPINE standard sample pins and the sample changer developed at the ESRF. As the EMBL Hamburg sample changer is based upon an industrial platform it possesses a degree of flexibility and can be adapted to other tasks, such as crystal flash cooling and *in situ* crystallization plate screening. Such techniques are currently under development and will be reported elsewhere.

7. CRYSTAL CENTERING

In order to operate the sample changer in a purely hands-off manner it is necessary to develop routines that will allow crystals to be automatically centered once mounted. The software package XREC is a novel method for automated recognition of flash cooled protein crystals[7]. It uses a number of texture-based image-processing algorithms and is able to cope with a variety of crystal morphologies and illumination conditions characteristic of beamlines at different sources around the world. The results from processing of the series of images in different orientations of the crystal, together with their estimated standard uncertainties, provide the crystal location and allow an internal assessment of their reliability. The functionality of the software is being extended for automated scoring of crystallization conditions. (http://www.embl-hamburg.de/XREC/).

8. AUTOMATED DATA COLLECTION STRATEGIES

The software program BEST was developed in-house[8] for the optimal planning of X-ray data collection experiments. From a few initial diffraction images, BEST estimates the statistical characteristics of the data set for different combinations of data-collection parameters and suggests the most optimal ones. The anisotropy in diffraction and the permitted width of oscillation without spatially overlapping reflections are taken into account. According to the option chosen, the optimal set of parameters provides a given average signal-to-noise ratio at a given resolution either in the shortest time or with the minimum total radiation dose. The software has proved to be extremely useful in using the available data-collection time in the most efficient way.

9. AUTOMATED STRUCTURE SOLUTION

The EMBL-Hamburg Automated Crystal Structure Determination Platform (Autorickshaw) is a system that combines a number of existing macromolecular crystallographic computer programs and several decision-makers into a software pipeline for automated and efficient crystal structure determination[9]. The pipeline can be invoked as soon as X-ray data from derivatized protein crystals have been collected and processed. It is controlled by a web-based graphical user interface for data and parameter input, and for monitoring the progress of structure determination. A large number of possible structure-solution paths are encoded in the system and the optimal path is selected by the decision-makers as the structure solution evolves. The processes have been optimized for speed so that the pipeline can be used effectively for validating the X-ray experiment at a synchrotron beamline.

10.   AUTOMATED STRUCTURE REFINEMENT

Once a structure solution has been found, it is then necessary to refine that solution to produce the final model. Over the last four years the EMBL Outstation has been running a 16 processor cluster, which accepts computational jobs to run ARP/wARP[10]. This package uses "dummy" atoms in order to improve the fit between the experimental data and the current model. Model (re)building and ligand identification routines are also incorporated.

11.   SUMMARY

Attempts have been made to automate the major steps involved in a protein structure pipeline. However, while it is possible to perform sample crystallization, data collection and processing with minimal manual intervention, these stages are not fully integrated. Techniques are still required in order to automatically produce optimized crystallization screens once initial crystallization conditions have been obtained and to robustly harvest crystals. Both of these techniques will rely strongly on visualizing protein crystals under a variety of conditions. Crystal optimization will likely require identification of initial small (<1μm) crystals in crystallization plates. Once identified databases may be used to generate optimization screens. Our work in florescence dye co-crystallization may be an important first step in this procedure. Sample mounting for flash-cooling will require tracking of the protein crystals from the crystallization drop to their final location within the sample mount. Us yet no mechanism is commercialized which can visualize and handle such small objects. The development of systems are some of our primary focuses.

REFERENCES

1. H.M. Berman, J. Westbrook, Z. Feng, G. Gilliland, T.N. Bhat, H. Weissig, I.N. Shindyalov, and P.E. Bourne (2000): The Protein Data Bank. Nucleic Acids Research, 28 pp. 235-242.
2. W.A. Hendrickson and C.M. Ogata (1997): "Phase Determination from Multiwavelength Anomalous Diffraction Measurements". Methods in Enzymology 276, 494-523.
3. J. Lodge, and M. Groves (2006) submitted to JMB.
4. Jochen Müller-Dieckmann et al. manuscript in preparation.
5. J. Wilson (2002) "Towards the automated evaluation of crystallization trials" Acta Cryst. (D58), 1907-1914.
6. J. Müller-Dieckmann, I.B. Müller, C. Müller-Dieckmann, X. Kreplin and M. Groves (2006) submitted to Acta Cryst D.
7. S. B. Pothineni, T. Strutz and V. S. Lamzin (2006) "Automated detection and centring of flash-cooled protein crystals" Acta Cryst D, in press.
8. A.N. Popov and G.P Bourenkov. (2006) "Choice of data-collection parameters based on statistic modelling." Acta Cryst D59(Pt 7):1145-53.
9. S. Panjikar, V. Parthasarathy, V. S. Lamzin, M. S. Weiss and P. A. Tucker (2005) "Auto-Rickshaw: an automated crystal structure determination platform as an efficient tool for the validation of an X-ray diffraction experiment" Acta Cryst. D61, 449-457.
10. A. Perrakis, T. K. Sixma, K. S. Wilson and V. S. Lamzin (1997) "wARP: Improvement and Extension of Crystallographic Phases by Weighted Averaging of Multiple-Refined Dummy Atomic Models" Acta Cryst. D53, 448-455.

Matthew.Groves@embl-hamburg.de

3D MACROMOLECULAR STRUCTURE ANALYSES: APPLICATIONS IN PLANT PROTEINS

Filiz Dede, Gizem Dinler and Zehra Sayers
Faculty of Engineering and Natural Sciences, Sabanci University, Orhanli, Tuzla 34956 Istanbul, Turkey

Abstract: Attempts to relate the function to the structural features of biological macromolecules have intensified through the use of synchrotron radiation (SR) based techniques. Small Angle X-ray Scattering (SXAS) is a technique that has become readily accessible through several bamlines on SR sources all over the world. SAXS is used for obtaining low resolution structural information and is particularly useful for macromolecules that do not easily crystallize or alternatively when time-resolved structural information is required. In this paper use of SAXS for determination of structural parameters of a wheat metallothionein as a fusion protein with glutathione-s-transferase (GST) is presented together with low resolution models. Results are discussed in the framework of functional roles of wheat metallothionein.

Key words: metallothionein, SAXS, synchrotron radiation

1. INTRODUCTION

In the last decades structure analyses of biological macromolecules have accelerated with the availability of synchrotron radiation (SR) sources and the possibility of using several SR based techniques to obtain information at different structural resolution. SR sources have facilitated studies where the structure of the molecule is probed by complementary techniques not only to gain information at different levels of organization in the macromolecule but also to follow changes in the structure due to perturbations introduced

Vasili Tsakanov and Helmut Wiedemann (eds.), Brilliant Light in Life and Material Sciences, 141–151.
© 2007 *Springer.*

in the sample environment. A technique particularly useful in this area is small angle X-ray scattering (SAXS), which depending on the experimental conditions and data analysis methods, may yield information about the structure in the resolution range between one and few hundred nm with a time resolution of a few milliseconds. SAXS may be used to study the structure of noncrystalline systems including solutions of proteins, large macromolecular complexes, fibrous systems, lipid micelles and viruses.

In small angle solution scattering the scattered intensity I(s) is recorded as a function of s, the momentum transfer vector and is dependent on the inhomogeneities in the electron density due to the macromolecules dispersed in the uniform electron density of the buffer solution. It is assumed that all solute particles are identical and randomly positioned and oriented in the solvent and the scattering pattern thus contains information about the spherically averaged structure of the macromolecule. The scattering curve, described by a distance probability function, contains a region at small angle corresponding to the long range organization of the macromolecule (i.e. its shape) and a large angle region where the internal structure of the solute dominates. It is therefore possible to extend the measurements to the wide angle range to probe the internal structural features of the particles in solution. The specific features of the scattering pattern are determined by the symmetries of the macromolecules i.e. frequency of occurrence of given distances. The simplest interpretation of SAXS curves from globular particles is based on the Guinier approximation (Guinier, 1939), through which overall parameters such as the Rg, asymmetries in the shape and the relative molecular mass may be determined (Sayers, 1988). Availability of powerful SR sources and development of *ab initio* data interpretation methods based on spherical harmonics, global minimization algorithms and rigid-body refinement (through improved computing power) have resulted in full exploitation of the SAXS method leading to time-resolved studies on macromolecular assembly and folding. Details of experimental set-ups, data collection and evaluations methods and various applications can be found in recent reviews including Koch (2006) and Koch et al (2003).

A protein family where SAXS technique can be readily applied for structural analyses is metallothioneins (MTs). MTs, present in a wide range of species from fungi to humans, are low molecular weight (6-8 kDa) cystein rich proteins that do not readily crystallize and exist in solution in a range of oligomeric forms (Zangger et al., 2001). MTs, coordinate metal ions (e.g. copper, cadmium, zinc, silver, mercury) through thiol groups of the cystein residues that are localized in the N- and C-termini of the protein in specific motifs (reviewed in Vasak and Hasler, 2000). Classification of the members of the MT family has been based on the distribution of the cystein residue motifs in the amino acid sequences of the known proteins and 15 subfamilies have been identified (Binz and Kagi, 1997). Several functional roles including heavy metal detoxification, zinc and copper homeostasis, free radical scavenging, regulation of metalloenzyme and transcription factors have been reported for mammalian MTs (see Bilecen et al., 2005 for a comprehensive list of references). Plant MT family has been further subdivided into 4 types according to the cystein motifs (Cobbett and Goldsbrough, 2002). The presence of the all of the four types of MTs has been shown in *A. thaliana* (Zhou and Goldsbrough, 1995) and their tissue specific expression is being investigated (Guo et al., 2003). Metal binding propensity of purified plant recombinant MTs (Kille et al., 1991; Bilecen et al., 2005, Domenech et al., 2005) and up-regulation of the *mt* genes upon exposure of plants to heavy metals (Guo etal., 2003) has lead to the conclusion that MTs play a role in metal homeostasis and heay metal tolerance mechanisms. Involvement of plant MTs in metal chaperoning, scavenging of reactive oxygen species (Akashi et al., 2004) stress response (Zimeri et al., 2005), senescence and fruit ripening (Cobbett and Goldsbrough, 2002) have also been reported. Investigations on some natural metal accumulator plants (Ma et al., 2003) have also pointed towards the possibility of using plant MTs for environmental purposes either as a possible indicator of accumulation of heavy metals in soil (Dallinger et al., 2004) or as a potential agent for phytoremediation (Eapen and Souza, 2005).

The 3D structure of MTs originating from mammalian and fungal sources has been extensively investigated using NMR and X-ray crystallography (Vasak and Hasler, 2000). In type 1 MTs, the most common type, a total of seven metals are bound through cysteins at the N- and C-termini forming ((Me(II)3Cys9) and ((Me(II)4Cys11) clusters which are connected by a short hinge region devoid of cysteins. The protein appears to have little secondary structure and folds through metal binding (Rigby and Stillman, 2004). These observations are confirmed by a recent report of the crystal structure of the single domain yeast metallothionein (Calderone et al. 2005). The 3D structure of plant MTs also appear to consist of two metal-binding domains connected by a hinge region with the difference that in most type 1 plant MTs the hinge region is unusually long, containing up to about 50 amino acids as compared to 2-10 amino acids in their mammalian counterpart.

We have been interested in investigating the involvement of MTs in tolerance to Cd toxicity in wheat, and we have previously identified a type 1 MT from a Cd tolerant durum wheat (dMT) cultivar. This *mt* gene (*dmt*) was cloned and overexpressed in *E. coli* as a glutathione-S-transferase (GST) fusion protein (GSTdMT). According to the amino acid sequence of dMT it contains the two metal binding domains with a total of 12 cysteins and the hinge region consists of 48 amino acids. Preliminary X-ray solution scattering measurements and homology modeling indicated that the hinge region extends out from the structure and may be readily available for interactions with other components in solution. (Bilecen et al.,2005). Following the preliminary studies a procedure was developed for isolation of highly homogeneous monodisperse fusion protein for SAXS measurements and biochemical and biophysical characterization. In this report recent results are presented together with the low resolution models based on the SAXS measurements and implications for functional roles of MT are discussed.

2.   MATERIALS AND METHODS

2.1   Purification and biochemical characterization of GSTdMT

Durum wheat MT was expressed as a fusion protein with GST in *E. coli* using the pGEX-4T2 vector (Amersham) and the recombinant protein (GSTdMT) was purified basically as described in Bilecen et al. (2005). The procedure involves isolation of GSTdMT from the cellular lysate by affinity chromatography using a 5ml GSTrap FF affinity column (Amersham) followed by size exclusion fractionation using a HiLoad 26/60 Superdex 75 column (Amersham) connected to an Akta FPLC system (Amersham). Modifications to this procedure for improving stability and for obtaining a well defined oligomeric state of the purified protein are discussed below. Purified protein was immediately flash frozen in liquid N_2 and stored at -80 °C until measurements. Proteolytic degradation and presence of aggregated protein were monitored by SDS and native PAGE analysis (Bilecen et al. 2005). Fraction of different oligomeric states of recombinant GSTdMT was determined by dynamic light scattering using a Zeta-sizer Nano ZS (Malvern Instruments).

2.2   SAXS measurements and data analysis

The SR X-ray scattering data were collected on the X33 camera (Koch and Bordas, 1983, Boulin et al., 1988) of the European Molecular Biology Laboratory (EMBL) on the storage ring DORIS III of the Deutsches

Elektronen Synchrotron (DESY) in Hamburg, using a mar345 Image Plate detector (Marresearch GmbH). Scattering patterns were recorded at a sample-detector distance of 2.4 m covering the range of momentum transfer $0.15 < s < 3.5$ nm^{-1} where $s = 4\pi \sin(\theta)/\lambda$, 2θ is the scattering angle and $\lambda = 0.15$ nm is the X-ray wavelength.

Samples were taken from Istanbul to the synchrotron on dry ice and were defrosted immediately before measurements. Solutions of GSTdMT fusion protein in buffer A (50 mM Tris-HCl pH 7.5, 2.5 mM MgCl$_2$, 100 mM NaCl and 0,5 mM PMSF) were measured at concentrations between 1.2 and 3.5 mg/ml. Protein concentrations were determined by measuring the absorbance at 280 nm as described. Due to aggregation problems at higher concentrations, only the scattering curves measured at concentrations below 3 mg/ml were taken for structural analyses. Protein solutions were made 2 mM in DTT prior to measurements. Bovine serum albumin (BSA) was measured as a molecular mass standard at 7 mg/ml in a buffer with 50 mM Hepes, pH 8.0 and 150 mM NaCl. Radiation damage during the scattering experiments was monitored by collecting data in 2 successive 2-minute runs. Data reduction, background subtraction and correction for detector response followed standard procedures using the program PRIMUS (Konarev et al., 2003).

The forward scattering *I(0)* and the radius of gyration R$_g$ were evaluated using the Guinier approximation assuming that at very small angles ($s<1.3/R_g$) the intensity is represented as $I(s) = I(0) \exp(-(sR_g)^2/3)$. These parameters were also computed from the entire scattering pattern using the indirect transform package GNOM (Svergun, 1992), which also provides the distance distribution function *p(r)* of the particle. The molecular mass (MM) of the solute was evaluated by comparison of the forward scattering with that from a reference solution of bovine serum albumin (MM = 66 kDa). The accuracy of this method is limited by the uncertainty in the measured protein concentration used for data normalization. The excluded volume of the hydrated particle (Porod volume) was computed without model assumptions and independently of normalization from the shape of the scattering curve as described in Bilecen et al. (2005).

3.   RESULTS AND DISCUSSION

Structural studies on plant MTs have been impaired due to proteolytic susceptibility and oxidation sensitivity of the purified native protein. Furthermore, oxidized forms of MTs lose a number of the bound metals and form higher molecular mass aggregates which are difficult to fractionate

(Sayers et al., 1993, Sayers et al., 1999, Bilecen et al., 2005, Domenech et al., 2005). It has also been hard to fractionate and quantify MTs due to their small molecular weight and the lack of aromatic amino acids (Sayers et al., 1999). Recently different laboratories have attempted to express plant MTs in *E. coli* as GST fusion proteins to circumvent some of the problems (e.g. Kille et al., 1991, Bilecen et al., 2005; Domenech et al., 2005; Akashi et al., 2004). The general procedure involves growth of bacteria in media supplemented with metals and isolation of MT-GST fusion by GST affinity purification under stringent anaerobic conditions to facilitate preservation of the integrity of the holo-protein. Expression of MTs as a fusion protein serves to speed up the purification process, stabilizes the protein and detection becomes easier at different stages of the isolation procedure. One draw back is that GST is also prone to aggregation forming dimers and higher order structures (Schroeder and de Marco, 2005). Special care needs to be taken to modify the standard procedures for the isolation of the recombinant fusion protein and to determine the oligomeric state of the isolated species.

3.1   Determination of homogeneity, stability and the oligomeric state of GSTdMT

In order to improve the quality of the SAXS data, we investigated conditions for increasing the stability of the purified fusion protein and those that allow characterization of the oligomeric state of the recombinant fusion protein. Modifications to the purification procedure reported previously (Bilecen et al., 2005) included the use of 0.1 mM $CdCl_2$ in the lysis buffer for bacteria and the use of 1 mM DTT in all buffers.

As can be seen in figure 1 (a), these modifications resulted in clear separation of the dimeric form of GSTdMT from other high molecular weight aggregates by size exclusion chromatography.

The dimer eluted as the dominant form and the protein yield increased to 30-50 mg per 1.5 l *E. coli* culture when compared to 10 to 15 mg in the earlier preparations. The elution position of the major peak corresponded to about 70 kDa molecular mass which is the value estimated from the known amino acid compositions of GST and dMT including the thrombin cleavage site introduced through the vector. Integrity of the purified protein was analyzed by SDS PAGE as shown in figure 1 (b). Here a single band at about 32 kDa corresponding to the intact GSTdMT is observed. Further monitoring of the homogeneity of the dimer solutions was carried out by native PAGE analysis and by dynamic light scattering. The lack of high molecular weight bands on the native gel given in figure 1 (b) and the presence of a single peak in volume fraction plot for dynamic light scattering shown in figure 1 (c) indicate that the solutions of purified GSTdMT are monodisperse and that the major species in solution is in dimeric form.

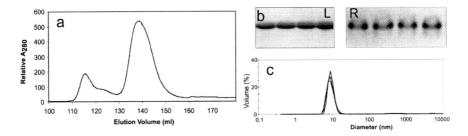

Figure 1. Fractionation and characterization of GSTdMT. (a) GSTdMT is eluted from the size exclusion column in a major peak at a position corresponding to the dimer. (b) SDS-PAGE Analysis of the top peak fractions show that the fusion protein has a molecular weight of about 32 kDa (L) and native gel analysis indicates a single species in the isolated fractions (R). (c) Dynamic light scattering results presented as volume fraction of particles in solution yield a single peak corresponding particles with a diameter less than 10 nm.

A second parameter which needs to be determined accurately for SAXS measurements is the concentration of protein in solutions (Bilecen et al., 2005). MTs are difficult to quantify by spectrophotometric methods due the lack of aromatic amino acids (Sayers et al., 1999). For data handling procedures given below the protein concentration was determined by using the extinction coefficient 41,960 calculated for the fusion protein from the amino acid sequence of the whole protein (Gill and von Hipple, 1989). Since the presence of metals bound to the protein can be inferred from the charge transfer band between 240 and 260 nm the ratio of A_{280} to A_{254} was taken as a measure of the metal content of the protein, a ratio of about 1.7 corresponds to the holo-protein (Dinler, G. et al. unpublished results).

3.2 SAXS measurements, determination of structural parameters of GSTdMT fusion protein and model calculations

The X-ray scattering pattern from a 1.2 mg/ml GSTdMT solution after corrections for buffer scattering, detector response and normalization for the X-ray beam intensity is shown in figure 2 (a) together with a plot of the data as $\ln(I(s))$ vs s^2 in the range $0.57 < sR_g < 1.3$ shown in Figure 2 (b).

The latter represents the Guinier plot of the scattering curve and the radius of gyration (R_g) of the particles in solution can be calculated from the slope of the straight line fitted to the data in the given s range. Extrapolation of the Guinier plot to s = 0 gives the scattered intensity at zero angle (I(0)) which can be used in estimations of the molecular mass of the species in solution relative to a known standard.

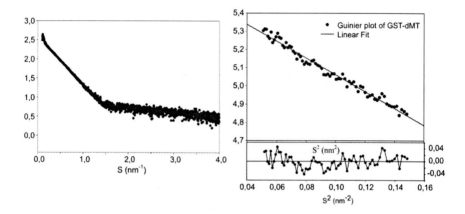

Figure 2. SAXS measurements and data analysis from GSTdMT. (a) SAXS curve of a 1.3 mg/ml GSTdMT in buffer A after data reduction. Scattering data was collected between 0.15<s<0.35 nm-1. (b) Guinier plot from the scattering curve in (a). At the lower part of the figure residuals for the linear fit are shown.

For this set of measurements R_g for GSTdMT is estimated to be 3.2 ± 0.04 nm and the molecular mass obtained from the data is about 55 ± 10 kDa. This value is less than 68.07 kDa, which is the value calculated on the basis of the amino acid sequence of the components in the dimer. The estimate of the molecular mass from the Guinier plot is dependent on the concentration of the protein in solution and any inaccuracy in this value is reflected to molecular mass calculations. It is likely that the lower value obtained from the Guinier plot for the molecular mass of GSTdMT is due to the problems in determination of the concentration of the protein. The molecular mass of the particles may also be calculated from the Porod volume, this calculation is independent of protein concentration (Porod, 1982). The molecular mass calculated from the Porod volume for is about 62 kDa which is close to that estimated from the amino acid composition. The overall parameters and the distance distribution function can also be calculated from scattering data using the indirect transformation program GNOM (Svergun, 1992). Results obtained with this approach were consistent with the Guinier analysis and showed that the GSTdMT dimer has an elongated shape in solution (data not shown).

Low resolution shapes were calculated for the GSTdMT using the software DAMMIN (Svergun, 1999). Models were constructed assuming P2 symmetry as the protein was shown to be in dimeric form and overall structural parameters were taken from the GNOM analyses run on the scattering curves. Alignment of the shapes obtained from 12 independent DAMMIN runs performed on one of the measurements is shown in Fig. 3(a).

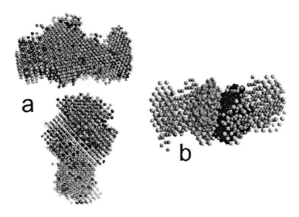

Figure 3. ab initio shape determination of GSTdMT. (a) Alignment of low-resolution GSTdMT structural models from 12 DAMMIN runs. The bottom aligned GSTdMT models represent the same models rotated clockwise by 90° around the y-axis. (b) Alignment of GST-dMT structural models with the GST dimer crystal structure. Ab-initio models of two GST-dMT dimers (greean and yellow) obtained from two independent DAMMIN runs overlaid on to a GST dimer (PDB ID:11GS; chain A and B of GST dimer colored navy and blue, respectively).

As can be seen from this figure, all models lead to an elongated shape consistent with the previous calculations (Bilecen et al., 2005). As the GST component is known to constitute the larger mass fraction of the fusion protein and it is prone to formation of dimers the known structure for GST was overlaid on the low resolution model of the GSTdMT dimer. It can be seen in Figure 3 (b) that the GST dimer fits in the central concentration of mass observed in the models shown in Figure 3 (a) and the two dMT components extend out from the bulk of the molecule in opposite directions not in contact with one another. This feature of dMT is supported by observations that the dMT component of the fusion protein is particularly susceptible proteolytic attack when the earlier isolation procedure was used.

It appears that the components in the fusion protein fold independently and the dMT structure in the fusion complex is close to that of the native protein.

In conclusion purification of the well characterized dimeric state of GSTdMT fusion protein has made it possible to carry out SASX measurements on monodisperse solutions of the recombinant protein. The low resolution structural model obtained based on the recent SAXS measurements support an extended structure for the dMT component. Results also indicate that some functional features attributed to dMT e.g. possibility of interactions with DNA, may be tested using the using the GSTdMT fusion because the proteins appear to be folding independently and that the dMT component would be structurally available for interactions with other macromolecules. Currently studies are underway to clarify the problems with concentration determination of the fusion protein as well as for testing interactions of GSTdMT (and cleaved dMT) with components such as DNA to gain insights into functional roles of plant MTs.

ACKNOWLEDGEMENTS

We would like to thank Drs. M. Koch, D. Svergun, M. Roessle and P. Konarev and other members of staff at EMBL Hamburg Outstation for their help with this worsk. We are grateful to Dr. F. Yagci from Koc University, Istanbul for her hel with the dynamic light scattering measurements. The work carried out at the EMBL Hamburg Outstation was made possible by a grant from the bilateral program between TUBITAK, Turkey and Juelich Research Center, Germany.

REFERENCES

Akashi, K. Noriyuki, N., Ishida, Y. and Yokota, A. (2004) Biochem. Biophys. Res. Comm. 323:72-78.

Bilecen, K., Öztürk, Ü.H., Duru, A.D., Sütlü, T., Petoukhov, M., Svergun, D. I. , Koch, M.H.J., Sezerman, U., Cakmak, I. and Sayers, Z. (2005) J. Biol. Chem. 280(14):13701-13711.

Binz, P.A., and Kagi, H.R. (1997). Metallothionein: Molecular evolution and classification. In Metallothionein IV, C.D. Klaassen, ed (Basel Boston Berlin: Birkhauser Verlag), pp. 7-21.

Boulin, C.J., Kempf, R., Gabriel, A., and Koch, M.H.J. (1988) Nucl. Instrum. Meth. A 269:312-320.

Calderone, V., Dolderer, B., Hartmann, H-J., Echner, H., Luchinat, C., Del Bianco, C., Mangani, S. and Weser, U. (2005) Proc. Natl. Acad. Sci. U. S. A. 102(1):51-56.

Cobbett, C., and Goldsbrough, P. (2002) Annu. Rev. Plant Biol. 53:159-182.

Dallinger, R., Lagg, B., Egg, M., Scipflinger, R. and Chabicovsky, M. (2004) Ecotoxicology 13:757-772.

Domenech, J., Mir, G., Huguet, G., Capdevila, M., Molinas, M. and Atrian, S. (2005) Biochimie 88(6):583-593.

Eapen, S. and D'Souza, S. (2005) Biotechn. Adv. 23:97-114.

Gill, S. C. and von Hipple, P. H. (1989) Annal. Biochem. 182(2):319-326.

Guinier, A. (1939) Ann. Physique 12:161-237.

Guo, W.-J., Bundithya, W. and Goldsbrough, P. B. (2003) New Phytologist 159:369-381.

Kille, P., Winge, D.R., Harwood, J.L., and Kay, J. (1991) FEBS Lett. 295:171-175.

Koch, M. H. J. (2006) Chem. Soc. Rev. 35(2):123-133.

Koch, M. H. J., Vachette, P. and Svergun, D. I. (2003) Quart. Rev. Biophys. 36:147-227.

Koch, M.H.J., and Bordas, J. (1983) Nucl. Instrum. and Methods 208:461-469.

Konarev, P. V., Volkv, V. V., Sokolova, A. V., Koch, M. H. J. and Svergun, D. I. (2003) J. Appl. Cryst. 36:1277-1282.

Ma, M., Lau, P.S., Jia Y.T., Tsang, W.K., Lam, S.K.S., Tam, N.F.Y., Wong, Y.S. (2003) Plant Science 164:51-60.

Porod, G. (1982) in "Small-angle X-ray scattering" (Glatter, O., and Kratky, O. eds) pp. 17-51, Academic Press, London.

Rigby, K. E. and Stillman, M. J. (2004) Biochem. Biophys. Res. Commun. 325:1271-1278.

Sayers, Z. (1988) in "Topics in Curent Chemistry" (E. Mandelkow, ed.) pp. 204-232, Springer Verlag, Berlin.

Sayers, Z., Brouillon, P., Vorgias, C., Nolting, H. F., Hermes, C. and Koch, M. H. J. (1993). Eur. J. Biochem. 212:521-528.

Sayers, Z., Brouillon, P., Svergun, D.I., Zielenkiewicz, P., and Koch, M.H.J. (1999) Eur. J. Biochem. 262(3):858-865.

Schroedel, A.and de Marco, A. (2005) BMC Biochemistry 6:10.

Svergun, D.I. (1992) J. Appl. Crystallogr. 25:495–503.

Svergun, D.I. (1999) Biophys. J. 76: 2879-2886.

Vasak, M., and Hasler, D.W. (2000) *Curr. Opin. Chem. Biol.* 4:177-183.

Zangger, K., Shen, G., Oz, G., Otvos, J. and Armitage, I. (2001) Biochem. J. 359:353-360.

Zimeri, A. M., Dhanker, O. P., McCaig, B. and Meagher, R.B. (2005) Plant. Mol. Biol. 58:839-855.

Zhou, J. M. and Goldsbrough, P.B. (1995) Mol. Gen. Genetics 248:318-328.

ALIGNMENT OF MOLECULS IN SOLUTION
NMR Spectroscopy Applications

Aleksan Shahkhatuni
Molecule Structure Research Center, Azatutian 26, 375014, Yerevan, Armenia

Abstract: Various physical methods for achieving partial alignment of dissolved molecules have been briefly reviewed. They include the use of external magnetic or electric fields, electromagnetic or acoustic waves, flow, pressure, electric current, stretching or straining of microporous matrices, various liquid crystalline phases, and solutions of anisotropic disc- or rod-like particles. SR sources can be used in all stages of this research; for the study of partially aligning medium itself, the study of structure and properties of aligned solute molecules, and finally, SR radiation can be used as aligning agent.

Key words: liquid crystals; weakly aligning media; NMR spectroscopy; molecular structure determination; anisotropic media; alignment of molecules;

1.   INTRODUCTION

During the past 10 years the methods to control the alignment and orientation of solute molecules in liquids are in the focus of intensive experimental activity. Although controlling the alignment always was one of the most exciting goals of physical chemistry, this research was expanded rapidly, when its potentials for the 3D structure determination of solute molecules, including biomacromolecules, had been demonstrated by NMR spectroscopy[1,2]. NMR spectroscopy in isotropic solution was already established as the unique method for the study of structure and dynamics of proteins and nucleic acids. The introduction of weak alignment makes measurable the important anisotropic parameters lost in conventional NMR. These parameters represent the anisotropic interactions such as magnetic dipole-dipole interaction, chemical shift anisotropy, or electric quadrupole

153

Vasili Tsakanov and Helmut Wiedemann (eds.), Brilliant Light in Life and Material Sciences, 153–163.
© 2007 *Springer*.

interaction and contain structural information. In isotropic solution they are fully averaged to zero due to rapid isotropic reorientation of molecules. When molecules are partially aligned, the averaging of anisotropic interaction becomes incomplete and residual dipolar couplings (RDCs), residual chemical shift anisotropies and residual electric quadrupole interactions appear in NMR spectra. Residual anisotropic parameters, especially RDCs, are very sensitive to structural changes, can be measured quite accurately, and are extremely useful for molecular structure and dynamics determination.

The orientation of molecules in solutions can be achieved by two approaches; using already oriented medium (liquid crystals, solution of aligned nanorods or nanodiscs) as a solvent, where dissolved molecules of interest align by the physical interaction (steric, electrostatic, etc.) with the elements of media; or by directly aligning molecules, if they have large anisotropy of one of their own physical properties (magnetic, electric, steric, etc.). The combination of these two approaches is also possible.

Here we will consider briefly some aspects of alignment of molecules of interest in liquid and liquid-crystalline (LC) phases.

The alignment of molecules in gas phase is also a topic of intensive study (see, for example. review[3]). Here we will consider only the alignment of molecules in solution by the means of NMR spectroscopy.

2.   ALIGNMENT OF MOLECULES IN SOLUTION

The central dogma for alignment is the following:

- The anisotropy of certain molecular properties can cause an interaction with the corresponding anisotropic external field.
- As a result of such interaction, the molecules with different orientation of their axis relative to the direction of external field have different energies.
- The difference in energies provides a rotational moment, which will turn molecules and induce their alignment.
- An alignment can be detected by observation of changes in NMR parameters (RDCs, etc.).
- The induced alignment is in competition with Brownian motion.
- The orientation of the molecules obeys a Boltzmann distribution.

In optics the partial orientation of the molecules lead to birefringence, which can be detected by appropriate techniques.

In NMR the partial orientation of molecules causes incomplete rotational averaging of dipolar and quadrupolar couplings, which leads to additional or new splittings in high-resolution NMR spectra, and produce anisotropies of chemical shifts leading to the shifts of the lines.

The electric or magnetic fields, pressure, mechanical stretching or straining of microporous matrices, electric current, flow, electromagnetic or acoustic waves, can be used as an aligning anisotropic agent. Liquid crystals (LCs), solution of anisotropic particles, polymer gels, zeolites, etc. can be used as an aligning environment. Many of these factors had been already mentioned in brilliant review of Buckingham and McLauchlan[4] in 1967. In Table 1 the different anisotropic interactions inducing alignment are represented.

Table 1. Anisotropic interactions and phenomena inducing molecular alignment.

Orienting agent	Orienting phenomenon	Molecular property
magnetic field	diamagnetic anisotropy	magnetic dipole moment
electric field	dielectric anisotropy	dipole moment and/or anisotropy of polarizability
electric field	electric field gradient	quadrupole moment
electromagnetic wave	dielectric anisotropy	anisotropy of polarizability
acoustic wave	density gradient	shape anisotropy
pressure	density gradient	shape anisotropy
electric current	ions velocity gradient	shape anisotropy
flow	velocity gradient	shape anisotropy
anisotropic particles	anisotropy of rotational viscosity	shape anisotropy
anisotropic cavities	anisotropy of rotational viscosity	shape anisotropy
liquid crystals	anisotropy of rotational viscosity	shape anisotropy

3. STRONG ALIGNMENT IN LIQUID CRYSTALS

In NMR experiments LC medium usually has fixed orientation of LC director, depending on the sign of anisotropy of diamagnetic susceptibility of LC, $\Delta\chi$ (director of LC is parallel to magnetic field direction for positive $\Delta\chi$ and perpendicular for negative ones). It is possible to control the angle between the magnetic field and the LC director by spinning sample at various angles. The orientation of the solute molecule in LC solvents depends on the mechanisms of solute orientation and the alignment of orienting LC medium in the magnetic field itself. The mechanism of solute orientation in LC media can be steric and/or electrostatic. Steric mechanism relies on the anisotropy of the shape of molecule, whereas electrostatic one depends on the charge distribution in the medium and molecule.

The basic property of LC medium is a **strong orientation** of solute molecules achieved in it except for a few molecules of tetrahedral symmetry. Strong orientation leads to dramatic growth of number of lines in NMR spectrum, and the interpretation becomes extremely complex due to overlap of lines. This phenomenon limits the number of magnetic nuclei in molecule, which can be successfully investigated, to 10-12. Up today 700 molecules containing 4-9 magnetic nuclei were studied in different LC media.

Thermotropic nematic liquid crystals are traditional and widely used media for the structural studies of small organic compounds. A great number of LCs and LC mixtures had been proposed. Due to theoretical estimations, the precision of determined structural parameters was expected to be very high, but in practice so called "solvent effects", i.e. wide spread of values of calculated structural parameter, observed in different thermotropic LCs at different experimental conditions (temperature, concentration, etc.), exceeds the estimated standard error of each experiment by 2-3 orders. Although many attempts to interpret and correct the observed solvent effects have been done, but none of them is universally applicable. The reason of these effects is not clear till now[5]. Recently we have shown that the large spread of structural data do not occur in all types of LCs and is not random, but inversely proportional to the order parameter of molecule in certain thermotropic LCs[6]. We also found the correlation of solvent effects with the signs and values of diamagnetic and dielectric anisotropies of LCs[7].

Lyotropic LCs were relatively rarely used as solvents for structure determination of small organic molecules. Recently the use of lyotropic LCs increased rapidly due to the idea of weak alignment, which can be easily done in lyotropic systems. A number of such systems have been developed. Our studies of some model molecules (halomethanes, benzene, furan, thiophene) in different weakly aligning lyotropic LCs, based on charged or nonpolar surfactants show their relevance as solvents for molecular structure determination[6-8].

4.   WEAK ALIGNMENT

Weak alignment of molecules can be attained by a greater number of methods than strong alignment. The type of alignment depends on the specific interactions of molecule with aligning forces or particles present in solution. The orienting properties of various proposed weakly aligning media for biomacromolecules have been described in detail in many reviews[9-11]. Currently, intensive search for new methods and new media for orientation of biomolecules and large organic molecules is in progress.

4.1 Alignment by physical factors

Molecules with appropriate anisotropic property under anisotropic physical exposures can be partially aligned. The simplest exposures are magnetic or electric fields, electromagnetic or acoustic waves, flow gradient.

4.1.1 Magnetic field

Molecules with non-zero anisotropy of magnetic susceptibility can be aligned by the applied magnetic field. The orientation of molecules in solution obeys Boltzmann's law with the degree of alignment being proportional to the square of magnetic field strength. In optics the phenomenon is known as magnetooptical (or Cotton-Mouton) effect. In NMR experiments the strong magnetic field always exists, but in practice, the degree of alignment of molecules is slight and as a rule can not be observed under normal conditions, except for complexes with paramagnetic metals. The partial alignment of molecules induced by magnetic field was first observed by Bothner-By and coworkers[12]. With the progress in NMR hardware, the magnetic fields in the spectrometers became stronger; therefore, the orientation effects are observed more and more frequently[13] and they should be taken into account for precise determination of spin-spin coupling constants[14].

Large and anisotropic molecules can be oriented by magnetic field easier. For instance, it was shown[15] that biological molecules, for example DNA duplexes and quadruplexes, are aligned in the magnetic field of the spectrometer with the measurable RDCs.

To enhance the alignment of molecules in magnetic field, it is necessary to increase the $\Delta\chi$ of understudy molecules or orienting system (membranes, media, etc.). This can be done, for example, by adding paramagnetic metal ions (mainly, lanthanide ions), which can form a paramagnetic complex with molecule, better aligned in the magnetic field.

4.1.2 Electric field

Alignment of molecules in electric field is well known phenomenon (so called electrooptical or Kerr effect). NMR spectroscopy using DC electric field for alignment was proposed and observed by Buckingham[16-17]. DC electric field has many disadvantages connected with accompanying processes (possible electrochemical and electroconvective processes, etc.), therefore, pulsed or variable AC field is preferable. Despite the certain difficulties, alignment by electric field is a very promising method. Especially it is connected with the possibility to develop two-dimensional (2D) NMR spectroscopy to examine correlations between isotropic and electrically induced anisotropic spectral effects, which can be detected in NMR experiments and used for molecular structure investigations [18-19].

In LCs an important mechanism of solute ordering based on anisotropic interaction between solute molecular quadrupole moment and the average solvent electric field gradient has been presented by Burnell and de Lange[5].

4.1.3 Polarized electromagnetic wave

The alignment of molecules in electric field of electromagnetic wave was also theoretically predicted by Buckingham in 1955[20], and experimentally was confirmed by Mayer and Gires. In optics the alignment manifests in the birefringence resulting from the application of an intense beam of polarized light. The mentioned phenomenon is known as an optical Kerr effect, in contrast to electrooptical Kerr effect, discussed above.

Light induced alignment depends on the strength of the wave electric field and rotational energy of molecule. Molecules align with their axis of largest polarizability along the electromagnetic wave polarization axes.

As a source of intense polarized electromagnetic wave, a laser or synchrotron radiation (SR) can be used. The later is preferable, because alignment effects depend on $\mathbf{E}^2/\lambda$, where $\mathbf{E}$ is an electric field and λ is a wavelength of radiation. Using soft X-Ray SR instead of laser we need 3 orders lower intensity of radiation for achieving the same aligning effect.

In gases, by using a strong linearly polarized laser pulse in a seeded supersonic beam, a very high degree of alignment of molecule can be achieved[3]. Even three dimensional alignment is possible, if elliptically polarized laser fields are used[3].

In solution the achieved alignment of molecules is very small due to their fast reorientation and tumbling. For weak alignment approach it was perfect. In liquid crystals the effect of irradiation is stronger due to interaction with whole media, sometimes leading to realignment of director.

By changing the linear polarization plane of electromagnetic wave we can manipulate with solute molecules, for example, can force them to spin. In combination with the real sample spinning along one axes we can simultaneously rotate molecules in sample along another axes by light.

4.1.4 Flow gradient

Alignment of molecules also can be induced by flow gradient (so called Maxwell effect). In laminar flow with the different velocities of shear the orientation of molecules with anisotropic shape can be produced. In simplest case, when the gradient of velocities is perpendicular to flow direction, molecules align with their long axis at angle 45° with respect to flow direction. This phenomenon now is successfully used for visualization of flow by reological NMR[21].

The similar alignment can be produced by electric current in liquids, when the movement of ions occurs in presence of velocity gradient.

4.1.5 Acoustic wave

It is known that acoustic wave can produce the birefringence in liquids (acoustooptic effect). The interaction between acoustic (ultrasonic) wave and a molecule depends on anisotropic molecular shape, frequency and intensity of acoustic wave. The potential energy of this interaction has minimum, when the long molecular axes is aligned along or perpendicular to the wave direction depending from the wave phase; along the wave propagation direction in stretching phase and perpendicular in compressing phase.

4.2 Weak aligning lyotropic systems

A number of lyotropic LC media allow achievement of weak alignment of macromolecules. First it was convincingly demonstrated in a mixture of two phosphoplipides, forming so called bicellar system[1]. Many other weakly aligning surfactant based lyotropic LC systems were proposed after that[9-11]. The exact morphology of these systems is still under debate. Among possible constitution the following ones are considered: lamellae, strongly perfored lamellae, wormlike structures, anisotropic disc-like or rod-like micelles, etc. Irrespective of the morphology of the phase, these media orient in magnetic field and provide weak alignment of dissolved molecules.

4.3   Alignment by particles

Therefore, instead of anisotropic micelles it is more preferable to use colloidal solutions of more stable particles able to be oriented and to orient dissolved molecules. Some suspensions of relatively rigid, charged or neutral, rod- or disc-like organic or inorganic nano-size moieties can be self-assembled or oriented by magnetic field. A number of different aqueous mixtures of such particles were proposed as solvents for biomolecular structure determination. Among them we can mention rod-shaped viruses or filamentous phages, purple membrane fragments, inorganic filamentary crystallites, polymeric crystallites and other relatively rigid filamentary polymeric structures, cellulose crystallites, natural or artificial nano- or micro- tubes, mineral liquid crystals[22], etc.

4.4   Alignment in porous media

The effect of orientation of small molecules incorporated in rubber-like polymers was known long ago[23]. The pore sizes and the diffusion properties of the gels can be changed either by controlling the concentration of ingredients during the preparation, or by mechanical treatment (compression or stretching) after preparation. A distinctive feature of polymer gels is that the direction of their orientation is defined by the preparation procedure and does not depend on magnetic field. This allows one to change the gel orientation smoothly with respect to the direction of magnetic field.

Another option is to incorporate other particles into the gel during polymerization, for example, membrane fragments, phages and viruses able to orient proteins. Polymerization is carried out in magnetic field and, hence, the particles are incorporated in gel in the oriented state. The orientation of particles immobilized in the gel during polymerization can be attained also using other types of external macroscopic, for example electric or mechanical, treatment. The orientation of molecules introduced into the gel is due to the interaction with the oriented particles.

There is a large number of substances which form cavities during crystallization or solidification, big enough to accommodate guest molecules or water solutions inside their crystal lattices or cavities. In some of them continuous unidirectional channels can be formed. The most interesting are natural or synthetic zeolites and molecular sieves[4].

5.   OTHER NMR SPECTROSCOPY APPLICATIONS

Except the trivial NMR spectroscopy in oriented molecules described above, the possibility of inducing, switching and tuning alignment of molecules allows to develop different types of multidimensional experiments based on correlation of isotropic and anisotropic parameters[24] (Table 2).

Table 2. Classification of anisotropic-isotropic correlation experiments.

	Sample status	Description of experiment
Type I	Sample is oriented only in one stage of NMR experiment	Orientation can be made by switching on/off different physical influences
Type II	Sample is always oriented	Anisotropic interactions can be removed by manipulation of rotations in space or spin space

Type I experiments (pulse synchronized molecule orientation) can be implemented using different fast switchable physical influences leading to alignment of molecules, for instance, applying AC or DC electric field[18,19].

Type II experiments (pulse synchronized sample or spin reorientation) can be implemented using modulations of anisotropic interactions by manipulations of sample (changing spinning axes, speed, or orientation) or by manipulations with spins in spin space (for example, by using multipulse sequence for coherent reduction of anisotropic interactions).

6.   NMR SPECTROSCOPY AND SR

Small angle X-ray scattering (SAXS) is a powerful technique to study molecular organization and dynamics of lyotropic LC systems. SR sources can be used to study partially aligning medium itself, the achieved alignment, the properties of aligned solute molecules, their crystal structure for the comparison with NMR solution structure, as well as aligning agent.

SR controlled NMR spectroscopy provides a great perspective for molecular structure determination in solution. SR can be used for manipulation of solute molecules or for reorientation of LC director. SR aligning effects can be expanded by using dyes or additional electric field.

7.   CONCLUSION

In the last years the search of the methods of weak alignment of molecules in solution is rapidly growing because of their necessity for precise 3D molecular structure determination by NMR spectroscopy. Despite many proposed solutions, the search of perfect weakly aligning media and methods, allowing to have any desired type of alignment and to manage it precisely in a very wide range, including the disordered (isotropic) state, is only at the beginning. The modern structural and functional genomics requires and initiates further studies, which will be very useful for a large variety of other applications in different fields of chemistry, physics, biology, material science and nanotechnology as well.

Here we have considered the basic methods of alignment. Indeed, the simultaneous use of two or more methods will increase the possibilities and quality of achieved alignment. It is especially important to get controlled and tunable switching of the direction and the degree of alignment.

REFERENCES

1.   N. Tjandra and A. Bax, Direct measurement of distances and angles in biomolecules by NMR in a dilute liquid crystalline medium. Science, 278, 1111-1114 (1997).
2.   J. H. Prestegard, New techniques in structural NMR – Anisotropic interactions, Nat. Struct. Biol., NMR Suppl., 5, 517-522 (1998).
3.   H. Stapelfeldt and T. Seideman, Aligning molecules with strong laser pulses. Rev. Mod. Phys., 75, 343-359 (2003).
4.   A. D. Buckingham and K. A. McLauchlan, in Progress in NMR Spectroscopy, edited by J. W. Emsley, J. Feeney, and L. H. Sutcliffe (Pergamon Press, Oxford, 1967), pp. 63-109.
5.   E. E. Burnell and C. A. de Lange, in NMR of Ordered Liquids, edited by E. E. Burnell and C. A. de Lange, (Kluwer, Norwell, MA, 2003), pp 221-240.
6.   A. G.Shahkhatuni, A. A. Shahkhatuni, H. A Panosyan, G. H. J. Park, R. W. Martin, and A. Pines, NMR studies of 13C-iodomethane: Comparison of spectral behaviors in thermotropic and lyotropic liquid crystals. J. Phys. Chem. A, 108(33), 6809-6813 (2004).
7.   A. A. Shahkhatuni, Solvent independence of structural parameters of furan and thiophene in some liquid crystalline solvents: J. Mol. Struct., 743, 217-222 (2005).
8.   A. A. Shahkhatuni, A. G. Shahkhatuni, H. A. Panosyan, A. B. Sahakyan, I.-J. L. Byeon, and A. M. Gronenborn, Assessment of solvent effects imposed by alignment media – Weak alignment does not perturb the structure of the solute, Magn. Reson. Chem., (submitted).
9.   J. H. Prestegard, C. M.Bougault, and A. I. Kishore, Residual dipolar couplings in structure determination of biomolecules. Chem. Rev., 104(8), 3519-3540 (2004).

10. A. A. Shahkhatuni. and A. G. Shahkhatuni, Determination of the three-dimensional structure for weakly aligned biomolecules by NMR spectroscopy. Rus. Chem. Rev., 71(12), 1005-1040 (2002).

11. A. M. Gronenborn, The importance of being ordered: improving NMR structures using residual dipolar couplings. C. R. Biol., 325(9), 957-966 (2002).

12. C. Gayathri, A. A. Bothner-By, P. C. M. van Zijl, and C. MacLean, Dipolar magnetic field effects in NMR spectra of liquids, Chem. Phys. Lett., 87(2), 192-196 (1982).

13. E. W. Bastiaan and C. MacLean, Molecular orientation in high-field high-resolution NMR. NMR Basic Princ. Progr., 25, 17-43 (1990).

14. A. V., Makarkina, S. S. Golotvin, and V. A. Chertkov, Orientation of benzofuran by magnetic field in isotropic liquid phase. Chem. Heterocycl. Comp., 31(9), 1214-1219 (1995).

15. H. C. Kung, K. Y. Wang, I. Goljer, and P. H. Bolton, Magnetic alignment of duplex and quadruplex DNAs. J. Magn. Reson. B, 109(3), 323-325 (1995).

16. A. D. Buckingham and K. A. McLauchlan, The absolute sign of the spin-spin coupling constant. Proc. Chem. Soc., 144 (1963).

17. C. W. Hilbers and C. MacLean, NMR of molecules oriented in electric fields. NMR Basic Princ. Progr., 7, 1-52 (1972).

18. A. Peshkovsky and A. E. McDermott, NMR spectroscopy in the presence of strong AC electric fields: The degree of alignment of polar molecules, J. Phys. Chem. A, 103(43), 8604-8611 (1999).

19. S. A. Riley and M. P. Augustine Extracting residual NMR coupling constants from electrically aligned liquids. J. Phys. Chem. A, 104(15), 3326-3331 (2000).

20. A. D. Buckingham and J. A. Pople, Theoretical studies of the Kerr effect. 1: Deviations from a linear polarization law. Proc. Phys. Soc. A, 68, 905-909 (1955).

21. P. T. Callaghan, Rheo-NMR: a new window on the rheology of complex fluids, Encyclopedia Magn. Reson., 9, 739–750 (2002).

22. J. C. P. Gabriel and P. Davidson, Mineral liquid crystals from from self-assembly of anisotropic nanosystems. Top. Curr. Chem., 226, 119-172 (2003).

23. B. Deloche and E. T. Samulski, Short-range nematic-like orientational order in strained elastomers: a deuterium magnetic resonance study. Macromolecules, 14(3), 575-581 (1981).

24. A. Bax, N. M. Szeverenyi, and G. E. Maciel, Correlation of isotropic shifts and chemical shift anisotropies by two-dimensional Fourier-transform magic-angle hopping NMR spectroscopy. J. Magn. Reson., 52(1), 147-152 (1983).

STRUCTURAL INVESTIGATION OF ORDERING IN BIOPOLYMERS

Vladimir F. Morozov, Yevgeni Sh. Mamasakhlisov, Anna V. Mkrtchyan, Artem V. Tsarukyan, Tatyana Yu. Buryakina, Shushanik Tonoyan, Sergey V. Mkrtchyan

Department of Molecular Physics, Yerevan State University, A. Manougian Str.1, 375025, Yerevan, Armenia.

Abstract: The Generalized Model of the Polypeptide Chain is developed for the conformational transitions in double-strand polynucleotides. The melting temperature and interval and the correlation length have been calculated. The inter-chain interaction, effects of heterogeneity and the stacking-interaction influence have been incorporated into the model and the corresponding parameters have been estimated.

Key words: helix-coil transition, DNA melting, correlation length, cooperativity

1.   INTRODUCTION

The helix - coil transition phenomenon in biopolymers has been known since the 1960's[1-7] and is still vigorously discussed[8-15]. Traditionally the theoretical models for the transition assume that each base pair can be in either the helical or the coil state. That is why it is convenient to use the Ising model[16-19] or to calculate the free energy directly as though the system were a dilute one-dimensional solution of helix and coil junctions[6]. Most traditional theories use the mean-field approximation, i.e., the Hamiltonians of these models include parameters that are averaged over all conformations of the molecule (e.g. the cooperativity parameter[16-19] or the junction free energy[6]). Some microscopic theories have been reported for polypeptides[20-22] and for DNA[9-13] that do not invoke the mean field

Vasili Tsakanov and Helmut Wiedemann (eds.), Brilliant Light in Life and Material Sciences, 165–174.
© 2007 *Springer.*

assumption and take into account many of the important structural features and interactions in biopolymers. Thus, many interesting models and approaches have been offered and much is known about the helix-coil transition. However, some important problems remain. In particular, the microscopical theory of heteropolymer melting, where the free energy is explicitly evaluated (i.e. not using a mean field approach) is interesting not only for analyzing heteropolymer melting, but is generally useful for the theory of disordered systems, with applications to many problems of principal interest[2].

Most heteropolymer approaches are based on the Zimm-Bragg model[1,16-19], a version of the 1-D Ising model. One exact and many approximate solutions for heteropolymer melting, within the scope of the Zimm-Bragg model for random and Markoff distributions of disorder[1], have been available since the 1960's.

In previous publications[23,24] we reported applications of the microscopic Potts-like model, with many particle interactions, to investigate the helix-coil transition in DNA's. Earlier we successfully took into account inhomogeneities for heteropolypeptides, in which the number of possible conformations of each repeated unit is different (each particular amino acid residue had a particular number of rotational isomers). It was shown, that the heterogeneity of polypeptides is not crucial for calculating the helix-coil transition and can be averaged out in the Hamiltonian with the help of redefinition of a number of possible conformations of each repeated unit[25].

2. THE BASIC MODEL

A microscopic Potts-like one-dimensional model with Δ-particle interactions describing the helix–coil transition in polypeptides was developed in[25, 26]. Subsequently, it was shown that the same approach could be applied to DNA if the large-scale loop factor is ignored[23]. The model was called as Generalized Model of Polypeptide Chain (GMPC). The Hamiltonian of GMPC has the form

$$-\beta H = J\sum_{i=1}^{N}\delta_i^{(\Delta)} , \qquad (1)$$

where the sum is over all repeated units; $\beta = T^{-1}$ is inverse temperature; N is the number of repeated units; $J = U/T$ is the temperature-reduced energy of interchain hydrogen bonding;

$$\delta_j^{(\Delta)} = \prod_{k=\Delta-1}^{0} \delta(\gamma_{j-k},1), \text{ with Kronecker } \delta(x,1);$$

γ_l is the spin that can take on values from 1 to Q, and describes the conformation of l-th repeated unit. The case when γ_l is equal to 1 denotes the helical state. The other $(Q-1)$ cases correspond to coil states. Q is the number of conformations of each repeated unit and thus corresponds to the conformational variability. The Kronecker delta inside the Hamiltonian ensures that energy J emerges only when all Δ neighboring repeated units are in the helical conformation. Thus, restrictions on backbone chain conformations imposed by hydrogen bond formation are taken into account[23].

The transfer - matrix, corresponding to the Hamiltonian Eq. (1) has the form:

$$\hat{G}(\Delta \times \Delta) = \begin{pmatrix} W & 1 & 0 & \dots & 0 & 0 & 0 \\ 0 & 0 & 1 & \dots & 0 & 0 & 0 \\ \dots & \dots & \dots & \dots & \dots & \dots & \dots \\ 0 & 0 & 0 & \dots & 0 & 0 & Q-1 \\ 1 & 1 & 1 & \dots & 1 & 1 & Q-1 \end{pmatrix}, \tag{2}$$

where all elements of first row are equal to $W = \exp(J)$; all elements of first lower pseudodiagonal are 1; the (Δ, Δ) element is Q; all other elements are zero. The secular equation for this matrix is:

$$\lambda^{\Delta-1}(\lambda - W)(\lambda - Q) = (W-1)(Q-1) \tag{3}$$

Within the scope of the proposed model the correlation length of the system can be evaluated as:

$$\xi = \left[\ln \lambda_1 / \lambda_2\right]^{-1} \tag{4}$$

Where λ_1 is the largest eigenvalue and λ_2 is the second largest. Previously, the transition point was found[23-26] via the condition $W \approx Q$ as $T_m = U / \ln Q$. At this point the correlation length ξ passes through a maximum:

$$\xi_{max} \sim Q^{\frac{\Delta-1}{2}} \tag{5}$$

The parameters s and σ of Zimm-Bragg theory can be obtained from our model[26] as

$$\sigma = \xi_{max}^{-2} \quad \text{and} \quad s = \frac{W}{Q}$$

Using this model it was possible to find expressions for the degree of helicity, for the average number of junctions between helix and coil sections, and therefore for the mean length of helical sections[25]. Within the scope of this model it is also possible to describe the influence of solvent[26]. The topological restrictions, imposed on the helix-coil transition in circular closed DNA may be taken into account as well[27]. Important stacking interactions were recently included[24].

3.   INTERACTION BETWEEN THE TWO GMPC CHAINS

Let us consider the two chains, describing by the GMPC, which are interacting as follows

$$-\beta H = J\sum_{i=1}^{N}\delta_i^{(\Delta)} + J\sum_{j=1}^{N}\delta_j^{(\Delta)} +$$

$$+I_1\sum_{i=j=1}^{N}\delta_i^{(\Delta)}\delta_j^{(\Delta)} + I_2\sum_{i=j=1}^{N}\delta_i^{(\Delta)}(1-\delta_j^{(\Delta)}) + I_3\sum_{i=j=1}^{N}(1-\delta_i^{(\Delta)})\delta_j^{(\Delta)} + I_4\sum_{i=j=1}^{N}(1-\delta_i^{(\Delta)})(1-\delta_j^{(\Delta)}) = \tag{6}$$

$$= (J+I_2-I_4)\sum_{i=1}^{N}\delta_i^{(\Delta)} + (J+I_3-I_4)\sum_{j=1}^{N}\delta_j^{(\Delta)} + (I_1-I_2-I_3+I_4)\sum_{i=j=1}^{N}\delta_i^{(\Delta)}\delta_j^{(\Delta)} + NI_4$$

The inter-chain interaction includes the four types of contacts: hh, hc, ch, cc.

Because of symmetry of problem $I_2=I_3$ and $Const = NI_4$ let us introduce the following notations

$$J \equiv (J+I_2-I_4) = (J+I_3-I_4)$$

$$I_1 \equiv (I_1-I_2-I_3+I_4)$$

Using these notations the Hamiltonian of the system can be presented as

$$-\beta H = J \sum_{i=1}^{N} \delta_i^{(\Delta)} + J \sum_{j=1}^{N} \delta_j^{(\Delta)} + I \sum_{i=j=1}^{N} \delta_i^{(\Delta)} \delta_j^{(\Delta)} \tag{7}$$

The effective energy of the inter-chain interaction can be presented as $I = \alpha \cdot J$, where we need to consider the following cases: $\alpha > 0$, $\alpha < 0$ and $\alpha = 0$.

The transfer-matrix, corresponding to Hamiltonian (7) is

$$M = g \otimes g + e^{2J}\left(e^{\alpha} - 1\right)R$$

Where R is Δ^2range matrix, where only 11 element is 1, and other elements are 0.

The helicity degree is defined by equation

$$\theta = \frac{1}{2NZ} \frac{\partial Z}{\partial J}$$

The dependence for the helicity degree on the reduced temperature t at the different values of α presented on the Fig.1(left). It is easy to see that the growth of the positive values of α leads to the increasing of the melting temperature and decreasing the melting interval. So, the DNA-DNA interaction additionally stabilized the secondary structure and increased the cooperativity. These results are confirmed by the behavior of correlation length (Fig.1 right).

For the negative values of α the melting curve become two-steps with the shifting left the first step (see Fig. 2 left), forming the plato $\theta = 1/2$. The investigation of the correlation length (Fig. 2 right) shows that on the plato $\theta = 1/2$ the correlation length reach its maximum value. Thus, the sharpness of the melting transition is now the same as cooperativity which is naturally to measure by the correlation length.

Effect of random Heterogeneity: The heterogeneous DNA modeling by Hamiltonian

$$-\beta H = \sum_{i=1}^{n} J_i \delta_i^{(\Delta)}, \tag{8}$$

where the energy of hydrogen bond formation is dependent on base pair number along the chain. In other words:

$$J_i = \begin{cases} J_{AT}, \text{if } i-th \text{ base pair is of } A-T \text{ type}; \\ J_{GC}, \text{if } i-th \text{ base pair is of } G-C \text{ type} \end{cases}$$

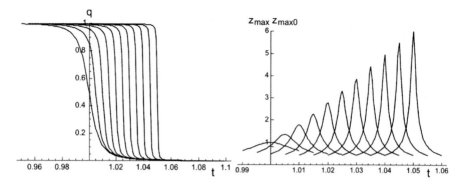

Figure 1. left)Helicity degree vs. reduced temperature at *the* $\alpha(0 \div 0.1)$ *and* $\Delta = 3, Q = 60$; right) Correlation length vs. reduced temperature at the $\alpha(-0.1 \div 0)$ *and* $\Delta = 3, Q = 60$

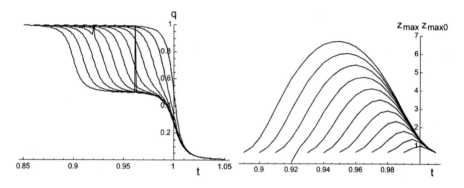

Figure 2. left)Helicity degree vs .reduced temperature at the $\alpha(-0.1 \div 0)$ *and* $\Delta = 3, Q = 60$; right) *Correlation length* ξ *vs. reduced temperature at the* $\alpha(-0.1 \div 0)$ *and* $\Delta = 3, Q = 60$

Using the transfer-matrix approach, the partition function for a given sequence of base pairs (given the realization of disorder) may be written as:

$$Z_{seq} = \text{Tr} \prod_{i=1}^{n} \hat{G}_i; \tag{9}$$

Using the microcanonical approach we obtain the expression for the free energy[28, 29]:

$$\begin{cases} f(y) = -q \ln q - (1-q) \ln(1-q) - \ln\left[\dfrac{(1+y)\lambda_1(y)}{y^{1-q}}\right]. \\[2mm] \dfrac{d}{d \ln y}\left[\ln(1+y)\lambda_1(y)\right] = 1-q \end{cases} \tag{10}$$

Here $\lambda_1(y)$ is the largest eigenvalue of the transfer matrix of the basic model with the redefinition

$$W(y) = \frac{W_{GC} + yW_{AT}}{1+y}.$$

Thus the heteropolymer reduces to the homopolymeric case.

To obtain the transition point, we have investigated the behavior of the free energy (10) and the following results have been obtained. The melting temperature as:

$$T_m = qT_{GC} + (1-q)T_{AT} \tag{11}$$

and an expression for the heteropolymer melting interval as:

$$\Delta T = 2q(1-q)T_m \left(\frac{T_{GC} - T_{AT}}{T_m}\right)^2 \ln Q \tag{12}$$

4. HOMOGENEOUS STACKING

The basic model considers cooperativity as conditioned by chain rigidity brought about by hydrogen bonding. In other approaches the cooperativity is considered to also arise from so-called stacking interactions[1,2,5]. Stacking interactions can be taken into account by analogy with Hamiltonian (1),

$$-\beta H = J\sum_{i=1}^{N}\delta_i^{(\Delta)} + I\sum_{i=1}^{N}\delta_i^{(2)} \tag{13}$$

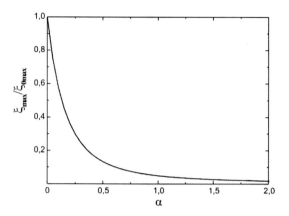

Figure 3. Dependence of reduced maximum correlation length , $\dfrac{\xi_{max}}{\xi_{0\,max}}$, on α for values

$\Delta = 15$, $Q = 3$, characteristic for DNA [23].

The first term in Eq. (13) is the same Hamiltonian in Eq. (1), that describes longitudinal Δ- range interactions. The second term describes nearest-neighbor-range stacking interactions. Stacking fixes two nearest-neighbor repeated units in the helical conformation. Here $I = E/T$ is the reduced energy of stacking interactions. The Kronecker $\delta_i^{(2)}$ ensures that the reduced energy I emerges when two nearest - neighboring repeated units are in the same, i.e. helical conformation. The transfer matrix for the model with the Hamiltonian Eq. (13) is given by,

$$\hat{G}(\Delta) = \begin{pmatrix} VR & VR & VR & \dots & VR & VR \\ R & 0 & 0 & \dots & 0 & 0 \\ 0 & R & 0 & \dots & 0 & 0 \\ \dots & \dots & \dots & \dots & \dots & \dots \\ 0 & 0 & 0 & \dots & R-1 & R-1 \\ 0 & 0 & 0 & \dots & 1 & Q \end{pmatrix}, \tag{7}$$

where $R = \exp[I]$; $VR = W - R$; $W = \exp[J + I]$.

The behavior of the maximum correlation length is presented on the Fig. 3. It can be see on the Fig. 3 that the reduced maximum correlation length decreases with increased stacking energy (α). Thus, increased stacking energy relative to hydrogen bonding energy results in a decreased correlation range.

5. HETEROGENEOUS STACKING IN REGULAR SEQUENCE

To construct a heterogeneous sequence DNA model, consider that the energies of hydrogen bonding J_i and stacking interaction I_i depend on the base pair number along the chain.

The Hamiltonian for this system is

$$-\beta H = \sum_{i=1}^{N} J_i \delta_i^{(\Delta)} + \sum_{i=1}^{N} I_i \delta_i^{(2)} \tag{14}$$

We have considered the behavior of regular heterogeneous sequences and the following results have been obtained.

In case of homogeneous hydrogen bonding the stacking heterogeneity results to increasing cooperativity in some cases. Proper inclusion of stacking interaction heterogeneity and investigations of the influence on the cooperativity of the duplex DNA melting produced somewhat unexpected results. There are heterogeneous systems that are more cooperative than the homogeneous system. In general, our results indicate melting cooperativity of duplex DNA, that is heterogeneous in both hydrogen bonding and stacking interactions depends on both types of interactions which, in several different ways compete with each other; and in other cases they both provide parallel effects on melting cooperativity.

REFERENCES

1. Poland, D.C.; Scheraga, H.A. The Theory of Helix-Coil Transition; Academic Press: New York, 1970.
2. Grosberg, A.Yu.; Khokhlov, A.R. Statistical Physics of Macromolecules; AIP Press: New York, 1994; Chapter 7.
3. Cantor, C.R.; Shimmel, T.R. Biophysical Chemistry; Freeman and Co.: San-Francisco, 1980; Chapter 20.
4. Flory, P.J. Statistical Mechanics of Chain Molecules; Interscience: New York, 1969; Chapter 7.
5. Wartell, R.M.; Benight, A.S. Phys Rep 1985, 126 (2), 67-107.
6. Vedenov, A.A.; Dykhne, A.M.; Frank-Kamenetskii, M.D. Usp Phys Nauk (in Russian) 1971, 105, 479-519.
7. Wada, A.; Suyama, A. Prog Biophys Mol Biol 1986, 47, 113-157.
8. Chalikian, T. Biopolymers 2003, 70, 492-496.
9. Garel, T.; Monthus, C.; Orland, H. Europhysics Lett 2001, 55 (1), 132-138.
10. Cule, D.; Hwa, T. Phys Rev Lett 1997, 79 (12), 2375-2378.
11. Baiesi, M.; Carlon, E.; Orlandini, E; Stella, A. L. Cond-Mat/0207122 2002.
12. Peyrard, M. Nonlinearity 2004, 17 R1.

174 *V. Morozov, Y. Mamasakhlisov, A. Mkrtchyan, A. Tsarukyan et al.*

13. Barbi, M.; Lepri, S.; Peyrard, M.; Theodorakopoulos, N. Phys Rev E 2003, 68, 061909.
14. Takano, M.; Nagayama, K.; Suyama, A. Journal Chem Phys 2002, 116(5), 2219-2228.
15. Munoz, V.; Serrno, L. Biopolymers 1997, 41, 495-509.
16. Zimm, B.H.; Doty, P.; Iso, K. Proc Natl Acad Sci USA 1959, 45, 1601-1607.
17. Zimm, B. H.; Bragg, J.K. Journal Chem Phys 1959, 31, 526-535.
18. Zimm, B.H. Journal Chem Phys 1960, 33 (5), 1349-1356.
19. Zimm, B.H.; Rice, N. Mol Phys 1960, 3 (4), 391-407.
20. Lifson, S.; Roig, A. Journal Chem Phys 1961, 34, 1963-1974.
21. Lifson, S.; Zimm, B.H. Biopolymers 1963, 1, 15-23.
22. Lifson, S.; Allegra, J. Biopolymers 1964, 2, 65-68.
23. Morozov, V.F.; Mamasakhlisov, E.Sh.; Hayryan, Sh.A.; Chin-Kun Hu Physica A 2000, 281 (1-4), 51-59.
24. Morozov, V.F.; Badasyan, A.V.; Grigoryan, A.V.; Sahakyan, M.A.; Mamasakhlisov, E.Sh. Biopolymers 2004, 75, 434-439.
25. Ananikyan, N.S.; Hayryan, Sh.A.; Mamasakhlisov, E.Sh.; Morozov, V.F. Biopolymers 1990, 30, 357-367.
26. Hayryan, Sh.A.; Mamasakhlisov, E.Sh.; Morozov, V.F. Biopolymers 1995, 35, 75-84.
27. Morozov, V.F.; Mamasakhlisov, E.Sh.; Grigoryan, A.V.; Badasyan, A.V.; Hayryan, Sh.; Chin-Kun Hu Physica A 2005, 348C, 327-338.
28. Crisanti, A.; Paladin, G.; Vulpiani, A. Products of random matrices in statistical physics; Springer-Verlag: Berlin, 1993.
29. Badasyan A.V., Grigoryan A.V., Mamasakhlisov E.Sh., Benight A.S., and Morozov V.F. Journal of Chemical Physics 2005, 123, 194701.

RECONSTRUCTION OF THE ELECTRON TRANSPORT CHAIN OF SECRETORY GRANULE MEMBRANE

Anna Boyajyan
Institute of Molecular Bilogy

Abstract: In the present study we identified full molecular assembly of the electron transport chain of chromaffin granule membrane and demonstrated a modulating effect of phospholipid microenvironment on this chain.

Key words: chromaffin granules, electron transport chain, phospholipids.

1.   INTRODUCTION

A number of key oxidoreduction reactions of vital cell take place in its membrane subdomains with participation of electron transport chains. These reactions have a broad spectrum of actions, beginning from biosynthesis of macroergic and physiologically active compounds to detoxification of xenobiotics, thus participating in realization of many essential functions of living organisms.

Numerous studies suggest about the important regulatory role of membrane lipids, particularly phospholipids, in functioning of the electron transport chains through influence on the components of these chains, electron transporters and enzymes. This is achieved due to the ability of lipid microenvironment to modulate structural and catalytic properties of membrane proteins, both integral and peripheral[1-8].

Chromaffin granules (Cgs) are specialized subcellular organelles localized in adrenal medulla and responsible for accumulation, storage and secretion of hormones (catecholamines). In these organelles one of the

Vasili Tsakanov and Helmut Wiedemann (eds.), Brilliant Light in Life and Material Sciences, 175–184.
© 2007 *Springer.*

catecholamines' biosynthesis reactions also takes place, namely. hydroxylation of dopamine to noradrenaline, which is catalyzed by the enzyme dopamine-beta-monooxygenase (DBM; EC 1.14.17.1). The enzyme is also capable to catalyze the hydroxylation of tyramine to oxytyramine. The basic principles of structural and functional organization of Cgs are identical to those of other secretory vesicles localized in different tissues of the organism, including brain. On the other hand, the method developed for isolation of Cgs is in a sufficient degree easier and more advanced than those developed for isolation of other secretory vesicles. Thus, Cgs are considered as an ideal model for in vitro study of the general principles of the molecular organization and functional activity of the secretory vesicles[9, 10].

Before we started our study, there was a considerable evidence to suggest the existence of an electron transport chain on Cgs membrane; and it was proposed that this chain is probably supplying membrane form of DBM (mDBM) with electrons for noradrenaline biosynthesis according to the following scheme:

NADH → NADH: (acceptor) reductase →?   → Cytochrome b561 →? → mDBM → O2 → O2-2; Dopamine + O2-2 → Noradrenaline +H2O

The present work was designed to identify the nature of the electron transporters functioning in the Cgs membrane electron transport chain between NADH: (acceptor) reductase and cytochrome b561, on the one hand, and between cytochrome b561 and mDBM, on the other, and to study the role of lipid surrounding in the functional activity of this chain.

The experiments were performed in vitro using electrophoretically homogeneous preparations of Cgs proteins and vesicles made from the Cgs membrane phospholipids.

2.   MATERIALS AND METHODS

Cgs were isolated from bovine adrenal medulla according to previously described procedure[11] and were found to be pure by electron microscopy.

Procedures of isolation of the Cgs membranes and electrophoretically pure preparations of mDBM, acidic copper containing protein (ACP), cytochrome b561 (b561), and flavoprotein (FP) were performed as described earlier[12-16].

Isolation of the total phospholipid fraction and individual phospholipids from the Cgs membrane and preparation of liposomes and micelles from these fractions were performed as described earlier[17-19].

To determine the specific activity of mDBM and to register kinetics of the enzyme catalyzing reactions (hydroxylation of dopamine or tyramine) earlier developed spectrophotometric procedures were applied[20, 21]. The oxygen consumption during mDBM catalyzed reactions was measured on a Beckman M-0226 oxygen analyzer, using a Clark electrode.

Visible spectroscopy detection on a Beckman-DU7 spectrophotometer was applied to follow oxidation/reduction state of hem in b561 and flavine-adenine dinucleotide in FP, and to register kinetics of the electron transfer reactions with the participation of these proteins. ESR spectroscopy detection on a Varian E-6 ESR spectrometer was used to determine the oxidation/reduction state of copper localized in active sites of both DBM and ACP and responsible for electron transfer reactions catalyzed by these proteins.

Fluorescent labeling of mDBM with fluorescein isothiocyanate-Celite (FITC; Sigma) was performed in 20mM potassium phosphate, pH 7.5, as described earlier[22]. All fluorimetric measurements with FITC-labeled enzyme were carried out using an Applied Photophysics SP3 spectrofluorimeter at excitation and emission wavelengths of 280 and 320 nm, respectively. Fluorescence anisotropy (FA) was determined as described previously[23].

Ultracentrifugation (100,000g x 60 min) was performed using Beckman Spinco L-2 ultracentrifuge.

Data analysis was performed using GraphPad Prism software.

The data presented here were obtained in 3 separate experiments (the variability coefficient did not typically exceed 10%). Each experimental point is an average of 3 measurements.

3. RESULTS AND DISCUSSION

As it was mentioned in the introduction, the nature of the electron transporters functioning in the Cgs membrane electron transport chain between NADH: (acceptor) reductase and cytochrome b561, on the one hand, and between cytochrome b561 and mDBM, on the other, was unclear. To clear this issue, in the first part of our study we examined whether ACP and/or FP, the Cgs proteins with unknown function, which were earlier isolated and characterized by us,[14,16] might be involved in the electron transfer reactions of the Cgs membrane electron transport chain. Related types of proteins, notably containing copper or FMF/FAD as a prosthetic group, are known to function as electron transporters in many electron transport chains of vital cell.

First, we compared the kinetics of product formation (for two different substrates, tyramine and dopamine), oxidation of electron donor, and oxygen consumption in the course of the reaction catalyzed by mDBM in the presence of the reduced form of ACP, ascorbate and ferrocyanide. The two latest are known as the most potent electron donors of the enzyme[24]. The results obtained demonstrated the ability of ACP to act as an electron donor in the mDBM catalyzed reactions. Moreover, it was shown that ACP is the most efficient electron donor for mDBM than ferrocyanide or ascorbate. Notably, the latest is considered as a natural electron donor of the soluble form of DBM[24]. In Table 1 the kinetic parameters of the reaction of hydroxylation of tyramine catalyzing by mDBM in the presence of the reduced form of ACP, ascorbate and ferrocyanide, as the electron donors, are compared. The same results were obtained when dopamine was used as a substrate.

Table 1. Kinetic parameters of the reaction of hydroxylation of tyramine catalyzed by mDBM in the presence of the reduced form of ACP, ascorbate and ferrocyanide.

Electron donor (Ed)	Km for Ed, mM (M±m)	Km for tyramine, mM (M±m)	Vmax of O2 consumption, mM/min (M±m)
ACP	0.41±0.04	0.13±0.01	0.51±0.05
ascorbate	0.50±0.05	0.21±0.02	0.33±0.04
ferrocyanide	0.63±0.06	0.37±0.04	0.23±0.02

Experiments with the use of 3-component system including b561, ACP and mDBM demonstrated the ability of ACP to act as an intermediate chain between b561 and mDBM, transferring electrons from b561 to mDBM.

Further, we studied interaction of the reduced form of FP with a number of natural oxidizers, including oxidized form of b561 (Table 2). The results obtained demonstrated the ability of FP to transfer electrons to b561.

Table 2. Influence of some natural oxidizers on the reduced form of FP.

Oxidizer	Ability to oxidize reduced form of FP
ACP, oxidized form	-
mDBM, oxidized form	-
b561, oxidized form	+
Cytochrome c, oxidized form	-
Cytochrome b5, oxidized form	-

Based upon the results obtained, we suggest that the electron transport chain of the Cgs membrane represents the following sequence of electron transporters and relevant electron transfer reactions:

NADH → NADH: (acceptor) reductase → FP → b561 → ACP →
→ mDBM → O2 → O2-2; Dopamine + O2-2 → noradrenaline + H2O

In the second part of our study we investigated the interaction of the components of the electron transport chain of the Cgs membrane with lipids, using vesicles made from the individual phospholipids (bilamellar liposomes and micelles) as well as from the total phospholipid extract of the Cgs membrane (multilamellar liposomes).

First, we compared the kinetics of product formation (for 2 different substrates, tyramine and dopamine), oxidation of electron donor (for 3 different electron donors, ACP, ascorbate and ferrocyanide), and oxygen consumption in the course of the reaction catalyzed by mDBM in the presence of vesicles made from the individual phospholipids of the Cgs membrane. The results obtained indicated that the enzyme is activated in the presence of either phosphatidylcholine (PC) or lysophosphatidylcholine (lPC), but is inhibited by phosphatidylserine (PS), phosphatidyleth-anolamine (PE), and phosphatidic acid (PA). The data obtained with ACP, as an electron donor, and with tyramine, as a substrate, are shown on Table 3. The same results were obtained when other electron donors or substrate were used.

Table 3. Kinetic parameters of the reaction of hydroxylation of tyramine catalyzed by mDBM (with the reduced form of ACP as an electron donor), in the presence of phospholipid vesicles made from lPC, PC, PS, PA, PE and in the absence of phospholipid vesicles.

Lipid	Km for ACP, mM (M±m)	Km for tyramine, mM (M±m)	Vmax of O_2 consumption, mM/min (M±m)
-	0.41±0.04	0.130±0.010	0.029±0.003
lPC	0.16±0.02	0.050±0.100	0.029±0.003
PC	0.24±0.03	0.076±0.010	0.030±0.003
PS	0.41±0.04	0.130±0.020	0.021±0.002
PA	0.41±0.04	0.130±0.010	0.016±0.002
PE	0.27±0.03	0.086±0.020	0.018±0.002

Next, we studied the association of mDBM to liposomes and micelles made from phosphatidylcholine and lysophosphatidylcholine, respectively, using the FA technique, and the association of mDBM to liposomes made from the total extract of the Cgs phospholipids, using ultracentrifugation technique. The complex formation between the enzyme and lipid vesicles was studied as a function of lipid, pH, and salt concentration. The efficiencies of complex formation for the apo- and holo-forms of mDBM were also compared. According to the results obtained, the most efficient association of mDBM to the phospholipid vesicles occurs in the pH range 5.0-6.0 (Figure. 1-2) and at a low ionic strength (Figure.3-4).

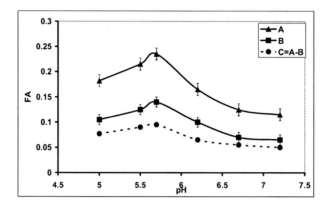

Figure 1. Dependence of FA of FITC-labeled mDBM on pH in the presence (A) and absence (B) of lPC micelles. C=A-B - differential curve demonstrating pH dependence of the association of mDPM to lPC micelles.

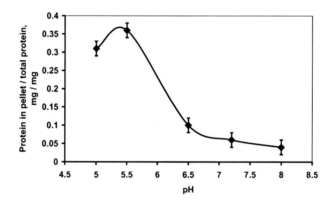

Figure 2. pH dependence of the association of mDBM to multilamellar liposomes prepared from the total extract of Cgs phospholipids.

In connection with this, it is important to note that mDBM undergoes reversible tetramer to dimer dissociation, and this dissociation is pH dependent. At pH 5.5-5.7 the tetrameric form dominantly exists in the enzyme population, whereas further increase in pH affects the decrease in tetramer to dimer ratio[24, 25]. Thus, our data suggest that the affinity of phospholipid vesicles to tetramer is greater than to dimer. It is interesting that mDBM, being peripheral membrane protein, is localized in the interior of the Cgs, which is at pH 5.5-5.7[25]. In this pH range the enzyme is more active, because the tetrameric form has a lower Kd for substrate than

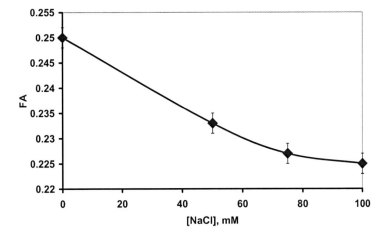

Figure 3. Dependence of FA of FITC-labeled mDBM on NaCl concentration in the presence of IPC micelles at pH 5.7

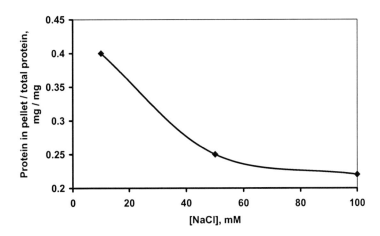

Figure 4. Dependence of the association of mDBM to multilamellar liposomes prepared from the total extract of Cgs phospholipids on NaCl concentration at pH 5.5.

dimeric does[24]. So far the results of our experiments suggest that the association of mDBM to vesicles formed from the Cgs membrane phospholipids is favored under the pH, characteristic to the in vivo environment.

It seems likely that, apart from the charge, other factors such as conformational changes can modify the binding properties of mDBM. This follows from the pH dependence of the association, and besides, from the inability of apoenzyme to associate with liposomes or micelles. According to the previous data, the conformation of the apo-DBM differs from that of the native enzyme[26].

Further we studied the influence of the individual phospholipids on the electron transfer reaction between the reduced form of b561 and oxidized form of ACP. During the study we revealed the inhibitory effect of PE and lysophosphatidylethanolamine (lPE) and the activation effects of PS and PA on this reaction. The data obtained are presented on Table 4.

Table 4. Kinetic parameters of the reaction of the oxidation of b561 by ACP in the presence and in the absence of phospholipid vesicles. Coefficient of variability for all calculated parameters did not exceed 15%.

Lipid	Effect	Kd, mM	Hill's coefficient	Vmax, μM/min
-	N/A	1.85	6	5.7
PE	activation	1.85	6	11.6
lPE	activation	1.85	6	37.5
PS	inhibition, $K_i = 0.33$ mM	1.85	6	0.16
PA	inhibition, $K_i = 1.77$ mM	1.85	6	0.08

The results obtained suggest about the important physiological role of phospholipid environment in binding of mDBM to membrane, modulation of the reaction catalyzed by this enzyme as well as the electron transfer reactions between other components of the electron transport chain of the Cgs membrane.

REFERENCES

1. Green D.E., Tzagoloff A. Role of lipids in the structure and function of biological membranes. J Lipid Res., 1966, 7(5), 587-602.
2. Jolliot A., Demandre C., Mazliak P. Role of lipids in the function of microsomal electron transport chains of potato. Biochimie, 1978, 60(8), 767-775.
3. Dmitriev L. F. The role of lipids in enzyme reactions with charge transfer. Mol. Biol. (Mosk.), 1983, 17(5), 1060-1067.
4. Archakov A. I., Borodin E.A., Dobretsov G.E., Karasevich E.I., Karyakin A.V. The influence of cholesterol incorporation and removal on lipid-bilayer viscosity and electron transfer in rat-liver microsomes. Eur. J. Biochem., 1983, 134(1), 89-95.
5. Lambeth J.D., Seybert D.W., Lancaster J.R. Jr., Salerno J.C., Kamin H. Steroidogenic electron transport in adrenal cortex mitochondria. Mol. Cell. Biochem., 1982, 45(1), 13-31.
6. Gwak S. H., Yu L., Yu C. A. Studies of protein-phospholipid interaction in isolated mitochondrial ubiquinone-cytochrome c reductase. Biochim. Biophys. Acta, 1985, 809(2), 187-198.

7. Fattal D.R., Ben-Shaul A. A molecular model for lipid-protein interaction in membranes: the role of hydrophobic mismatch. Biophysical Journal, 1993, 65, 1795-1809.

8. Lee A.G. Lipid-protein interactions in biological membranes: a structural perspective. Biochim. Biophys. Acta, 2003, 1612(1), 1-40.

9. Lee A.G. How lipids affect the activities of integral membrane proteins. Biochim. Biophys. Acta, 2004, 1666(1-2), 62-87.

10. Winkler H. Membrane composition of adrenergic large and small dense cored vesicles and of synaptic vesicles: consequences for their biogenesis. Neurochem. Res., 1997, 22(8), 921-932.

11. Apps D. K. Membrane and soluble proteins of adrenal chromaffin granules. Semin. Cell. Dev. Biol., 1997, 8(2), 121-131.

12. Hoffman P.G., Zinder O., Bonner W.M., Pollard H.B. Role of ATP and beta, gamma-iminoadenosinetriphosphate in the stimulation of epinephrine and protein release from isolated adrenal secretory vesicles. Arch. Biochem. Biophys., 1976, 176 (1), 375-388.

13. Boyadjian A.S., Nalbandyan R.M., Buniatyan H. Ch. Comparative investigation of physico-chemical properties of two forms of dopamine-beta-hydroxylase from chromaffin granules. Biochemistry (Moscow), 1981, 46(4), 635-641.

14. A.S. Boiadzhian. Purification of dopamine-beta-monooxygenase and extremely acidic, copper-containing proteins from the adrenal medulla. Extremely acidic, copper-containing proteins as electron donors for dopamine-beta-monooxygenase. Biokhimiia, 1985, 50(1), 84-90.

15. Petrosian S.A., Boiadzhian A.S., Karagezian K.G. Purification of cytochrome B-561 from chromaffin granules and its physico-chemical properties. Ukr. Biokhim. Zh., 1991, 63(2), 39-45.

16. Petrosian S.A., Boyajian A.S., Karageosian K.G., Avakian S.A., Abramov R.E. The purification of a new flavoprotein from chromaffin granules and investigation of its properties. Neurochemistry, 1992, 11 (1), 29-40

17. Pogosyan A.G., Boyajian A.S., Mkrtchyan M.Y., Karagezyan K.G. Interaction of dopamine-beta-monooxygenase with chromaffin granule membrane lipids. Biochem. Biophys. Res. Commun., 1992, 188(2), 678-683.

18. Boyadzhyan A.S, Karagezyan K.G. Interaction of dopamine-beta-monooxygenase with lipids: 1. Lipids as modulators of the catalytic activity of the enzyme. Biomedical Science, 1990, 1(1), 379-383.

19. Pogosyan A.G., Boyajian A.S., Mailyan K.R., Aivazyan V.A. Fluorescence spectroscopy studies of the association of membrane and soluble dopamine-beta-monooxygenase to bipolar phospholipid liposomes and micelles. Membr. Cell Biol., 2000, 13(5), 657-665.

20. Kuzuya H., Nagatsu T. A simple assay of dopamine beta-hydroxylase activity in the homogenate of the adrenal medulla. Enzymologia, 1969, 36(1), 31–38.

21. Skotland T., Ljones T. Direct spectrophotometric detection of ascorbate free radical formed by dopamine beta-monooxygenase and by ascorbate oxidase. Biochim. Biophys. Acta, 1980, 630(1), 30–35.

22. Tompa P., Bar J., Batke J. Interaction of enzymes involved in triosephosphate metabolism. Comparison of yeast and rabbit muscle cytoplasmic systems. Eur. J. Biochem., 1986, 159 (1), 117-124.

23. Lacowicz J.R. Principles of fluorescence spectroscopy, 1983, 1-496, N.Y. & London: Plenum Press.

24. Skotland T., Ljones T. Dopamine-beta-monooxygenase: structure, mechanism, and properties of the enzyme-bound copper. Inorg. Persp. Biol. Med., 1979, 2, 151-180.
25. Saxena A., Hensley P., Osborne J.C. Jr., Fleming P.J. The pH-dependent subunit dissociation and catalytic activity of bovine dopamine beta-hydroxylase. J. Biol. Chem., 1985, 260(6), 3386-3392.
26. Dhawan S., Hensly P., Osborne J.C. Jr., Fleming J., Fleming P.J. Adenosine 5'-diphosphate-dependent subunit dissociation of bovine dopamine beta-hydroxylase. J. Biol. Chem., 1986, 261 (17), 7680-7684.
27. Stewart L.C., Klinman J.P. Dopamine-beta-hydroxylase of adrenal chromaffin granules: structure and function. Ann. Rev. Biochem., 1988, 57 (3), 551-592.

EMISSION SPECTROSCOPY IN BIOPHYSICAL MACROMOLECULAR RESEARCH

I. From Stokes shift to multiphoton induced fluorescence

BORIS KIERDASZUK

University of Warsaw, Institute of Experimental Physics, Department of Biophysics, 93 Zwirki i Wigury St., 02-089 Warsaw, Poland; e-mail: borys@biogeo.uw.edu.pl

Abstract: The purpose of this review is twofold: first, to recall some modern emission methods in biochemistry and biophysics and second, to furnish their scope and limitations. Methodology of this part is based on the shift between excitation and emission spectra (Stokes' shift) and fluorescence anisotropy, including advantages of multi-photon excitations. Their applications in confocal microscopy led to higher spatial and temporal resolution, and lower photodamages, thus permit dipper penetration of the cells and tissues, and a ultraprecise nanosurgical applications. Higher selectivity of two-photon excitation (TPE) of tryptophan vs tyrosine residues is attainable due to short-wavelength shift of the TPE spectrum of tyrosine, and zero or slightly negative values of its TPE fundamental anisotropy.

Key words: emission spectroscopy; fluorescence; one- and two-photon excitation; multi-photon excitation

Abbreviations: NADH, β-nicotinamide adenine dinucleotide; NAMN, β-nicotinamide mononucleotide; NATA, N-acetyl-L-tryptophanamide; NATyrA, N-acetyl-L-tyrosinamide; OPE and TPE, one-photon and two-photon excitation; OPIF and TPIF, one-photon and two-photon induced fluorescence

1. INTRODUCTION

There are two categories of luminescence emission of light (photons) from electronically excited states depending on their nature (Fig. 1), i.e. fluorescence emission from excited singlet state $(S_1 \rightarrow S_0)$, and phosphorescence emission from triplet excited state $(T_1 \rightarrow S_0)$. Singlet-

Vasili Tsakanov and Helmut Wiedemann (eds.), Brilliant Light in Life and Material Sciences, 185–197.
© 2007 *Springer.*

singlet transition is spin allowed, and occurs with the emission rates which are in the $10^{12} - 10^{8}$ s^{-1} range, resulting in picosecond and nanosecond fluorescence lifetimes. In contrast, triplet-singlet transition is spin forbidden and the emission rates are slow (10^{6}-10^{-2} s^{-1}), and phosphorescence lifetimes are usually much longer, typically microseconds to seconds. They both are subjected to many deactivation processes, e.g. non-emissive decays and quenching, which compete with emission. It is expected that emission intensity decays reflect intra- and intermolecular interactions, structure and dynamics of biological molecules, and thus the increased information about system, as compared with steady-state parameters.

1.1   Phenomenon of fluorescence emission

First observation of fluorescence from a quinine solution in sunlight was reported by John F.W. Herschel (1845), who observed it in a homogeneous quinine solution internally colourless. Following Herschel's discovery, few years later John G.G. Stokes in Cambridge first observed that quinine emission is red shifted relative to excitation (Stokes, 1852). Now it is evident that Herschel and Stokes discovered very basic phenomenon which was explained a century later, but to this day, the fluorescence of quinine, which may be easy observed in the Indian tonic water, remains one of the most used example of fluorescence. It is worth noticing that Stokes experiment was an early example of simplest homemade spectrofluorometer, where UV excitation (< 400 nm) was provided by sunlight and blue glass filter (part of the stained glass window), bottle containing a solution of quinine was used as a cuvette, and blue fluorescence of quinine (~450 nm) free of the exciting light filtrated by a yellow glass (of wine) filter, was easily visible by detector (Stokes eye). One century later quinine fluorophore was also responsible for stimulating the development of the first commercial spectrofluorometer in the 1950s. This was based on a successful fluorometric assays of quinine as one of the antimalaria drugs employed during World War II by US Department of Defence, subsequently followed by NIH program on the development of the spectrofluorometer and its commercialization (Undenfriend, 1995).

The phenomena that occur after absorption of light (photons) are usually described using some form of Jabłoński diagram (1935). Typical Jabłoński diagram (Figure 1) contains horizontal lines which denote singlet ground (S_0) and excited (S_1), and triplet (T_1) electronic states, where molecules may exist in a number of vibrational energy levels. Transitions between states are denoted by vertical solid arrows and reflect the fact that they occur in about 10^{-15} s, i.e. in the absence of displacement of nuclei (the Frank-Condon principle). At room temperature, thermal energy is not adequate to significantly populate the excited vibrational states, therefore absorption starts at lowest vibrational states of the molecule. Absorption transition energy $E_0 \rightarrow E_i$ requires photons of frequency $\nu_i = (E_i - E_0)/h$, where h is the Planck constant. Strength of the absorption depends on the strength of the electric and magnetic filed vectors (light intensity), their mutual orientation *vs* molecule, and the molecular transition moments. Since contribution of the quadruple transition moments and the magnetic dipole transition moments are negligible low, the electric dipole transition moments $\hat{\mu}_{fi}$ between initial ψ_i and final ψ_f electronic states (Eq. 1) sufficiently well describes the light absorption (Michl and Thulstrup, 1986).

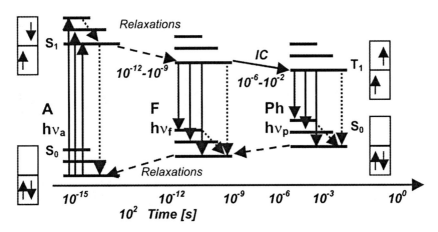

Figure 1. One form of Jabloński diagram of singlet (S0, S1) and triplet (T1) electronic states of molecules and possible transitions between them: radiative transitions (solid vertical arrows), i.e. absorption (A, hna), emission of fluorescence (F, hnf) and phosphorescence (Ph, hnp), and nonradiative transitions, i.e. intersystem crossing (IC, solid arrow), relaxations (dashed arrows) and internal conversions (dotted arrows). Energy of the electronic states increases vertically, while horizontal arrow in the bottom denotes tentative time scale of particular transitions, as indicated.

$$\hat{\mu}_{fi} = \langle f | \hat{\mu} | i \rangle = \int \psi_f^* \hat{\mu} \psi_i d\tau \qquad (1)$$

where $\hat{\mu}$ is the operator of electric dipole moment of the molecule.

After light (photon) absorption several processes occur like internal conversions within electronic singlet (S_1 and S_2) states or vibrational states, or relaxation of environment (Figure 1). The latter is caused by the fact that absorption of photon is usually accompanied by significant displacement of the electron density within molecules, and thus by an increase of the excited state dipole moment. Because of rapid relaxation, molecules, initially excited to higher electronic and vibrational levels, quickly dissipate ($\sim 10^{-12}$ s) excess of energy via internal processes and interactions with the solvent. Prior to emission, molecule is in the lowest vibrational level of S_1, so called the fluorescent electronic state, independently of excitation wavelength (Kasha's rule, 1950). This is actually one of the main reason for long wavelength shifts of fluorescence emission relative excitation spectrum (Stokes shifts), described by several empirical formula, e.g. by well known Lippert equation (Lakowicz, 1999), which relates electronic properties of the system (dielectric constant, refractive coefficient, dipole moment in the ground and excited electronic states) and above mentioned shifts between emission end excitation spectra. Fluorescence emission spectrum is measured for particular excitation wavelength and showed as a plot of the fluorescence intensity *vs* wavelength or wavenumber, while fluorescence excitation spectrum is measured for particular emission wavelength and showed as a plot of the fluorescence intensity *vs* excitation wavelength (wavenumber). Consequently, emission spectra may show individual vibrational levels of the ground state, while excitation spectra show that of the excited state, which are specifically pronounced for rigid and planar molecules, e.g. perylene. The emission spectra are dependent on the chemical (electronic) structure of fluorophore and the environment, e.g. solvent, in which fluorophore is dissolved.

1.2   Fluorescence polarization (anisotropy)

Interaction of the linearly (vertically, along the z-axis) polarized light with the isotropic collection of fluorophore molecules leads to selective excitation of the molecules, which have a dipole of absorption transition moments owing to the polarized absorption (photoselection) described by the probability distribution $f_1(\omega)d\omega = \cos^2 \omega \cdot \sin \omega \cdot d\omega$ of molecules in the angle between ω and $\omega + d\omega$, where ω is the angle between dipole of absorption transition moment and the z-axis. Photoselective excitation is the origin of fluorescence anisotropy in isotropic media. Exited fluorophores are symmetrically distributed around z-axis. For anisotropy determinations, fluorescence excited by the vertical polarized photons (I_v) should pass through a monochromator (and/or filters) and a polarizer set at the vertical (F_{vv}), horizontal (F_{vh}) or the magic angle (54.7°) orientation relative to the excitation (Figure 2).

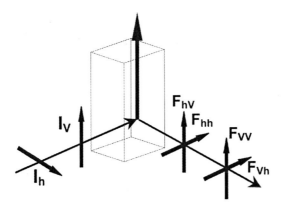

Figure 2. Schematic diagram for measurements of fluorescence anisotropy, see text for further details.

After Jabloński (1960, 1961, 1967) and Kawski (1993), concept of emission anisotropy is based on (i) the lack of correlation between the photons of excitation ($\sim 10^{-15}$ s) and emission ($\sim 10^{-9}$ s), which is true for low light intensity, i.e. in the absence of stimulated emission, (ii) emission is a random event, and each fluorophore has the same probability of emitting in a given period of time, and (iii) the directions of the absorption and emission moments are fixed within the molecular coordinates. Fluorescence anisotropy at given wavelength (λ) is defined as r = $(F_{vv} - gF_{vh})/(F_{vv} + 2gF_{vh})$, after subtraction of background fluorescence. Fluorescence anisotropy, r(λ), spectra were determined from polarized fluorescence spectra as r(λ) = $(F_{vv}(\lambda) - gF_{vh}(\lambda))/(F_{vv}(\lambda) + 2gF_{vh}(\lambda))$, where $F_{vv}(\lambda)$ is vertically polarized emission intensity (spectrum) with vertically polarized excitation, and $F_{vh}(\lambda)$ is horizontally polarized emission intensity (spectrum) with vertically polarized excitation. The instrument correction factor, g, is equal to g = $F_{hv}(\lambda)/F_{hh}(\lambda)$, where $F_{hv}(\lambda)$ is vertically polarized emission intensity (spectrum) with horizontally polarized excitation, and $F_{hh}(\lambda)$ is horizontally polarized emission intensity (spectrum) with horizontally polarized excitation. This g factor corrects for the small polarization bias of the optical system and photomultiplier.

Fundamental anisotropy (r_0), i.e. anisotropy free of the depolarising processes such as rotational diffusion or energy transfer is determined for vitrified solutions, usually in aqueous medium containing at least 80% of glycerol or propylene glycol (v/v) at -60 °C. Such solutions form glassy solution where molecules are immobile during the lifetime of excited state. Fundamental anisotropy may be determined as the time-zero anisotropy value, sometimes called as limiting (maximal) anisotropy, $r_0 = r(t = 0)$, in time resolved anisotropy measurements in liquid media. Resulted fluorescence anisotropy values depend on the angular displacement (β) of the absorption and emission dipoles, known as a *Perrin product*,

$$r_0 = \frac{2}{5}\left(\frac{3\cos^2\beta - 1}{2} \right)$$

where a factor of 2/5 reflects a loss of anisotropy due to photoselection (see above), and β angle is accounted by similar formula as that for displacement of the emission dipole by an angle ω from the z-axis. Accordingly, maximal value of fundamental anisotropy (r_0=0.4) is achieved for β = 0, while r_0 = 0 for β = 54.7°, and the minimum value r_0 =-0.2 for β=90°. Only few fluorophores show r_0=0.4, while for most of them r_0 values are lower than 0.4. Since the orientation of absorption dipoles differs for each absorption bands, r_0 depends on excitation wavelength. Resulted excitation anisotropy spectra enable determination of β and absorption bands for identified transitions, and thus resolution of electronic states, e.g. one-photon (Valeur and Weber, 1977; Eftink et al., 1990) and two-photon absorption (Kierdaszuk et al., 1997) of indole (tryptophan). Generally fundamental anisotropy value is independent on the emission wavelength, with the exception of that for overlapping emission bands occurred from more than one excited state. Photoselection process, and thus fundamental anisotropy, are strongly affected by multiphoton excitation (Sect. 2).

2. ONE-PHOTON *VS.* MULTI-PHOTON INDUCED FLUORESCENCE

Multi-photon absorption is a relatively old idea in quantum mechanics – first proposed as theoretical prediction of two-photon absorption by Maria Göppert-Mayer (Nobel laureate) in her doctoral dissertation (Göppert-Mayer, 1931), and verified experimentally shortly after the invention of the laser (Kaiser and Garrett, 1960). Although quantum-mechanical theory of two-photon absorption is rather complex (Göppert-Mayer, 1931; Mazurenko, 1971; McClain and Harris, 1977), it is understandable phenomenologically taking into account a virtual state produced by interaction of one photon with a ground state (ψ_0), which is the energetically perturb nonstationary state of the molecule. Consequently, incoming low energy second photon will have a small but finite probability of being absorbed to the virtual state. Both photons are absorbed simultaneously, and they are not necessarily of the same energy (colour), but sum of their energy values should be equal to the energy of photon required for one-photon excitation. In this case "simultaneous" means within about 10^{-18} s.

Quantum-mechanical considerations showed that probability of such a process is very low, and consequently experimentally determined absorption cross-sections are extremely low. To achieve this one should use photons highly spatially (in space) and temporarily (in the time space) overlapped. Resulted fluorescence intensity depends on the squared laser power for two-photon excitation, cubed (third) power for three-photon excitation, and the fourth power for four-photon excitation, frequently used to control the mode of excitation. Due to low cross section values for multiphase absorptions, the photon density must be at least 10^6 times higher than that required to generate the same number of one-photon absorptions. Such high density of photons are easily achieved using mode-locked lasers, where power of each peak (typically from ~100 fs to ~1 ps fwhm) is high enough (~TW/cm^2) to generate two-photon excitation, while average power (~10 mW/ cm^2) is in the same range as for one-photon excitation. This gives rise to significant advantages associated with multi-photon excitation leading to broad applications of modern fluorescence spectroscopy in biology, biochemistry and biomedical sciences, including analytical assays.

2.1 Advantages of multi-photon excitation

First of all, red and infrared photons (690-1050 nm), which fit well to the therapeutic window of the cells and tissues, my be used for three-photon excitation of fluorophores absorbing in a range from deep UV (~ 230 nm) to near UV (~350 nm). Simultaneously, two-photon excitation of fluorophores absorbing in 345-525 nm may also occur. This may reduce photodamages of the cells and tissues as compare as with one-photon excitations, which at the wavelengths below 300 nm are harmful.

Another powerful advantages of using multi-photon excitation arise from the physical principle that the absorption depends on the i-th power of the excitation intensity, where i is the number of absorbed photons. In practice, the only place where photon density is high enough to be absorbed efficiently is the focal point. It means that in a confocal microscope two-photon excitation induces fluorescence only at the focal plane, free of background fluorescence usually present in the classical one-photon confocal microscopy. It means that three-dimensional resolution of a two-photon excitation microscopy (without pinhole!) is the same as ideal confocal microscopy, and lack of out-of-focus absorption allow dipper penetration of the sample (Denk et al., 1995, 1990), as well as application of the femtosecond NIR laser microscopy as a ultra precise Nan surgical tools with cut sizes 100-300 nm (König et al., 1999). Moreover, photobleaching in conventional microscope is nearly through the sample while it is limited to the focal plane in two-photon confocal microscopy, and thus minimizes photobleaching and photodamages of the cells and tissues (Feijó and Moreno, 2004; König, 2000; Piston, 1999).

Promising advantages of multi-photon excitations arise from its strong effect on the fundamental anisotropy values due to polarized photoselection with collinear transitions. Maximal anisotropy values (r_{mi}) of fluorescence (in the absence of depolarising processes), $r_{mi} = 2i / (2i + 3)$, where i - number of absorbed photons, are 2/5, 4/7, 2/3 and 8/11 for the one-, two-, three or four-photon transitions, respectively. They originate with $\cos^{2i}\omega$ (i=1,2,3 and 4) photoselection, where ω is the angle between the excitation polarization and the transition moment of the molecule. The fundamental anisotropy is given by

$$r_{0i} = r_{mi}\left(\frac{3\cos^2\beta_i - 1}{2}\right) = \left(2i/(2i+3)\right)\left(\frac{3\cos^2\beta_i - 1}{2}\right)$$

where the angle β_i is the angular displacement between the absorption and emission transition, and need not to be identical for each mode of excitation (*i*). More specifically, if the effective β values are identical, then the fundamental anisotropies are related by above equations. This is observed for some fluorophores, e.g. POPOP (Lakowicz et al., 1995) and β-nicotinamide adenine dinucleotide (NADH) (Kierdaszuk et al., 1996). Moreover, excitation anisotropy spectrum of NADH was remarkably affected by complex formation with liver alcohol dehydrogenase (LADH), especially at wavelengths below 310 nm, where absorption is dominated by tryptophan residues and adenine (Lakowicz et al., 1996). There is no doubt that fluorescence probes exhibiting higher values of fundamental (time-zero) anisotropy are better for detection of depolarising effects, e.g. rotational diffusion reflecting dynamical nature of biomolecules, or effect of the enzyme-ligand interaction, such as LADH-NADH complex formation (Lakowicz et al., 1996). However, tryptophan and tyrosine derivatives, and proteins serve as good examples that fluorophores can display more complex behaviour (Kierdaszuk et al., 1997; Lakowicz et al., 1995; Lakowicz and Gryczynski, 1992; Lakowicz et al., 1992), see also Sect. 2.2 .

2.2   One-, two- and three-photon induced fluorescence of tryptophan and tyrosine

The most important for potential applications of emission spectroscopy in biochemical and biophysical studies are availability of appropriate fluorescence probes. There are only few natural (intrinsic) biomolecules which enable luminescence studies without addition of an extra molecules, called as synthetic fluorescence probes. Natural fluorescence of proteins is observable due to the aromatic amino acids, tryptophan (Trp), tyrosine (Tyr) and phenylalanine (Phe). As fluorescence quantum yield of Phe is very low (~0.02), protein absorption and fluorescence spectra are dominated by the indole and phenol moieties of Trp and Tyr, respectively. However, there are several factors in the native protein environments which, in spite of the same values of fluorescence quantum yields (~0.14), lead to a substantial decrease of tyrosine emission relative to that of tryptophan. First of all, tyrosine has 5-fold lower molar extinction coefficient, its singlet excited state is strongly quenched by hydroxyl groups of carboxylic amino acids and peptide chain or excitation energy transfer to Trp. This is further confirmed by an increase of Tyr fluorescence intensity after denaturation of proteins, but with practically no effect on the position of the excitation (~275 nm) and emission (~304 nm) maxima. Fortunately, the emission of Trp is highly sensitive to local environment. Spectral shift and changes of indole fluorescence intensity may be observed as a result of several phenomena, such as protein-ligand or protein-protein association, conformational changes and an average exposure of indol to the aqueous phase and/or nonpolar environment and/or quenching by halogens, acrylamide or nearby disulfides or other electron-deficient groups like – NH_3^+, $-CO_2H$, and protonated histidine residues. The presence of Trp residue(s) in different environments are often a source of non-exponential fluorescence decays, described in subsequent chapter.

Two-photon induced fluorescence (TPIF) of aromatic amino acids was studied using amide or *N*-acetyl derivatives and their fluorescent counterparts such as indole and its derivatives, and phenol. Fluorescence emission spectra of *N*-acetyl-L-tryptophanamide (NATA) (Lakowicz et al., 1992; Gryczynski et al., 1996) and *N*-acetyl-L-tyrosinamide (NATyrA) (Lakowicz et al., 1995; Gryczynski et al., 1999) are similar for one-photon, two-photon and three-photon excitation, and indicate that emission starts from essentially the same electronic states for all modes of excitation. Linear, quadratic and cubic dependence of fluorescence intensity on the intensity of incidence light was observed due to single- bi- and tri-photonic absorption, respectively. With TPIF similar results were observed for indole and phenol, as well as for NADH and its fluorescent counterpart, β-nicotinamide mononucleotide (NAMN). Two-photon cross section (relative to that of p-bis(O-methylstyryl)benzene standard) for NATA is very low (~0.7%), and for NATyrA (<0.01%) and NADH (0.001-0.01%) is lower (Kierdaszuk et al., 1997). As expected, three-photon excitations of NATA (Gryczynski et al., 1996) and NATyrA (Gryczynski et al., 1999) exhibited even lower cross section values.

It was shown unequivocally by independent determinations in two laboratories that two-photon absorption spectra of tyrosine and tryptophan residues free in aqueous solutions (Kierdaszuk et al., 1995; Rehms and Callis, 1993) and in proteins (Kierdaszuk et al., 1995) allow elevated (better) selectivity of excitation of tryptophan vs tyrosine residue, as compare as with their one-photon absorption spectra (Figure 3). It was shown that the two-photon absorption spectrum of tyrosine (λ_{max} 517 nm) is shifted (~33 nm) to shorter wavelengths relative to the one-photon absorption spectrum (λ_{max} 275 nm) at two-photon energy (550 nm). This was unequivocally confirmed from studies of OPIF and TPIF of tyrosine-tryptophan mixture. For a mixture of NATyrA and NATA, OPE results in a dominant tyrosine emission, while in TPE, the emission from the same mixture is dominantly from tryptophan residues (Kierdaszuk et al., 1995). It is rather surprising that for NATyrA, essentially the same excitation spectrum was observed with three-photon excitation and one-photon excitation (Gryczynski et al., 1999), and similar observation was made for NATA (Gryczynski et al., 1996).

Remarkably, for indole or tryptophan an increase of two-photon (Lakowicz et al., 1992) and three-photon excitation (Gryczynski et al., 1996) fundamental anisotropy (r_0) values could not be explained by a $\cos^4\omega$ and $\cos^6\omega$ photoselection factor, respectively, which is due to different cross sections of the ^1L_a and ^1L_b states for one-, two- and three-photon absorption. Unexpectedly, the TPE anisotropy spectrum of NATA is lower than OPE anisotropy spectrum at all excitation wavelengths, and indicate that for TPE the ^1L_b absorption spectrum is red-shifted, which results in initial absorption of photons by the ^1L_b state, followed by emission from ^1L_a state, for which the transition moment is rotated by 90°. This was confirmed by the resolved ^1L_a and ^1L_b spectra (Kierdaszuk et al., 1997), and further interpreted by Callis (1997) in terms of two-photon tensors and molecular orbital theory.

Unexpectedly, for the anisotropy spectrum of L-tyrosine and NATyrA (Lakowicz et al., 1995), the TPE anisotropy was zero or slightly negative (Figure 4) as compare as to near 0.3 for one-photon, and three-photon excitation (Gryczynski et al., 1999). Theoretical considerations (Lakowicz et al., 1995) do not negate the existence of such low anisotropy.

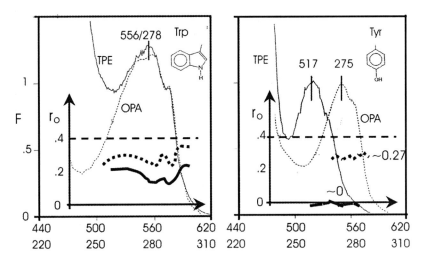

Figure 3. One-photon absorption (OPA) spectra (- - -) and two-photon (TPE) excitation spectra at one-photon energy (_____), and one-photon (dotted line) and two-photon (solid line) excitation anisotropy spectra (r0) for tryptophan and tyrosine residues (as indicated). Absorption and excitation spectra are for in 0.01 M neutral aqueous solution at 20 oC (Adapted from Rhems and Callis, 1993), and excitation anisotropy spectra are for 0.01 M neutral aqueous solution containing 80% glycerol (v/v) at –60 oC (Adapted from Lakowicz et al., 1992, 1995).

Three-photon anisotropy values were essentially similar to that in one-photon anisotropy for both NATA (Gryczynski et al., 1996) as well as for NATyrA (Gryczynski et al., 1999), which may arise from the fact that for one-photon and three-photon transition the ground and excited state are of a different evenness, while two-photon transition goes to the state of the same evenness as the ground state. Furthermore, different vibration couplings might be involved in these transitions. Further experimental and theoretical studies on the model compounds are needed to demonstrate how these electronic transitions are related to the photophysical characteristic of the system.

ACKNOWLEDGEMENTS

This research was supported by the Polish Ministry of Science and Higher Education (grant No 3P04A02425).

REFERENCES

Callis, P.R., 1997, ^1L_a and ^1L_b transitions of tryptophan: applications of theory and experimental observations to fluorescence of proteins, Methods Enzymol. **278**:113-50.

Denk, W., Piston, D.W. and Webb, W.W., 1995, Two-photon molecular excitation in laser-scanning microscopy, in: *The Handbook of Biological Confocal Microscopy, 2nd edition* (Pawley, J.B. Ed.), Plenum Press, New York, pp. 445-458.

Denk, W., Strickler, J.H. and Webb, W.W., 1990, Two-photon laser scanning fluorescence microscopy, Science, **248**:73-76.

Eftink, M.R., Selvidge, L.A., Callis, P.R., and Rehms, A.A., 1990, Photophysics of indole derivatives: Experimental resolution of ^1L_a and ^1L_b transitions and comparison with theory, J. Phys. Chem., **94**:3469-79.

Feijó, J.A. and Moreno, N., 2004, Imaging plant cells by two-photon excitation, Protoplasma, **223**:1-23.

Gryczynski, I., Malak, H. and Lakowicz, J.R., 1999, Three-photon excitation of N-acetyl-L-tyrosinamide, Biophys. Chem., **79**:25-32.

Gryczynski, I., Malak, H. and Lakowicz, J.R., 1996, Three-photon excitation of a tryptophan derivative using a fs Ti:Saphire laser, Biospectroscopy, **2**:9-15.

Herschel, J.F.W., 1845, On a case of superficial colour presented by a homogeneous liquid internally colourless, Phil. Trans. R. Soc. London **135**:143-145.

Jabłoński, A., 1935, Über den mechanisms des photolumineszenz von farbstoffphosphoren, Z. Phys., **94**:3846.

Jabłoński, A., 1960, On the notion of emission anisotropy, Bull. Acad. Pol. Sci. Ser. A **8**:259-264.

Jabłoński, A., 1961, Über die abklingungsvorgänge polarisierter photolumineszenz, Z. Naturforsch., 16a:1-4.

Jabłoński, A., 1967, Emission anisotropy, Bull. Acad. Pol. Sci. Ser. A **15**:885-892.

Jabłoński, A., 1970, Anisotropy of fluorescence of molecules excited by excitation transfer, Acta Phys. Polon., A **38**:453-458.

Kaiser, W. and Garrett, C.G., 1961, Two-photon excitation in $CaF_2:Eu^{2+}$, Phys. Rev. Lett. **7**:229-231.

Kasha, M., 1950, Characterisation of electronic transitions in complex molecules, Disc. Faraday Soc., **9**:14-19.

Kawski, A., 1993, Fluorescence anisotropy: theory and application of rotational depolarization, Crit. Rev. Anal. Chem., **23**:459-529.

Kierdaszuk, B., Gryczynski I., and Lakowicz, J.R., 1997, Two-photon induced fluorescence of proteins. In: *Topics in Fluorescence Spectroscopy, Volume 5: Nonlinear and Two-Photon-Induced Fluorescence,* (Lakowicz, J. Ed.) Plenum Publishing Co., New York, pp. 187-209.

Kierdaszuk, B., Gryczynski, I., Modrak-Wojcik, A., Bzowska, A., Shugar, D. and Lakowicz, J.R., 1995, Fluorescence of tyrosine and tryptophan in proteins using one- and two-photon excitation. Photochem. Photobiol. **61**:319-324.

Kierdaszuk, B., Malak H., Gryczynski, I, Calis, P. and Lakowicz, J.R., 1996, Fluorescence of reduced nicotinamides using one- and two-photon excitation, Biophys. Chem. **62**:1-13.

König, K., 2000, Multiphoton microscopy in life sciences, J. Micsosc. **200**:83-104.

König, K., Riemann, I., Fischer, P. and Halbhuber, K.J., 1999, Intracellular nanosurgery with infrared femtosecond laser pulses, Cell. Mol. Biol. **45**:195-201.

Lakowicz, J.R., 1999, Principles of fluorescence spectroscopy, 2nd edition. Kluwer Academic/Plenum Press.

Lakowicz, J.R. and Gryczynski, I., 1992, Tryptophan fluorescence intensity and anisotropy decay of human serum albumin resulting from one-photon and two-photon excitation, Biophys. Chem., **45**:1-6.

Lakowicz, J.R., Gryczynski, I., Danielsen, E. and Frisoli, J., 1992, Anisotropy spectra of indole and *N*-acetyl-L-tryptophanamide observed for two-photon excitation of fluorescence, Che. Phys. Lett., **194**:282-287.

Lakowicz, J.R., Kierdaszuk, B., Callis, P., Malak, H.   and Gryczynski, I., 1995, Fluorescence anisotropy of tyrosine using one- and two-photon excitation, Biophys. Chem. **56**:263-271.

Lakowicz, J.R., Kierdaszuk, B., Malak H. and Gryczynski I., 1996, Fluorescence of horse liver alcohol dehydrogenase using one- and two-photon excitation, J. Fluorescence **6**:51-59.

Mazurenko, Y.T., 1971, Polarization of luminescence from complex molecules with two-photon excitation. Dichroism of two-photon light absorption. Opt. Spectrosc. (USSR) **31**:413-414.

McClain, W.M. and Harris, R.A., 1977, Two-photon spectroscopy in liquid and gases. In: Excited Sates 3, (R.C. Lim Ed.) Academic Press, New York, pp. 2-77.

Michl, J. and Thulstrup, B.W., 1986, Spectroscopy with Polarized Light. Solute Alignment by Photoselection, in Liquid Crystals, Polymers, and Membranes. VCH Publisher, New York.

Piston, D.W., 1999, Imaging living cells and tissues by two-photon excitation microscopy, Trends Cell Biol. **9**:66-69.

Rehms, A.R. and Callis, P.R., 1993, Two-photon fluorescence excitation of aromatic amino acids, Chem. Phys. Lett., **208**:276-282.

Stokes, J.G.G., 1852, On the Change of Refrangibility of Light, Phil. Trans. R. Soc. London **142**: 463-562.

Valeur, B. and Weber, G.,1977, Resolution of the fluorescence excitation spectrum of indole inro ^1L_a and ^1L_b excitation bands, Photochem. Photobiol. **25**:441-444.

Undenfriend,S., 1995, Development of the spectrofluorometer and its commercialization, Protein Sci. **4**:542-551.

INFLUENCE OF LOWINTENCITY ELECTROMAGNETIC IRRADIATION ON THE WHEAT SEEDLINGS CHROMATIN FRACTIONS

L. A. Minasbekyan, V.P. Kalantaryan, P. H. Vardevanyan
Department of Biophysics and Department of Radiophysics, Yerevan State University, str. A. Manougian, 1, Yerevan, 375025, Armenia

Abstract: Have been investigated thermal parameters of wheat seeds chromatin. Obtained profiles of the eu- and hetero-chromatin differential melting curves from wheat 3-day seedlings. The influences of low-intensively non-thermal coherent electromagnetic irradiation (EMI) in mm-diapason on the conformational state of chromatin compartments have been studied. Shifts in the profile of DMC of the euchromatin fractions are occurred during germination, but treatment seeds by EMI of mm-diapason led to the small changes and in the profile of DMC heterochromatin. The possibility of heterochromatin took part in the replay of organism on the influence of environment by the modulation of nucleoporin basket and change of nuclear membrane permeability are discussed. Have been suggested, that EMI of mm-diapason can influence through change of nuclear membrane permeability.

Key words: chromatin, differential melting curves, euchromatin, heterochromatin, mm-waves, non-thermal coherent electromagnetic irradiation, radio replay pathway, seeds embryos, wheat seedlings.

1.   INTRODUCTION

Artificially created by man high-monochromatic coherent generators of electromagnetic MM–waves have an anomalous high biological activity, including therapeutic action. So study of biological action by the physical fields gains all greater importance.

Vasili Tsakanov and Helmut Wiedemann (eds.), Brilliant Light in Life and Material Sciences, 199–203.
© 2007 *Springer.*

Most eukaryotic genomes are packed into two types of chromatin: euchromatin and heterochromatin[1].Until recently function of heterochromatin while it remains debatable. It is known that genome function is subordinated influence of the phenomena of an effect of the position, under which imply genome inactivation under its transposition from a euchromatin in the heterochromatin regions of chromosomes that, for instance, occurs under chromosome rearrangement, caused by irradiation [2]. The nature of these influences can exist the epigenetic in one cell, but otherwise for the other cells under same genetic base [3, 4].

The purpose of the work revealing difference in heterochromatin and euchromatin fractions thermal parameter of seedlings to the third day of the germination after influence by extremely high frequency electromagnetic irradiation. Obtained results have allowed to value influence of the factor of the physical nature on a conformation of chromatin compartments under different functional conditions of the cells and consider possible ways of the response of the organism on environmental factors.

2.   MATERIALS AND METHODS

Seed germination. In our experiments have been used seeds of hexaploid wheat (*Triticum Aestivum, c*vs. Voskehask). Watering on the night seeds was treated by nonthermal coherent waves. Source of irradiation was served high-frequency generator of EMI signals G4-141 (made in USSR) on the frequency 50.3 GHz, 40min. After that seeds growing in petri dishes in darkness at 26^0C for 48 h yet.

Nuclei isolation and fractionation. Nuclei were isolated from seedlings by the method of Blobel and Potter in our modification[5]. To this end seedlings frozen in liquid nitrogen were powdered with a mortar and pestle, and 0.25 M sucrose in TKM buffer (10 mM Tris--HCl, pH 7.4, containing 25 mM KCl, and 15 mM $MgCl_2$) was added [5]. The pellet obtained contained purified intact nuclei.

Isolation of heterochromatin and euchromatin. Obtained nuclei are suspended in enough 0.25 M sucrose to yield a concentration of approximately 1 OD_{425} unit/ml and allowed to swell by standing for 10min at 4^0C. They are then disrupted by sonication for 10-15 sec in an sonifier UZDN-1 (made in USSR) at 6.2amps and 20.000 cycles/sec. A small amount of broken nuclei is removed by centrifugation at 100g for 5min. The resulting suspension of chromatin is composed three fractions: a heavy fractions of heterochromatin which is isolated by centrifugation at 3500g for 20min, an intermediate fraction composed of euchromatin contaminated with small amounts of heterochromatin which is sedimented by centrifugation at 78000g for 1h, and a very light fraction of eucromatin. Isolated by such manner chromatin fractions using further for investigation of thermal parameters.

Chromatin melting. The sample before melting dialised against 0.1 SSC during night under 4°C, as is described[6]. The temperature of the melting (T_m) and differential curves of the melting got the way of the numerical differentiation normalized integral crooked melting [7].

3. RESULTS AND DISCUSSIONS

Active or inactive states of genome inherited in cellular generations, defined by interchromosomal interaction or by translocations of the gene in determined intranuclear space, carry epigenetic nature that is to say they are not connected with change in DNA nucleotide sequences and are reversible. The molecular mechanisms of epigenetic inheritances at present unknown, unlike example of the inherited switching to activities gene, where essential role play the processes of reversible DNA methylation[8,9]. So study of the change of chromatin fractions conformation under influence of extremely high frequency EMI in the collection with data of earlier study[5,10,11], can throw light upon the nature of the changes, where main regulation role can belong to as conformational changes of chromatin, so and DNA methylation.

Differential melting curves of chromatin fractions for 3-day seedlings without treatment are presented on the Fig.1a. Appearance the number of small peaks in the DMC of heterochromatin in the field of 40-50° C, which is corresponding to AT- rich area. In DMC of euchromatin several peaks are revealed in the field of 50-65° C, which corresponds to low melting AT– rich chromatin sequence and several small peaks in GC-rich area.

Wide applicable in homes and technology, electro-, radio-, tele- and the other equipment when working irradiates the electromagnetic waves of no coherent natures that calls as noise waves. In consequence of external influence own fluctuations molecules increase that as a result can bring about conformational change. The studies[12] have shown that resonance interaction MM-waves with water and biological ambience exists at frequency - 50,3 and 51,8 GHz.

We have shown earlier that under the action of EMI extremely radio frequency within the range of 49-53 GHz the most clear change on the factor of the wheat seedlings growing, fermentative activity, biochemical and phospholpid composition both in nuclear membrane and soluble nuclear fraction exist at frequency 50.3 GHz exactly [11, 13, 14]. The influence of EHF EMI of this frequency on seed chromatin during germination led to the changes in heterochromatin, about which witness got by us data, submitted on Fig. 1b.

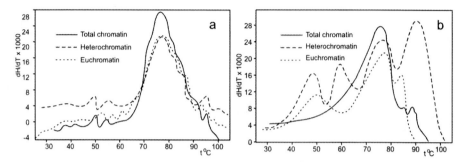

Figure 1. Differential melting curves of total chromatin, heterochromatin and euchromatin: a-isolated from intact nucleus of 3-day wheat seedlings, b- isolated from intact nucleus of 3-day wheat seedlings, treated by lowintecity coherent EMI mm-diapason of 50,3 GHz.

Change exists in DMC profile of total chromatin to within GC-rich area sequences. Have been revealed several peaks that is indicate about realignment or breaches in constitutive chromatin packing. A fraction of euchromatin during genome activation (Fig.1a), has a broad profile of the melting and are revealed two large peaks - in light melting 40–55°C and hard–melting 70-90°C areas. In contrast with euchromatin, the profile of the melting curve of heterochromatin has a complex structure. Probably, EHF EMI break a conformation of tightly packed, tandemly repetitive sequences of heterochromatin with the result that they are revealed peaks of cooperative melting condensed area of chromatin in area, adjoining to inner face of the nuclear membrane. Perhaps that exactly these high helical, compact DNA sequences check the replay reaction of the organism on external, occasionally stressful for organism, factors of environment. The changes in heterochromatin explain and throw light upon such inadequate changes in seedlings, as increasing fermentative activities and indistinctive increase of some classes of phospholipids in composition nucleus subfraction [14, 15].

Thereby, is shown that in genome under the action of EHF EMI of mm-range occur the conformational realignments of chromatin, touching in different degree changes in eu- and heterochromatin, which are expressed in change of melting parameter, enzymatic activities [13], composition of nuclear subfraction [11, 14] and growth factors of germination on the early stages [13]. However, such conformational changes carry temporary, epigenetic nature and raise the adaptation reaction of the organism, not infusing on the sequence of DNA and not inherited in the following generations chromosome realignments.

REFERENCES

1. Avramova Z.V. Heterochromatin in animals and plants. Similarities and differences. Plant Physiology, 129 (1), P. 40-49, (2002).
2. Pakhomov A.G.Current state and implication of research on biological effects of millimeter waves: a review of literature. http://www.rife.org
3. Gasser S.M., Gilson E.. Chromatin: a sticky silence. Curr.Biol., 6, P.1222-1225, (1996).
4. Marshall W.F., Fung J.C., Sedat J.W. Deconstructing the nucleus: global architecture from local interactions, Curr. Opinion Genet. Dev., 7. P.259-263, (1997).
5. Minasbekyan L.A., Yavroyan Zh.V., Darbinyan M.R., Vardevanyan P.O. The phospholipid composition of nuclear subfractions from germinating wheat embryos, Russian Journal of Plant Physiol., 51 (5), P.784-789, (2004).
6. Vardevanyan P.O., Vardapetyan R.R., Tiratsuyan S.G., Panosyan G.H.Changes in the content of histone fractions of wheat isolated embryos during germination. Russian Journal of Plant Physiol. , 33 (1), P. 121, (1986).
7. Minasbekyan L.A., Vardevanyan P.O., Simonyan A.G., Panosyan G.A. To the study of some structural peculiarity of cereals DNA ., Biol.Journal of Armenia, 42(6), P. 551-555, (1989).
8. Bergman Y., Mostoslavsky R. DNA demethylation: turning genes on. Biol.Chem.,379, P. 401-407, (1998).
9. Gvozdev V.A. regulation of gene activity due to chemi\cal modification (methylation) DNA . Soros educat.journal, 10, P.11-17, (1999).
10. Minasbekyan L.A., Parsadanyan M.A., Panosyan G.H., Vardevanyan P.O., Changes in th enucleotide composition and pattern of DNA methylation during the germination of cereals seeds, Russian Journal of Plant Physiol ,48 (2), P. 256-259, (2000).
11. Minasbekyan L.A., Nerkararyan A.V., Vardevanyan P.O. Study of the phospholipid composition of wheat seedlings nuclear subfraction under influence of electromagnetic radiation. The II Int.Conf. "Modern problems of genetics, radiobiology, radioecology and evolution",Yerevan , Sept., P.130, (2005).
12. Petrosyan V.I., Sinicyn N.I., Elkin V.A., et al. The role of rezonanse molecular-wave processes and its using for control and correction of ecological system state. Binedicinskaya electronika, 5-6, P. 62–112, (2001).
13. Nerkararyan A.V., Parsadanyan M.A., Minasbekyan L.A. et al. Influence of low-intecity nonthermal koheremt EMI mm-diapason on seedlings growth. VI International Symposium «New and nontraditional plants and perspectives utilisation», Pushino, 1, P.35, (2005).
14. Gevorgyan E.S., Yavroyan Zh.V Armenian Nuclear Power Plant // Proceedings on Int. Conf. «Unification and Optimization of Radiation., Minasbekyan L.A. et al. Phospholipid composition of chromatin preparations of wheat germinating seedlings grown on the waters from adjoing zones of the Monitoring on NPP Location Regions», Armenia, Yerevan, September 22-26, 2004, P.70-75.
15. Lohe A., Roberts P. Heterochromatin Molecular and Structural Aspects // Ed. R.S. Verma Camb. Univ. Press.1988. P. 148-186.

CORRELLATION BETWEEN AQUOES PORE PERMEABILITY AND SURFACE CHARGE OF WHEAT SEEDLINGS NUCLEI

[1]Liya Minasbekyan, [2]Hamlet Badalyan, [1]Poghos Vardevanyan
[1]Department of Biophysics and [2]Department of Physics, Yerevan State University, str. A. Manougian, 1, Yerevan, 375025, Armenia

Abstract: Nucleocytoplasmic transport is mediated by shuttling receptors that recognize specific signals on protein or RNA cargoes and translocate the cargoes through the nuclear pore complex. Transport receptors appear to move through the nuclear pore complex by facilitated diffusion. We discuss recent experimental data on the electrokinetic potential of nuclei and the possible role of phospholipids in regulation of this charge. Such possibility can be realized by generation of difference potential outside and inside of nuclear envelope due to asymmetric distribution and content of phospholipids in double membrane.

Key words: embryos, electrokinetic potential, intact nuclei, nuclear membrane, phospholipids, translocation model, wheat seedlings.

1. INTRODUCTION

The eukaryotic cells separate the nuclear synthesis of DNA and RNA from cytoplasm protein synthesis with barrier termed the nuclear envelope (NE). Macromolecular transport between the nucleus and cytoplasm mediated by large protein-aqueous assemblies is called nuclear pore complexes (NPCs), which spann the double membrane of the nuclear envelope and have mass of ~ 125 MDa[1, 2]. The detailed structure of the NPC is not well defined and its molecular mass remains controversial. Translocation through this dynamic supramolecular structure, that provides

Vasili Tsakanov and Helmut Wiedemann (eds.), Brilliant Light in Life and Material Sciences, 205–211.
© 2007 *Springer.*

an interaction interface for a large number of different receptor pathways, supports an enormous mass flow. Charge of phospholipids may to play significant role in such interactions and NPC stability support by formation of local surface charge around NPC. With this purpose we investigated electrokinetic potential of nuclei, phospolipid content of nuclear membrane and discuss the possibility of this charge role in the translocation cargoes through NPC.

2.   MATERIALS AND METHODS

Seeds Germination. In our experiments we used seeds of hexaploid wheat (Triticum Aestivum, cvs. Voskehask and Bezostaya-1) and tetraploid spelt (Tr.dicoccum, cv.. Arpi), which were grown in petri dishes in darkness at 26^0C for 72 h and 96 h. Embryos were isolated from dry seeds of abovementioned cereals[3].

Nuclei isolation and fractionation. Nuclei were isolated from embryos by the method of Blobel and Potter in our modification[4]. To this end, dry embryos or seedlings frozen in liquid nitrogen were powdered with a mortar and pestle, and 0.25 M sucrose in TKM buffer (10 mM Tris-HCl, pH 7.4, containing 25 mM KCl, and 15 mM $MgCl_2$) was added[4]. The pellet obtained contained purified intact nuclei.

Phospholipid extraction and quantification. Phospholipids were extracted from acetone powder of nuclear membrane with the mixture of chloroform and methanol 2:1 by the method of Folch, as described in work[3].

Determination of the nuclear electrokinetic potential. Nuclear electrophoretic mobility was measured by micro-electrophoresis. To this end, isolated intact nuclei from wheat embryos and seedlings were suspended in the medium[4].

3.   RESULTS AND DISCUSSION

Relationship between surface charge of the nuclear membrane and the genome functional activity was demonstrated in some studies performed with various tissues[3-7]. The surface charge of the nuclear membrane depends on the functional swelling or shrinkage of the nucleus[6, 7]. However, the mechanisms and significance of these changes, as well as nuclear transport through NPC are obscure. Thus, the electrostatic charge of the nuclear surface was shown to reflect the functional activity of the genetic material. The increase in the EKP values we obtained in our experiments was evidently related to the state of the genome and changes of its physico–chemical characteristics during germination as presented in Table1.

Table 1. The surface electrokinetic values of the nuclei isolated from dry embryos and germ of cereals.

Source	Elektrokinetic potential		Direction of move, %	
	embryos	seedlings	Anodic	Cathodic
Bezostaya	−3.54 ± 0.43	−11.6 ± 0.28	100%	0%
Voskeask	−5.47 ± 0.51	−7.7 ± 0.24	100%	0%
Spelt	−23.6 ± 0.27	−25.6 ± 0.2	49%	51%
		+ 26.0± 0.31		

The experimental data available permit conclusion that the value of EKP on the surface of the nuclear membrane can reflect some changes in the nuclear matrix, particularly in heterochromatin, which directly depend on the nucleus functional state. It is not inconceivable that these changes in the nuclear matrix reflect the shifts in the phospholipids composition of the nuclear membrane.

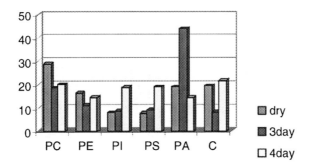

Figure 1. Proportion of phospholipids in the composition of the nuclear membrane from dry embryos and seedlings. . PC–phosphatidylcholine, PE–phosphatidylethanol, PI–phosphatidylinositol, PS–phosphatidylserin, PA–phosphatidic acid, CL–cardiolipin.

The changes of PL content in nuclear membrane reflect parallel morphological changes in a cell, which occur at the first stage of genome expression during germination. From the phospholipid content (see diagrams) we can calculate a proportion of neutral (PC and PE) and anionic (PI, PS, PA and CL) phospholipids in the nuclear membrane during germination have induced changes.

Such increase of anionic phospholipids can led to the increase of surface charge, which we observed and presented in Table 1. These changes evidence the significant role of phospholipids in forming the surface charge and permit us to judge of its inclusion in the translocation mechanism according to our suggested model.

Table 2. Proportion of neutral and anionic phospholipids in the composition of nuclear membrane from dry embryos and seedlings.

	Dry embryos	3-day seedlings	4 –day seedlings
Neutral PL	45,39 %	29,63%	18,37%
Anionic PL	54,56%	70,37%	81,71%

Nowadays two models have been proposed to explain the selectivity of the central channel. In one case, the NPC was postulated to contain a Brownian affinity gate made of a large number of flexible FG (phenylalanine-glycine) repeat-containing filaments. The second model for achieving selectivity of translocation through the NPC is the selective phase model[2]. However we have shown earlier, that there is a certain correlation between the activation of RNA-export through nuclear membrane and the value of electrokinetic potential of isolated nucleus during germination. This implies that the gradient of an electric field inside nuclear pore complexes can play an important role. On the other hand, hydrophobic interactions are carried out basically by Van–der–Vaals forces, which are weaker and can not provide so high speed of RNA-transport through NPC. By the most approximate calculations more than 1mln macromolecules passes through nuclear pore complexes in a minute. With the purpose to reveal the contribution of a gradient of an electric field potential inside nuclear pore complexes on RNA-export theoretical calculations have been carried out. As to a skeleton of NPC adjoin phospholipids, which consist of dipole "head" and hydrophobic neutral tails. Moreover the charge on the nuclear surface is distributed non-uniformly, that is inside the NPC there is a gradient of the potential of an electric field on one tail (outer) φ_1 and on the other end of the channel (inner) φ_2, big (outer) radius of NPC is R, and the distance from the certain point of a surface up to the given point inside of the channel r as it is shown in Fig. 2, then from equation of Laplas in cylindrical coordinates we have

$$\frac{\partial^2 \varphi}{\partial r^2} + \frac{\partial \varphi}{r \partial r} + \frac{\partial^2 \varphi}{r^2 \partial \theta^2} = 0 \tag{1}$$

Where θ angle between radius–vector up to the given point inside of NPC channel with an axis of the cylinder.

By solving (1) we have

$$\varphi(r = R\theta) = \sum \left[A_m \cos(m\theta) + B_m \sin(m\theta) \right] \qquad (2)$$

where A– is the amplitude on r, B – the amplitude on θ, and in the case when R=r.

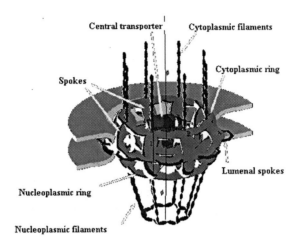

Figure 2. Framework of NPC and form electrical field gradient on two side of channel. This picture was worked out by R.W. Wozniak & C.P. Lusk, Quick guide, 2001, and modified by us with our purpose.

From (2) follows, that $B_m=0$, due to $\varphi(\theta) = \varphi(-\theta)$ in each point, what follows from cylindrical symmetry. Coefficient A_m can be obtained by integrating $\varphi_{(R,\theta)} \times \cos(m\theta)$ product on θ.

Consequently

$$\int_{-\pi/2}^{\pi/2} \varphi_1 \cos(n\theta) d\theta + \int_{\pi/2}^{3\pi/2} \varphi_2 \cos(n\theta) d\theta = \frac{2(\varphi - \varphi_2)}{n} \sin \frac{\pi n}{2} = \pi A_n$$

and $\pi(\varphi_2 + \varphi_1) = 2\pi A_0$ by this manner we receive potential in the given point of the NPC

$$\varphi = \frac{\varphi_1 + \varphi_2}{2} + \frac{2(\varphi_1 - \varphi_2)}{\pi} \sum_{n=1}^{\infty} \frac{(-1)^{n-1}}{2n-1} \left(\frac{\varphi}{R} \right)^{2n-1} x \cos\left[(2n-1)\theta \right] \qquad (3)$$

Numbers included in (3) can be summarized, if we use a condition

$$\sum_{n=1}^{\infty}(-1)^{n-1}\frac{x^{2n-1}}{2n-1}\cos[(2n-1)\theta] = \operatorname{Re}\int_{0}^{x}\frac{-i}{y}\sum_{n-1}^{\infty}\left(iye^{i\theta}\right)^{2n-1}dy \qquad (4)$$

By using more simple factorization and after integrating (4) we receive

$$\sum_{n=1}^{\infty}\frac{(-1)^{n-1}}{2n-1}x^{2n-1}\cos[(2n-1)\theta] = \frac{1}{2}I_m\ln\frac{1+ixe^{i\theta}}{1-ixe^{i\theta}} = \frac{1}{2}arctg\frac{2x\cos\theta}{1-x^2} \qquad (5)$$

Substituting (5) in (3) we receive

$$\varphi(r,\theta) = \frac{\varphi_1+\varphi_2}{2} + \frac{\varphi_1-\varphi_2}{2}arctg\frac{2rR\cos\theta}{R^2-r^2} \qquad (6)$$

Thus in any point inside the channel of NPC the potential is defined by the formula (6), that gives us possibility to determine a gradient of a field and force working on molecule RNA from the electric field.

The investigation of the reassembling forces acting among nucleoporins and phospholipids will be useful for understanding the mechanism of the translocation through these big protein aqueous complexes – NPC. This likely will require new approaches, including physical and biophysical methods with in vivo and in vitro systems for rapid time measurements of single events. This will be an exciting area of the future research in the nucleocytoplasmic transport field.

REFERENCES

1. Burger K.N. Greasing membrane fusion and fission machineries, Traffic, 1, P.605-613, (2002).
2. Bednenko, J., Cingoliani, G., and Gerace, L., Nucleocytoplasmic Transport: Navigating the Channel, Traffic, 4, pp. 127—135, (2003).
3. Minasbekyan L.A., Yavroyan Zh.V., Darbinyan M.R., Vardevanyan P.O. The phospholipid composition of nuclear subfractions from germinating wheat embryos, Russian Journal of Plant Physiol., 51(5), P.784-789, (2004).
4. Minasbekyan L.A., Parsadanyan M.A., Gonyan S.A., Vardevanyan P.O. RNA-export and Elektrokinetic Potential of Nuclei in Germinating Embryos of Cereal Crops, Russian Journal of Plant Physiology, 49(2), P.250-254, (2002).
5. Khesin E.Ya. The Size of Nuclei and Functional state of Cell (Meditcyna, Moscow, 1967), P.247

6. Vardevanyan P.O., Gonyan S.A., Tiratsuyan S.G., Relationship between electrokinetic percecularity isolated nuclei of rat hepatocytes with its functional activity, Biofizika, 46(2), P.271-274, (2001).

7. Vardevanyan P.O., Gonyan S.A., Minasbekyan L.A., Vardevanyan A.O. Electrokinetic Properties of Nuclei Isolated from Hepatocytes after Hormonal Induction, Biofizika, 42, P.1156-1157, (1997).

THERMODYNAMIC PROFILE OF INTERACTION OF PORPHYRINS WITH NUCLEIC ACIDS

Samvel Haroutiunian[1], Yeva Dalyan[1], Ara Ghazaryan[1] and Tigran Chalikian[2]
[1]Department of Molecular Physics, Yerevan State University, Yerevan Armenia,
[2]Department of Pharmaceutical Sciences, Leslie Dan Faculty of Pharmacy, University of Toronto, Canada

Abstract: The binding of a number of novel water soluble porphyrins to nucleic acids (NA) were studied monitoring the changes in Soret region of absorbance spectra, fluorescence spectra and circular dichroism (CD) spectra of formed complexes. The binding modes for all complexes were dtermined. The binding isotherms were used to calculate the binding constant, K_b, and binding free energy, $\Delta G_b = -RTlnK_b$. By performing these experiments as a function of temperature, we evaluated the van't Hoff binding enthalpies, ΔH_b and the binding entropies, ΔS_b. Generalization concerning the established correlation between the binding mode and thermodynamic profile was done.

Key words: porphyrin; nucleic acids; binding mode; thermodynamics.

1.   INTRODUCTION

Water-soluble cationic porphyrins bind to DNA. The pioneering work of Fiel (1979)[1] and subsequent publications of Pasternack and their coworkers (Pasternack & Gibbs, 1983)[2] established that the binding mechanism could be modulated by the nature of the metal ion and the size and location of the substituent groups on the periphery of the porphyrin. Generally, the free bases and square planar complexes intercalate between the base pairs of DNA[1]. For complexes having axially bound ligands or those with bulky

Vasili Tsakanov and Helmut Wiedemann (eds.), Brilliant Light in Life and Material Sciences, 213–218.
© 2007 *Springer.*

substituents located on the periphery of the structure, intercalation is blocked and "outside" binding occurs. The most extensively studied DNA binding porphyrin is *meso*-tetrakis(4*N*-methylpyridinium-4-yl)porphyrin [TMPyP] (Figure 1).

In addition to established chemical activity, porphyrins can be photochemicaly activated to attack biological molecules[3,4]. For example, photodynamic therapy (Berg et al., 1978; Diamond et al., 1977; Praseuth et al., 1986; Fiel et al., 1980) very likely involves damage to DNA and membranes through the photoinduced production of singlet oxygen.

The structure of the porphyrin-DNA complexes has been characterized using a variety of physical techniques. In an attempt to learn more about the manner in which cationic porphyrin complexes interact with nucleic acids (NA) and the forces that stabilize the complex we have focused on the binding thermodynamics. We assume that these data are essential for understanding their mode of action and for the development of design principles to guide the synthesis of new, improved compounds with enhanced or more selective activity.

2. MATERIALS AND METHODS

Meso-tetrakis(4*N*-hydroxyethylpyridinium-4-yl)porphyrin [TOEtPyP4], *meso*-tetrakis(4*N*-allylpyridinium-4-yl) porphyrin [TAlPyP4], *meso*-tetrakis (4*N*-metallylpyridinium-4-yl) porphyrin [TMetAlPyP4] and Ag-containing derivative of TAlPyP4 [AgTAlPyP4] (see Figure.1) were synthesized, purified, and kindly donated by Dr. Robert Ghazaryan (Faculty of Pharmaceutical Chemistry, Yerevan State Medical University, Armenia). The NA duplexes were obtained from Sigma-Aldrich Canada (Oakville, ON, Canada).

The concentrations of double-stranded polynucleotides were determined spectroscopically using molar extinction coefficients presented in our pervious papers [5,6].

All spectroscopic measurements were performed in 0.1 BPSE buffer (1BPSE = 6 mM Na_2HPO_4 + 2 mM NaH_2PO_4 + 185 mM NaCl + 1 mM Na_2EDTA), pH 7.0, $\mu = 0.02$.

All absorption and fluorescence titration experiments were carried out by stepwise addition of aliquots of a stock solution of NA to the porphyrin as previously described [5].

Visible absorption spectra of porphyrins in Soret region in the absence and presence of the RNA duplexes were collected at 18, 25, 35, and 45 °C using Perkin Elmer Lambda 800 UV/VIS spectrophotometer. Emission spectra were recorded at 18, 25, 35, and 45 °C using an AVIV model ATF 105 spectrofluorometer (Aviv Associates, Lakewood, NJ). CD spectra were recorded at 25 °C using an AVIV model 62 DS spectropolarimeter (Aviv Associates, Lakewood, NJ, USA).

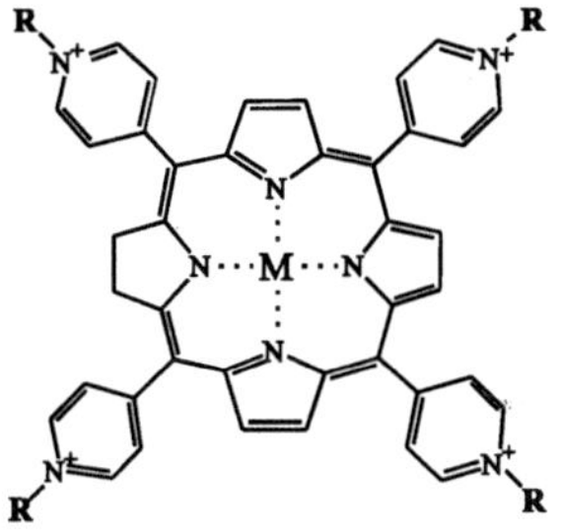

TMPyP(4)   R= CH₄

TOEtPyP(4)   R= CH₂-CH₂-OH

TAlPyP(4)   R= CH₂-CH=CH₂

TMetAlPyP(4)   R= CH₂-C(-CH₃)=CH₂

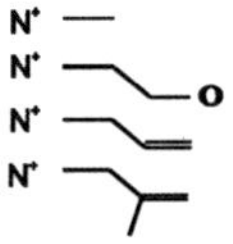

M = H₂, Ag

Figure 1. Porphyrin scheme.

3. RESULTS AND DISCUSSION

In our previous works we have characterized the binding of four novel water soluble porphyrins to nucleic RNA duplexes[5-8]. We estimated their binding modes and binding thermodynamics. Preliminary assumptions were made concerning the nature of porphyrin/nucleic acid interaction.

To investigate the possibility of extending the results and conclusions obtained for RNA/porphyrin interaction to the case of DNA we have studied the interaction of TAlPyP4 with homogeneous poly(dA)poly(dT) duplex. Positive peak of CD spectra in the Soret region obtained for this complex suggested that the binding mode is predominantly external which is rather expected since majority of porphyrins studied up-to-date prefer to bind to AT regions of DNA via this mode[2]. A further evidence for the CD-suggested mode of binding is provided by the fluorescence spectra. A splitting and significant increase in intensity of the emission spectrum of the dye was observed which is consistent with outside mode of binding[9].

Further, the absorbance titration data were assessed in terms of the noncooperative neighbor-exclusion model[5]. Our results show that TAlPyP4 binds to poly(dA)poly(dU) at 25°C with binding constant, K_b, equal to 9.9

$\pm 0.4 \times 10^5$ M^{-1} and binding stoichiometry, n, equal to 5.2 ± 0.27, numbers typical for porphyrin-nucleic acid interactions[2,10,11]. We determined the binding constants at 18, 25,35 and 45 °C. These experimental dependence was approximated by linear function (see figure 2); the slope of this function yield the binding enthalpy, $\Delta H_b = -R[\partial \ln K_b / \partial (1/T)]_P$. The binding free energy, ΔG_b, and entropies, ΔS_b, were calculated using $\Delta G_b = -RT \ln K_b$ and $\Delta S_b = (\Delta H_b - \Delta G_b)/T$. Our results show that the outside binding of TAlPyP4 to poly(dA)poly(dT) is accompanied by $\Delta G_b = -8.2 \pm 0.1$ kcal M^{-1}, highly unfavorable (positive) change in enthalpy of 14.4 ± 0.9 cal M^{-1} and favorable change in entropy , equal to 75.8 ± 3.2 cal M^{-1} K^{-1}.

Considering this result in a row with the data obtained by us before and data known from the literature[5-8,12-15], reveals a correlation between the thermodynamic profiles and the mode of porphyrin-NA association (see figure 3). For each porphyrin/NA complex the intercalation is accompanied by favorable change of enthalpy and rather modest or unfavorable change in entropy whereas external binding produces the large favorable values of ΔS_b and unfavorable ΔH_b. This correlation may reflect the differential nature of the molecular forces that are involved in stabilizing/destabilizing the two binding modes - intercalation versus outside stacking.

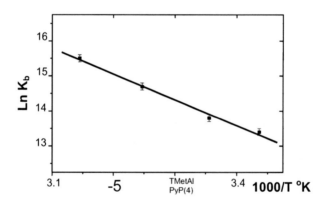

Figure 2. Temperature dependences of the binding constant for the associations of TAlPyP4 with poly(dA)poly(dT)

The majority of intercalating agents, including porphyrins, studied to date exhibit negative changes in enthalpy upon their association with nucleic acids[12-15]. For example, intercalation of TMPyP4 into three oligomeric G-tetraplexes differing in length and composition is accompanied by favorable changes in enthalpy[12]. Intercalation of this porphyrin and its metal derivative CuTMPyP4 into calf thymus DNA also causes favorable changes in enthalpy of -4.6 and -5.9 kcal M^{-1}

respectively[15]. On the other hand, the binding of ZnTMPyP4, an outside binder, to the same DNA causes an unfavorable change in enthalpy of 1.5 kcal M^{-1},[15]. A favorable change in enthalpy upon porphyrin intercalation is generally explained by the strong intermolecular interactions of intercalated porphyrin with the host DNA. Apparently, the favorable enthalpic contribution of strong electrostatic interactions between the positively charged pyridinium groups of the porphyrin and DNA phosphates prevails over the unfavorable contribution of disrupted and unformed interactions and results in a negative ΔH_b.

Unfavorable changes in enthalpy, ΔH_b, accompanying external porphyrin binding may, in part, reflect dehydration of some polar and charged groups of DNA without adequately supplying them by hydrogen bonds from the external porphyrin stacks.

Analogous to binding enthalpy, ΔH_b, binding entropy, ΔS_b, appears to correlate with the binding mode. As a working hypothesis, we propose that the major favorable contribution to the binding entropy comes from dehydration of porphyrin and NA groups upon their association. Clearly, external stacking of porphyrins along the duplex helices should be accompanied by more extensive burial of previously solvent-exposed atomic group and, consequently, more extensive dehydration than the intercalation. This rationalization is consistent with more favorable ΔS_b values observed for outside binders compared to those observed for intercalators.

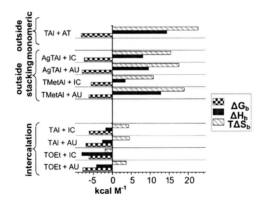

Figure 3. Thermodynamic parameters of porphyrins binding to nucleic acids at 25 °C.

This discussion and emerging thermodynamic picture are essentially consistent with our data presented in previous papers and in this work. We assume that a generalization can be made concerning the correlation between the thermodynamic profile and the binding mode of porphyrin to nucleic acids which can be used also as a footprint for binding mode determination.

This work was supported by NATO Collaborative Linkage Grant (LST.CLG.979777) and ISTC A301.2.

REFERENCES

1. R.J. Fiel, J.C. Howard, E.H. Mark, N. Datta Gupta, Interaction of DNA with a porphyrin ligand: evidence for intercalation, Nucl. Acids Res. 6, 3093–3118 (1979)
2. R. F. Pasternack, E. J. Gibbs, J. J. Villafranca, Interactions of Porphyrins with Nucleic Acids, Biochemistry, 22, 2406-2414 (1983)
3. I.E. Kochevar, D.A. Dunn, Photosensitized reactions of DNA: cleavage and addition. Bioorg. Photochem., Photochem. and the Nucl. Acids. (New York, 1990)
4. L.G. Marzilli, Medical aspects of DNA-porphyrin interactions, New J. Chem. 14, 409-420 (1990)
5. A.A. Ghazaryan, Y.B. Dalyan, S.G. Haroutiunian, A.Tikhomirova, N.Taulier, J.W. Wells, and T.V. Chalikian, Thermodynamics of Interactions of Water-Soluble Porphyrins with RNA Duplexes, J. Am. Chem. Soc. 128, 1914-1921 (2006)
6. A.A. Ghazaryan, Y.B. Dalyan, S.G. Haroutiunian, V.I. Vardanyan, R.K. Ghazaryan, T.V. Chalikian, Thermodynamics of Interactions of TAlPyP4 and AgTAlPyP4 Porphyrins with Poly(rA)poly(rU) and Poly(rI)poly(rC) Duplexes, J. Biomol. Struct. Dynam. 24(1), 66-71 (2006)
7. Y. B. Dalyan, The interaction of meso- tetra-(4N-oxyethylpyridyl)porphyrins with DNA. Effect of side group position , Biophysics translated from Biofizika, Russian Academy of Sciences, 47, 253 -258 (2002)
8. A. A. Ghazaryan, Ye. B. Dalyan, S. G. Haroutiunian, Spectral studies of interaction of meso-tetra-(4N-allylpyridyl)porphyrins with poly(rG)poly(rC), poly(rI)poly(rC) and poly(rA)poly(rU), Scientific Reports of YSU, 3, 73-78 (2005)
9. J.M. Kelly, M.J. Murphy, D.J. McConnell, C. OhUigin, A comperative study of the interaction of 5,10,15,20-tetrakis(N-methylpyridinium-4-yl) porphyrin and its zinc complex with DNA using fluorescence spectroscopy and topoisomerisation. Nucl. Acids Res., 13(1), 167-184 (1985).
10. T. Uno, K. Hamasaki, M. Tanigawa, S. Shimabayashi, Binding of meso-Tetrakis(N-methylpyridinium-4-yl)porphyrin to Double Helical RNA and DNA·RNA Hybrids, Inorg. Chem. 36, 1676-1683 (1997)
11. T. Uno, K. Aoki, T. Shikimi, Y. Hiranuma, Y. Tomisugi, Y. Ishikawa, Copper Insertion Facilitates Water-Soluble Porphyrin Binding to rA-rU and rA-dT Base Pairs in Duplex RNA and RNA-DNA Hybrids, Biochemistry, 41, 13059-13066 (2002)
12. I. Haq, J. O. Trent, B. Z. Chowdhry, T. C. Jenkins, Intercalative G-tetraplex stabilization of telomeric DNA by a cationic porphyrin. J. Am. Chem. Soc. 121, 1768-1779 (1999)
13. J.B. Chaires, Energetics of Drug–DNA Interactions, Biopolymers, 44, 201-215 (1997).
14. J.S. Ren, T.C. Jenkins, J.B. Chaires, Energetics of intercalation reactions, Biochemistry, 39, 8439-8447 (2000).
15. R.F. Pasternack, P. Garrity, B. Ehrlich,C. B. Davis, E. J. Gibbs, G. Orloff, A. Giartosio, C. Turano, The influence of ionic strength on the binding of a water soluble porphyrin to nucleic acids, Nucl. Acids Res. 14, 5919-5931 (1986)

SOME ASPECTS OF SYNCHROTRON RADIATION EFFECTS IN RADIOBIOLOGY

Ts.M. Avakian, A.S. Karaguesyan
Yerevan Physical Institute

Abstract: The modern electron accelerators and storage rings are unique source of synchrotron radiation(SR) Having   wide   spectrum   of radiation waves,minimal beam divergence ,high intensity and polarization SR is an interesting source of radiation for researches in the field of molecular biology and medicine. The main works on biological action of SR and perspectives of their use in radiation medicine,and genetics are presented.

Key words: Synchrotron radiation, ionization, chromosome aberrations.

Several published reports deal with   synchrotron radiation (SR) in biology and medicine[1]. One of the tasks of biology is the investigation of functioning mechanisms of biosystems, and dynamic and kinetic aspects of biological processes. From this point of view, SR is a unique technique. But SR has very strong damaged effect in those works. In reference[2] on accelerator SPIAR, it was found that during radiation structural analysis the stability of enzyme crystals to SR depended on time. In other investigations of Vasina[3] , it was showed the damaging effect of monochromatic SR on muscular fibers. These data along, with a wider diffusion   of this new source of radiation made radiobiologists more interested in investigating the biological action of SR .

Some characteristics of SR are very attractive for biological sciences, such as wide range spectrum, high intensity, small beam divergence, particularly high level of polarization. The last parameter very important for some experiments. For example, effects in DNA induced by radiation, can be investigated under conditions of conformational changes, which might be registered by circular dichroism[4]. Data about dependence of corner distribution of SR, its brightness and intensity from energy might be

Vasili Tsakanov and Helmut Wiedemann (eds.), Brilliant Light in Life and Material Sciences, 219–222.
© 2007 *Springer.*

found at Kunz`s article[5]. Therefore, SR can be superior to any alternative sources in vacuum ultraviolet and X-regions.

What can SR do for radiobiologists as an instrument for investigating radiation effects in bio-molecules and cells? Usually molecules are irradiated in a chaotic state, under the stochastic fluctuations of energy in atoms. Therefore, dependence of radiation damage in bio-molecules from the exact molecular energy configuration at the time of exposure cannot be studied. For such an approach, we should be able to select the wave-length of incident photons. Besides, beam intensity should be high enough to produce the damage, SR corresponds to these requirements.

By knowing K-side of absorption of element, it is possible to determine the sort of atom-absorbent[6]. Halpern and Motze[7] irradiated the microorganisms *Micrococcus denitrificans*, with partially bromized DNA, with monochromatic X-rays.They found that photons at 14,4 keV (above K-edge for Br), were more lethal per unit dose than 12.1keV (below the that K-edge). The results suggest that radiation effect depends on the localization of primarily absorbed energy events in the molecular system.

Using conventional sources of ionizing radiation it is not possible to change excitation energy for the formation of primarily radicals (PR) This is only possible using SR, and it will clarify chemical action of radiation in DAN and mechanisms of action of PR[8].

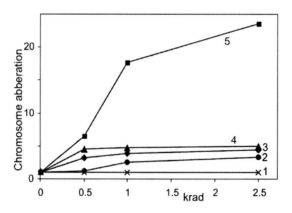

Figure 1. The action of X-rays (2, 3) and nonmonochromatic SR (4, 5) in the He (3) and oxygen (5) atmosphere; control (1).

We measured the effects of continuous and monochromatized SR (0.3-4.0 Å)[15]. We found oxygen effect (Fig. 1), noticeable mitotic delay of in plants rootlets, Besides, mitotic activity was lower than signals during action of 240GV X-rays (Fig. 2), and ESR signals were higher than control

(Fig. 3)[10-12]. These experiments showed that biological action of SR is very different from action of other sources (VEPP-3, Novosibirsk, Osippova L.P., 2001).

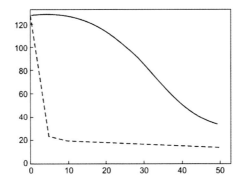

Figure 2. Dependence of the mitotic activity from X-ray(-)248 KeV,and SR(--).

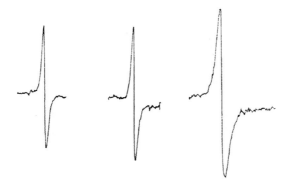

Figure 3. Spectrum analysis ESR a) dry seeds of Nic.tabacum-L b) after action of the X-ray c) after action of the SR.

REFERENCES

1. 1.Sturman H.B. Uses of SR in Biology, Academic Press, London, 1982
2. 2.Phillips J.C, at all, PNAS, 73, P.128-132, 1976
3. 3.Vasina A.A. Molecular Biology 8, 2, P., 243-284. Moscow, 1976 (in Russia)
4. 4.Poletaev A..Molecular Biology. 8,2.Moscow, 1972 (in Russia)
5. 5.Kunz C., Synchrotron Radiation Techniques and Application Springer-Verlag, N.Y, 1979
6. 6.Bocki.G.B. Roengen Structore analysis. Moscow, 1951 (in Russia)
7. 7.Halpern A.,Mutze B., Int.J.Rad.Biology 34, 67-72, 1978

8.   8.Halpern A.,Uses of SR in Biology (Edited by Sturman),P.255-283.,London,1982
9.   9.Avakian Ts.M, Karabekov I.P., Martirossian M.A. Preprint EPI-256(49) -77. Yerevan, 1977
10.   10 Minassian M. A,Avakian,Ts.M, Semerdjyan. SP, Radiobiologia, 18, 5, P.779-784., Moscow 1978 (in Russia)
11.   AvakianTs.M., at.all. SR Implementation in biology Modern problem of Radibiology, Radioecology and Evolution Dedicated to centenary of N.W. Timofeeff- Ressovsky. 96-97, Dubna, 2000
12.   Avakian Ts M. Action of Radiation on living organisms P.,1-59, Yerevan, 2003

e-mail(tsovak@mail.yerphi.am)

PART 3

Biomedical Research

THE BIOMEDICAL PROGRAMS AT THE ID17 BEAMLINE OF THE EUROPEAN SYNCHROTRON RADIATION FACILITY

A. Bravin
European Synchrotron Radiation Facility, B.P. 220, 38043 Grenoble Cedex, France

Abstract: The application of synchrotron radiation in medical research has become a mature field of research at synchrotron facilities worldwide. The ability to tune intense monochromatic beams over wide energy ranges differentiates these sources from standard clinical and research tools. At the European Synchrotron Radiation Facility in Grenoble, France, a dedicated research facility is designed to carry out a broad range of research ranging from cell radiation biology to *in vivo* human studies. Medical imaging programs at the ID17 biomedical beamline include absorption, K-edge and phase contrast imaging. Radiation therapy programs, namely the Stereotactic Synchrotron Radiotherapy and the Microbeam Radiation Therapy are rapidly progressing towards the clinical phase. This paper will present a very brief overview of the techniques and of the latest results achieved at the ESRF.

Key words: synchrotron radiation; medical imaging; radiotherapy.

1. INTRODUCTION

The European Synchrotron Radiation Facility (ESRF) is a 3rd generation synchrotron radiation (SR) machine which is operated at 6 GeV and 200 mA with a lifetime of more than 50 hours in the 2/3 filling mode. About 40 beamlines are presently operational, using monochromatic or filtered synchrotron radiation in a wide energy range, from the infrared to the hard X-rays (> 100 keV). The ID17 beamline is dedicated to preclinical and clinical biomedical research. Its source is a 21-pole wiggler, 150 mm

Vasili Tsakanov and Helmut Wiedemann (eds.), Brilliant Light in Life and Material Sciences, 225–239.
© 2007 *Springer.*

period, with a 1.6 T maximum magnetic field. The beamline is one of the two long ESRF beamlines; one experimental hutch (mainly used for the Microbeam Radiation Therapy programs) is located at about 30 m from the source, while a second experimental hutch is hosted in a satellite building, outside the ESRF main experimental hall, at about 150 m from the source (Fig. 1).

The original goal of the ID17 biomedical beamline at the ESRF was the development of imaging and radiation therapy techniques for preclinical and potentially clinical application. The two pioneering programs, both aiming at the clinical application in human beings, have been the Transvenous Coronary Angiography using the dual energy digital subtraction imaging technique and the Microbeam Radiation Therapy applied to the palliation and cure of brain tumours.

However, in the past years, there has been a dramatic increase of the users' community that has determined a large diversification of techniques and applications. The beamline activities are now focussed on functional and anatomical imaging, and radiotherapy (including basic radiobiology).

The research at ID17 is carried out by several groups: ESRF scientists as part of their In-House research programs, the Grenoble Hospital (CHU) team in association with the INSERM U647 group, and by outside users selected by peer review processes.

An overview of the main techniques applied at ID17 and of a selected number of applications is here presented and discussed.

Figure 1. An external view of the ID17 biomedical beamline at the ESRF. On the left the satellite building.

2. IMAGING

Several intrinsic characteristics of synchrotron radiation light produced by second and third generation storage rings are particularly interesting in biomedical imaging.

The radiation is highly intense and naturally collimated, permitting to select a narrow energy band (down to a few electron volts, typically several hundreds of eV) in a wide energy range (typical range used in biomedical imaging: ~10-100 keV) by a crystal monochromator; in this way, X-ray energies can be optimized with the specific sample characteristics (size and composition). The possibility of "freely" choosing a monochromatic X-ray beam has several implications: the dose delivered to the sample can be reduced with respect to conventional multi-energy imaging while the image contrast can be optimized. In addition, the effects of beam hardening, always present in conventional X-ray imaging are largely avoided, and therefore absolute quantification of elements in the sample can be determined.

Instead of a single beam, the digital subtraction imaging technique uses two monochromatic beams to take advantage from the discontinuity in absorption at the K-edge of specific contrast agent elements (Fig.2); details are given in the Paragraph 2.1.

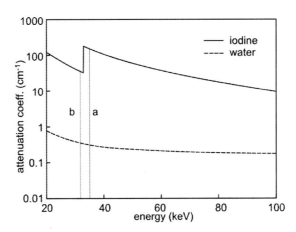

Figure 2. X-ray attenuation coefficient for iodine and water. "a": above the K-edge, "b": below the K-edge. In the K-edge digital subtraction technique, two images are simultaneously taken at these energies.

The very small source size (down to a few tents of microns in the vertical direction) and the large source to sample distances (typically 20-40 m in short bending magnet beamlines, but more than 100 m in long insertion device beamline like at the ESRF) allow good spatial resolution, limited only by the detector.

The high collimation (~ 0.1 mrad in the vertical direction, about one order of magnitude larger in the horizontal) permits to place the detector far from the object (up to several meters) so that scattering is effectively eliminated from the detected beam.

The small source size and the large source to sample distance determine the spatial coherence of the beam allowing for the application of the propagation phase contrast imaging technique (Baruchel et al., 2000).

2.1   K-edge digital subtraction imaging (KEDS)

The K-edge digital subtraction imaging method utilizes the sharp rise in the photoelectric component of the attenuation coefficient of a given element (typically iodine, xenon or gadolinium) at the binding energy of the K-electron. At ID17 it may be used in both planar and Computed Tomography (CT) mode. The single crystal bent Laue monochromator, which is one of the three X-ray monochromators presently installed at ID17, can simultaneously produce two beams with energies bracketing the K-absorption edge of the contrast agent to be visualized (e.g. 33.17 keV for I, 34.56 keV for Xe, 50.25 keV for Gd). Each beam has an energy bandwidth of about 125 eV and their mean energies are separated by about 300 eV at 33 keV.

The two beams transmitted by the sample (or by the patient) are then acquired by a dual line germanium detector or by a taper optics CCD camera (Bravin et al., 2003). Logarithmic subtraction of both measurements results in a contrast agent enhanced image (Sarnelli et al., 2005). In the case of planar imaging, the value of a pixel corresponds to the integral of the contrast agent concentration times the thickness of tissues containing contrast. In CT, as the tomography reconstruction integrates the concentration-thickness product, each pixel value corresponds to the concentration of the contrast agent in the sample's voxel. This technique also gives access to a complementary image, which is the X-ray map of the sample not containing the contrast agent (Elleaume et al., 2002).

2.2   Transvenous coronary angiography

The only program that has reached the clinical trial phase is transvenous coronary angiography, which has been carried out at ID17 in the period 2000-2003. Patients included in the protocol have previously undergone angioplasty with the implantation of a stent.

A research protocol was designed to evaluate the potential of this method, in comparison with the conventional gold-standard technique, in the visualization of intra-stent stenosis, which is the most common form of re-stenosis.

If a re-stenosis was suspected, the patient was imaged at the ESRF in the left anterior oblique (LAO) orientation, and within the next few days with the conventional technique at the hospital. The total irradiation dose was strictly limited by the French Ministry of Healthy to 200 mSv skin dose per patient (2 mSv effective dose) for the full series of images (Elleaume et al., 2000).

The imaging method included the insertion of a catheter into the brachial vein and advanced to the superior *vena cava* under fluoroscopy control; the patient was then installed in the scanning system in the imaging room.

By a remotely controlled system, 30–45 ml of iodine (350 mg/ml) were injected in a bolus. The image sequence was started a few seconds after the injection of the contrast agent depending on the transit time (time interval between the injection and the arrival of the bolus in the coronaries), previously measured. Each image line (0.35 mm thick) was acquired in 1.4 ms, and thus no motion artefact occurred. The complete image was obtained within 0.6 s. The dimension of the scan images was 15x15 cm^2, with a pixel size of 0.35×0.35 mm^2 (Peterzol et al., 2006; Bertrand et al., 2005; Elleaume et al., 2000). Images were quantitatively evaluated by post-processing (Sarnelli et al., 2005).

The accuracy of Synchrotron Radiation Angiography (SRA) was investigated in 57 men; the right coronary artery (RCA) in 27 patients, and the left anterior descending artery (LAD) in 30 patients. SRA and conventional Quantitative Coronary Angiography were performed within 2 days of each other. Image quality was good or excellent in most patients. Global sensitivity of SRA images was 64%, specificity was 95%, and positive and negative predictive values were ~85%. False negatives involved short eccentric lesions and superimposed segments, most frequently of the LAD. False positives occurred in intermediate stenoses slightly overestimated by SRA (Bertrand et al., 2005).

The SRA is currently the only minimally invasive technique allowing intra-stent stenosis follow-up which has been compared with the conventional gold standard technique. Because of concern about patients undergoing repeated techniques without benefit to themselves, the present study used a low X-ray dose of only 200 mSv to the skin for the entire imaging procedure. Use of higher radiation doses might improve image quality with regards to the signal-to-noise ratio.

SRA has the advantages of being unaffected by artefacts due to the stents, calcifications, or cardiac motion. Stenosis in any part of the RCA can be accurately visualized, but visualization of the LAD may cause problems due to superimposition of other vascular structures.

2.3 Bronchography

Small airways play a key role in the distribution of ventilation and in the matching of ventilation to perfusion. SR bronchography allows measurement of regional lung ventilation and evaluation of the function of airways with a small diameter. Monochromatic SR beams are used to obtain quantitative respiration-gated images of lungs and airways in anaesthetized and mechanically ventilated rabbits using inhaled stable xenon gas as a contrast agent. Two simultaneous images are acquired at two different energies, above and below the K-edge of Xe. Logarithmic subtraction of the two images yields absolute Xe concentrations. Two-dimensional planar and CT images are obtained showing spatial distribution of Xe concentrations within the airspaces, as well as the dynamics of filling with Xe (Bayat et al., 2001).

In a sequence of CT images it is possible to follow the evolution of Xe gas concentration; these data permit to deduct the local time constants of ventilation. In addition, CT imaging can also provide a quantitative tool with high spatial discrimination ability for assessment of changes in peripheral pulmonary gas distribution during mechanical ventilation (Porra et al., 2004).

At the ESRF, the effects of the tidal volume (VT) of inspiration on the ventilation distributions have been studied in detail.

The excellent spatial resolution of the KES method allows also studies of the fractal structure of the bronchial three leading to a deeper understanding of the factors affecting ventilation.

In addition, the application of spiral SR computed tomography (SRCT) technique has permitted for direct quantification of absolute regional lung volumes in the 3D rendering of the lungs (Monfraix et al., 2005).

2.4 Cerebral blood volume and blood flow measurement

Synchrotron radiation computed tomography applied in KEDS mode permits to quantitatively access the brain perfusion and permeability. In particular, this technique allows one to measure *in vivo* absolute contrast-agent concentrations with high accuracy and precision, and absolute Cerebral Blood Volume or flow can be derived from these measurements using tracer kinetic methods (Adam et al., 2003). An intravenous infusion of iodine is injected in rats while CT images are acquired to follow the temporal evolution of the contrast material in the blood circulation. As in the previous cases, two beams are used, of energies bracketing the K-edge of the iodine. The temporal evolution of the contrast agent concentration in each voxel was the basic information necessary to derive brain perfusion maps. The CBV and the Blood Brain Barrier permeability measurements were derived using well established perfusion models (Adam et al., 2005).

Measurements were obtained on a rat glioma model using SRCT and a high resolution CCD camera (47×47 μm^2 pixel size) (Coan et al., 2006). To date, this is technique is the only one permitting to access *in vivo* these brain parameters with such a spatial resolution, opening interesting perspectives for small-animal morphological and functional imaging. This technique is particularly useful for pathophysiologic brain studies involving hemodynamic changes, such as new treatment follow-up, where preclinical trials are usually performed on small rodents.

2.5 Analyser based imaging (diffraction enhanced imaging)

In absorption imaging, the contrast results from variation in X-ray absorption arising from density differences and from variation in the thickness and composition of the specimen or of the injected contrast agent.

Besides absorption, an X-ray beam traversing an object picks up information on its refraction properties that can also be utilized as a source of contrast for displaying the internal properties of the sample.

The behaviour of X-rays as they travel through an object can be described in terms of a complex index of refraction. In the X-ray region, it can be indicated as $n=1-\delta+i\beta$ where the real part δ corresponds to the phase shift (or refraction) and the imaginary part β to the absorption.

The real and imaginary parts have very different dependences on the photon energy; in the regime where the photoelectric effect dominates and far from absorption edges, $\beta \sim E^{-4}$ while $\delta \sim E^{-2}$. As a consequence, the values of δ in the X-ray energy range are orders of magnitude larger than β terms.

X-rays passing through regions of different δ values are subjected to phase shifts that correspond to being refracted. These changes, which can originate from the purely geometrical effect of the shape of the object or, for instance, from local homogeneity changes of the object, cannot often be visualized using absorption imaging techniques.

The analyser-based technique allows detecting the phase variations (Bravin, 2003); the X-rays transmitted through a sample are analysed by a

perfect crystal; only the X-ray satisfying the Bragg law for the diffraction
($2d \sin \theta = \lambda$, where d is the crystal d-spacing, θ is the grazing angle of
incidence to the crystal and λ is the radiation wavelength) can reach the
detector and then contribute to the image formation.

The angular resolution is provided by the choice of reflection from the
crystal, which acts as a spatial filter of the radiation refracted and scattered
inside the object. As an example, the full-width at half-maximum of the
reflectivity curve for the reflection (111) of the silicon spans the range
17.6-4.3 µrad for the energy interval 15–60 keV (Bravin, 2003).

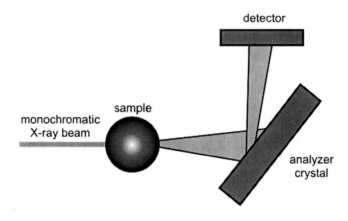

Figure 3. Scheme of experimental setup for analyzer-based imaging.

The small angular acceptance of the analyzer suppresses small-angle
scattered radiation and thereby increases the signal to noise ratio. The edge
enhancement, which is characteristic of this method, occurs at the interfaces
of regions with different refractive indices. It is achieved by slightly
detuning the analyzer away from the exact maximum of the reflectivity
curve. In that way images can be recorded at angles which lie somewhat
above and below the analyzer maximum, for example at the half values of
the maximum reflectivity where the reflectivity gradient is steepest. A ray
which is refracted by a tiny amount when passing through a zone of
changing refractive index is then reflected from the analyzer with a
considerably different reflectivity. From two images of the same object
taken on the low and on the high angle side of the rocking curve, an image
which represents apparent absorption (intensity behind the object affected
by absorption and suppression of scattering) and an image which represents
only refraction effects can be calculated (Chapman et al., 1997). This post-
processing technique is indicated in the literature as Diffraction Enhanced
Imaging (DEI).

At the ESRF, the DEI technique has been widely applied in *in vitro* mammography and cartilage and bone studies. In mammography, fine details of the structures such as strands of collagen and contours between glandular and adipose tissue, which are barely visible in the conventional absorption-based mammograms, are clearly visible in the diffraction-enhanced images. Images have been systematically compared with the stained histopathological sections; the correlation of the radiographic findings with the morphologic changes in specimens is unequivocally confirmed in several literature reports. Examples are found in (Keyriläinen et al., 2005; Pagot et al., 2005). An increased soft tissue contrast and a combination of information obtained with disparate diffraction-enhanced images provide better visibility of mammographically indistinguishable features (Pagot et al., 2005). This kind of additional structural information of the breast tissue is required to improve assessment accuracy and earlier detection of the breast lesions. These advances in image quality make the method a very promising candidate for clinical mammography.

DEI has been successfully applied in preclinical osteoarthritis studies. Osteoarthritis is a poorly understood disease that can affect the cartilage and other tissues in the joints of ageing people. Conventional radiography is sensitive only in cases of advanced disease in which there has been a loss of cartilage; structural abnormalities in the early stages of the degenerative process are generally not visualized in radiographs.

Measurements have been performed on human articular cartilage in disarticulated, as well as in intact joints. Gross cartilage defects, even at early stages of development, have been studied at 18 and 30 keV and compared with the absorption technique and show a clear, early visualization of the damage (Wagner et al., 2005). In addition, experiments were performed in rabbits and sheep sample with model implants to investigate the option for DEI as a tool in bone-implant research. DE images allow the identification of the quality of ingrowth of bone into the hydroxyapatite layer of the implant through the visualization of a highly refractive edge at the implant/bone border. Implants with bone fully grown onto the surface did not display a refractive signal. Therefore, the refractive signal could be utilized to diagnose implant healing and/or loosening (Wagner et al., 2006).

3. RADIATION THERAPY

The fundamental goal of radiation therapy is to deliver a high therapeutic dose of ionizing radiation to the tumour without exceeding normal tissue tolerance. This limitation is particularly severe in the case of brain tumors because of the high risk of adverse normal tissue morbidity. Ideally, radiation should be transported to the tumour with minimum damage to the surrounding tissue, and radiation should cause secondary processes that stop the tumour growth and eliminate the tumour.

Several approaches are being explored using synchrotron radiation to enhance the effectiveness of radiation therapy. At the ESRF, the research has focused on the treatment of high grade gliomas, which are some of the most radioresistant and aggressive human tumors. In brain tumours adjuvant chemotherapy generally fails to improve patient outcome, because of inadequate drug delivery and/or concentration inside the tumour. Hence, considerable efforts were provided to optimize chemo-radiotherapy treatments against brain tumors by increasing both chemotherapeutic drug concentration and radiation dose inside the tumour, while preserving healthy tissues (Behin et al., 2003).

Despite all efforts, presently used treatments are largely ineffective: the median survival time is 3 months without any treatment, and 8 to 12 months after surgery and conventional radiotherapy or radiosurgery (Black et al., 1998; de Crevoisier et al., 1997).

Innovative brain cancer treatments are under development at the ESRF: the Photon Activation Therapy (PAT) -in vitro- and the Synchrotron Stereotactic Radiotherapy (SSR) -in vivo, both using monochromatic X-rays and the Microbeam Radiation Therapy (MRT) using spatially fractionated filtered white X-ray beams.

3.1   The photon activation therapy (PAT) and the synchrotron stereotactic radiotherapy (SSR)

Both techniques are based on inserting a high Z material (associated or not with a chemotherapic drug which is targeting the DNA), either into the cells or in the tumour, and on irradiating the target with monochromatic X-rays at energies above the K-edge of the inserted element; the presence of the high Z material produces a dose enhancement. One possible explanation of this effect involves the release of high-LET Auger electrons following photoelectric absorption of an X-ray by an electronic shell of the incorporated heavy element. Such electrons have a very short mean free path and therefore have a high probability of creating Double Strand Breaks in the tumour DNA (Solberg et al., 1992).

One of the first works carried out by the INSERM-U647 team led by F. Esteve (Grenoble, France) consisted in estimating, in vitro, the cytotoxicity effect of the synchrotron X-rays on the cellular survival with or without iodinated contrast agent. Human cell lines were irradiated at different X-ray doses in the presence of various concentrations of iodine aiming at verifying the increase of the dose deposit (Corde et al., 2004). It has also been shown *in vitro* that the amount of double and single strand breaks is superior when the energy is tuned above the K-edge of the photo-activated heavy element (Corde et al., 2003).

In vivo tests in tumour bearing rats were performed using the SSR technique, in which the dose is continuously distributed over 360°, the tumour being at the centre of rotation. In addition, the beam is restricted to the tumour dimension leading to a peak dose at the target location (Boudou et al., 2005).

Preclinical trials using iodinated contrast agents have shown that it is possible to master the irradiation procedure with regards to the wavelength choice and to the ballistic, thus leading to an optimal energy deposit in the tumour while sparing surrounding tissues; life span was significantly extended in treated tumour bearing animals (Adam et al., 2006).

The same SSR procedure was then used to test the combination of chemotherapy (cisplatin) and radiotherapy in rats inoculated with F98 glioma cells. 3 µg of cisplatin were injected intra-cerebrally at the tumour site the day before irradiation with 78.8 keV photons. Median Survival Time (MeST) of untreated rats was 26 days. When cisplatin or synchrotron radiation alone was applied, the MeST was found to be 37 and 48 days, respectively. When both treatments were combined, a very large increase in life span was obtained (MeST = 206.5 days) compared to the controls. One year after treatment, 6 out of the 18 rats treated with CDDP and radiation, were still alive. This outstanding life span increase (694%) is the largest obtained to date with the F98 glioma model and demonstrate the interest of powerful monochromatic X-ray sources as new tools for cancer treatment.

3.2   The microbeam radiation therapy

Microbeam Radiation Therapy is based on the "dose-volume" effect: the smaller the irradiated macroscopic tissue volumes, the higher the threshold absorbed doses for damage to normal tissues (Curtis, 1967). Present-day clinical applications of this principle include stereotaxic radiosurgery and conformal radiotherapy, using photon beams collimated in millimetres. High-intensity SR X-ray wiggler beams with negligible divergence became available in the 1980s, and dose-volume relationships were tested in the microscopic range, first at Brookhaven National Laboratory's (BNL) National Synchrotron Light Source and then at the ESRF. In analogy to the model used by Curtis (1967), SR generated microbeams (≈25 µm wide) were delivered to normal rat brain tissues in a single exposure to achieve skin-entrance absorbed doses of 312 to 5000 Gy. Mature brain tissues displayed an extremely high resistance to tissue necrosis; the latter was only observed after entrance doses of 10000 Gy (Slatkin et al., 1995). The sparing effect has been attributed to rapid repair of microscopic lesions by minimally irradiated cells contiguous to the irradiated tissue slice, i.e., mainly by endothelia and perhaps by glial cells. Microplanar beams crossfired toward the target in parallel exposures at 100 μm intervals (entrance doses of 312 or 625 Gy) considerably extended the median survival time of young adult rats bearing advanced intracerebral gliosarcomas (Laissue et al., 1998). These results suggested the possibility of a differential effect between normal and tumour tissue for microbeam irradiation. This biological selectivity is likely to be related to differences between vasculature in tumours and in normal tissues even if the mechanisms are as yet unknown and under investigation.

MRT has the potential to treat infantile brain tumours when other kinds of radiotherapy would be excessively toxic to the developing normal brain. Preclinical research programs which were started and developed by a Swiss-based research team, led by Prof. J. Laissue (University of Bern), are presently carried out by several groups in collaboration with the ESRF. They are aiming to reach clinical trials, and are focused on different aspects of the problematic related to clinical application of MRT. They include the determination of the best irradiation parameters in rats and mice (Miura et al., 2006), studying the effect of MRT in highly radiation sensitive organs of large animals (Laissue et al., 2001), the combination of MRT and immunotherapy (Smilowitz et al., 2006), and the development of innovative irradiation geometry aiming at maximizing the dose delivered to the tumour while keeping it below the tolerance threshold of the healthy tissues (Bräuer-Krisch et al., 2005). Basic studies on vascular tissue damage by microbeams are giving extraordinary results on vascular reparation process for a better understanding of the mechanism involved in the tissues sparing effect (Serduc et al., 2006). Preparation of treatment planning requires detailed knowledge of locally deposited dose. With this aim, extensive theoretical (Monte Carlo) (Siegbahn et al., 2006) and experimental microdosimetric studies are also performed.

ACKNOWLEDGEMENTS

The ID17 team of the ESRF, the INSERM Unit U647 and ID17 users are kindly acknowledged for their support in the preparation of this manuscript.

REFERENCES

Adam, J. F., Elleaume, H., Le Duc, G., Corde, S., Charvet, A. M., Troprès, I., Le Bas, J. F., and Estève, F., 2003, Absolute cerebral blood volume and blood flow measurements based on synchrotron radiation quantitative computed tomography, *J. Cereb. Blood Flow Metab.* **23**:499-512.

Adam, J. F., Nemoz, C., Bravin, A., Fiedler, S., Bayat, S., Monfraix, S., Berruyer, G., Charvet, A.-M., Le Bas, J. F., Elleaume, H., Estève, F., 2005, High-resolution blood–brain barrier permeability and blood volume imaging using quantitative synchrotron radiation computed tomography: study on an F98 rat brain glioma, *J. Cereb. Blood Flow Metab.* **25**:145−153.

Adam, J. F., Joubert, A., Biston, M.-C., Charvet, A.-M., Peoc'h, M., Le Bas, J. F., Balosso, J., Esteve, F., and Elleaume, H., 2006, Prolonged survival of Fischer rats bearing F98 glioma after iodine-enhanced synchrotron stereotactic radiotherapy *Int. J. Radiat. Oncol. Biol. Phys.* **64**(2):603–611.

Bayat, S., Le Duc, G., Porra, L., Berruyer, G., Nemoz, C., Monfraix, S., Fiedler, S., Thomlinson, W., Suortti, P., Standertskjöld-Nordenstam, C. G., Sovijärvi, A. R. A., 2001, Quantitative functional lung imaging with synchrotron radiation using inhaled xenon as contrast agent, *Phys. Med. Biol.* **46**(12):3287-3299.

Baruchel, J., Cloetens, P., Härtwig, J., Ludwig, W., Mancini, L., Pernot, P., and Schlenker, M., 2000, Phase imaging using highly coherent X-rays: radiography, tomography, diffraction topography, *J. Sync. Rad.* **7**:196-201.

Behin, A., Hoang-Xuan, K., Carpentier, A. F., and Delattre, J. Y., 2003, Primary brain tumours in adults, *Lancet* **361**:323-31.

Bertrand, B., Estève, F., Elleaume, H., Nemoz, C., Fiedler, S., Bravin, A., Berruyer, G., Brochard, T., Renier, M., Machecourt, J., Thomlinson, W., and Le Bas J. F., 2005, Comparison of synchrotron radiation angiography with conventional angiography for the diagnosis of in-stent restenosis after percutaneous transluminal coronary angioplasty, *Eur. Heart J.* **26**:1284-1291.

Black, P., 1998, Management of malignant glioma: role of surgery in relation to multimodality therapy, *J Neurovirol.* **4**:227–236.

Boudou, C., Balosso, J., Esteve, F., and Elleaume, H., 2005, Monte Carlo dosimetry for synchrotron stereotactic radiotherapy of brain tumours, *Phys. Med. Biol.* **50**(20):4841-51.

Bräuer-Krisch, E., Requardt, H., Regnard, P., Corde, S., Siegbahn, E., Le Duc, G., Brochard, T., Blattmann, H., Laissue, J., and Bravin, A., 2005, New irradiation geometry for microbeam radiation therapy (MRT), *Phys. Med. Biol.* **50**:3103-3111.

Bravin, A., 2003, Exploiting the X-ray refraction contrast with an analyser: the state of the art, *J. Phys. D: Appl. Phys.* **36**:A24-A29.

Bravin, A., Fiedler, S., Coan, P., Labiche, J.-C., Ponchut, C., Peterzol, A., and Thomlinson, W., 2003, Comparison between a position sensitive germanium detector and a taper optics CCD "FRELON" camera for Diffraction Enhanced Imaging, *Nucl. Inst. Meth. A* **510**:35-40.

Chapman, D., Thomlinson, W., Johnston, R. E., Washburn, D., Pisano, E., Gmür, N., Zhong, Z., Menk, R., Arfelli, F., and Sayers, D., 1997, Diffraction enhanced x-ray imaging, *Phys. Med. Biol.* **42**:2015–2025.

Coan, P., Peterzol, A., Fiedler, S., Ponchut, C., Labiche J. C., and Bravin, A., 2006, Evaluation of imaging performance of a taper optics CCD `FReLoN' camera designed for medical imaging, *J. Synchrotron Radiat.* **13**:260-270.

Corde, S., Balosso, J., Elleaume, H., Renier, M., Joubert, A., Biston, M. C., Adam, J. F., Charvet, A. M., Brochard, T., Le Bas, J. F., Estève, F., and Foray, N., 2003, Synchrotron photoactivation of cis-platin (PAT-Plat) elicits an extra-number of DNA breaks that stimulate RAD51-mediated repair pathways, *Cancer Res.* **63**:3221-3227.

Corde, S., Joubert, A., Adam, J. F., Charvet, A. M., Le Bas, J. F., Esteve, F., Elleaume, H., and Balosso, J., 2004, Synchrotron radiation-based experimental determination of the optimal energy for cell radiotoxicity enhancement following photoelectric effect on stable iodinated compounds, *Br. J. Cancer* **91**:544–551.

Curtis, H. J., 1967, The interpretation of microbeam experiments for manned space flight, *Radiat. Res. Suppl.* **7**:258-264.

de Crevoisier, R, Pierga, J. Y., Dendale, R., Feuvret, L, Noel, G., Simon, J. M., and Mazeron, J. J., 1997, Radiotherapy of glioblastoma, *Cancer Radiother.* **1**(3):194 –207.

Elleaume, H., Fiedler, S., Estève, F., Bertrand, B., Charvet, A. M., Berkvens, P., Berruyer, G., Brochard, T., Le Duc, G., Nemoz, C., Renier, M., Suortti, P., Thomlinson, W., and Le Bas, J. F., 2000, First human transvenous coronary angiography at the European Synchrotron Radiation Facility, *Phys. Med. Biol.* **45**:L39-L43.

Elleaume, H., Charvet, A. M., Corde, S., Estève, F., and Le Bas J. F., 2002, Performance of computed tomography for contrast agent concentration measurement with monochromatic x-ray beam: comparison of K-edge versus temporal subtraction, *Phys. Med. Biol.* **47**:3369-3385.

Keyriläinen, J., Fernández, M., Fiedler, S., Bravin, A., Karjalainen-Lindsberg, M.-L., Virkkunen, P., Elo, E.-M., Tenhunen, M., Suortti, P., and Thomlinson, W., 2005, Visualisation of calcifications and thin collagen strands in human breast tumour specimens by the diffraction-enhanced imaging technique: a comparison with conventional mammography and histology, *Eur. J. Radiology* **53**:226-237.

Laissue, J. A., Geiser, G., Spanne, P. O., Dilmanian, F. A., Gebbers, J.-O., Geiser, M., Wu, X. Y., Makar, M. S., Micca, P. L., Nawrocky, M. M., Joel, D. D., and Slatkin, D. N., 1998, Neuropathology of ablation of rat gliosarcomas and contiguous brain tissues using a microplanar beam of synchrotron-wiggler-generated X rays, *Int. J. Cancer* **78**(5):654–660.

Laissue, J. A., Blattmann, H., Di Michiel, M., Slatkin, D. N., Lyubimova, N., Guzman, R., Werner Zimmermann, W., Jaggy, A., Smilowitz, H. M., Stettler, R., Kircher, P., Bley, T., Brauer, E., Stepanek, J., Bravin, A., Renier, M., Le Duc, G., Nemoz, C., Thomlinson, W. C., and Wagner, H. P., 2001, The weanling piglet cerebellum: a surrogate for tolerance to MRT (Microbeam Radiation Therapy) in pediatric neuro-oncology, *Proc. SPIE,* **4508**:65-73.

Miura, M., Blattmann, H., Brauer-Krisch, E., Bravin, A., Hanson, A. L., Nawrocky, M. M., Micca, P. L., Slatkin, D. N., and Laissue, J. A., 2006, Radiosurgical palliation of aggressive murine SCCVII squamous cell carcinomas using synchrotron-generated X-ray microbeams, *Br. J. Radiol.* **79**(937):71-75.

Monfraix, S., Bayat, S., Porra, L., Berruyer, G., Nemoz, C., Thomlinson, W., Suortti, P., and Sovijarvi, A. R. A., 2005, Quantitative measurement of regional lung gas volume by synchrotron radiation computed tomography, *Phys Med Biol* **50**(1):1-11.

Pagot, E., Fiedler, S., Cloetens, P., Bravin, A., Coan, P., Fezzaa, K., Baruchel, J., and Härtwig J., 2005, Quantitative comparison between two-phase contrast techniques: diffraction enhanced imaging and phase propagation imaging, *Phys. Med. Biol.* **50**:709-724.

Peterzol, A., Bravin, A., Coan, P., and Elleaume H., 2006, Performance of the K-Edge digital subtraction angiography imaging system at the European Synchrotron Radiation Facility, *Radiat. Prot. Dosimetry* 117(1-3):44-49.

Porra, L., Monfraix, S., Berruyer, G., Le Duc, G., Nemoz, C., Thomlinson, W., Suortti, P., Sovijarvi, A. R. A., and Bayat, S., 2004, Effect of tidal volume on distribution of ventilation assessed by synchrotron radiation CT in rabbit, *J. Appl. Physiol.* 96(5):1899-908.

Sarnelli, A., Nemoz, C., Elleaume, H., Estève, F., Bertrand, B., and Bravin, A., 2005, Quantitative analysis of synchrotron radiation intravenous angiographic images, *Phys. Med. Biol.* 50:725-740.

Serduc, R., Vérant, P., Vial, J.-C., Farion, R., Rocas, L., Rémy, C., Fadlallah, T., Brauer, E., Bravin, A., Laissue, J., Blattmann, H., and van der Sanden, B., 2006, In vivo two-photon microscopy study of short-term effects on normal mouse brain microvasculature after microbeam irradiation, *Int. J. Radiat. Oncol. Biol. Phys.* 64(5):1519-1527.

Siegbahn, E. A., Stepanek, J., Bräuer-Krisch, E., and Bravin, A., 2006, Determination of dosimetrical quantities used in microbeam radiation therapy (MRT) with Monte Carlo simulations, *Med. Phys.* 33(9): in press

Slatkin, D. N., Spanne, P., Dilmanian, F. A., Gebbers, J.-O., and Laissue, J. A., 1995, Subacute neuropathological effects of microplanar beams of x-rays from a synchrotron wiggler, *Proc. Natl. Acad. Sci. USA*, 92:8783-8787.

Smilowitz, H. M., Blattmann, H., Bräuer-Krisch, E., Bravin, A., Di Michiel, M., Gebbers, J.-O., Hanson, A. L., Lyubimova, N., Slatkin, D. N., Stepanek, J., and Laissue, J. A., 2006, Synergy of gene-mediated immunoprophylaxis and unidirectional Microbeam Radiation Therapy for advanced intracerebral rat 9L gliosarcomas, *J. Neurooncol.* 78:135–143.

Solberg, T. D., Iwamoto, K. S., and Norman, A., 1992, Calculation of radiation dose enhancement factors for dose enhancement therapy of brain tumours, *Phys. Med. Biol.* 37:439–443.

Wagner, A., Aurich, M., Sieber, N., Stoessel, M., Wetzel, W.-D., Schmuck, K., Lohmann, M., Metge, J., Reime, B., Coan, P., Bravin, A., Arfelli, F., Heitner, G., Menk, R., Irving, T., Zhong, Z., Muehleman, C., and Mollenhauer, J. A., 2005, Options and limitations of joint cartilage imaging: DEI in comparison to MRI and sonography, *Nucl. Instr. Meth. A* 548:47–53.

Wagner, A., Sachse, A., Keller, M., Wagner, O., Aurich, M., Wetzel, W.-D., Venbrocks, R. A., Wiederanders, B., Hortschansky, P., Horn, J., Schmuck, K., Lohmann, M., Reime, B., Metge, J., Arfelli, F., Menk, R., Rigon, L., Muehleman, C., Bravin, A., Coan, P., and Mollenhauer J., 2006, Evaluation of titanium implant ingrowth into bone by diffraction enhanced imaging (DEI), *Phys. Med. Biol.* 51:1313-1324.

NOVEL CHEMOTYPES IN PHARMACOCHEMICAL APPROACHES
Focus on Aldose Reductase Enzyme Inhibitors

Vassilis J. Demopoulos[1], Ioannis Nicolaou[1], Polyxeni Alexiou[1], Chariklia Zika[1], Katja Sturm[2], and Albin Kristl[2]

[1]Department of Pharmaceutical Chemistry, School of Pharmacy, Aristotle University of Thessaloniki, Thessaloniki 54124, Greece, [2]Faculty of Pharmacy, University of Ljubljana, Ljubljana 1000, Slovenia

Abstract: Diabetes mellitus exacts a huge toll in money and human suffering. At its present rate of increase, within few decades it will be one of the world's commonest diseases and biggest public-health problem, with an estimated minimum of half-a-billion cases. The diabetic individual is prone to late onset complications that are largely responsible for the morbidity and mortality observed in the patients. It has been demonstrated that the more severe and sustained the degree of hyperglycemia, the more likely it is that the chronic complications of diabetes will develop. Pharmaceutical intervention of hyperglycemia-induced diabetic complications is actively pursued since it is very difficult to maintain normoglycemia by any means in patients with diabetes mellitus. Aldose reductase enzyme (AR, ALR2, E.C. 1.1.1.21) of the polyol metabolic pathway was first found to be implicated in the etiology of secondary complications of diabetes. AR inhibitors (ARIs) have therefore been noted as possible pharmacotherapeutic agents. Although several ARIs have progressed to the clinical level, only one is currently on the market. However, the inhibition of the polyol pathway is considered to be a promising approach to control diabetes complications as well as a number of other pathological conditions like ischemia, abnormal vascular smooth muscle cell proliferation, cancers, and mood disorders. Thus, attention is currently targeted to discover ARIs of distinct chemical structures, being derivatives of neither hydantoin nor carboxylic acid, which are known to cause either toxicity, or posses' narrow spectrum of tissue activity. The focus of this brief overview will aim to support the notion that novel aldose reductase inhibitors with unique physicochemical properties could posses' pharmacotherapeutic potential.

Vasili Tsakanov and Helmut Wiedemann (eds.), Brilliant Light in Life and Material Sciences, 241–250.
© *2007 Springer.*

Key words: long term diabetes complications, pyrrole based aldose reductase inhibitors,
               intestinal absorption, synthetic methodologies

1.   INTRODUCTION[1-8]

The search for novel chemotypes as aldose reductase inhibitors started about ten years ago in the laboratory of Medicinal Chemistry at the University of Thessaloniki. Our main goal was the development of pharmacotherapeutic agents for the treatment of the long-term diabetic complications in a background of hyperglycemia.

Diabetes mellitus is a universal health problem, and type II diabetes is a disease fast approaching epidemic proportions throughout the world. At its present rate of increase, within few decades it will be one of the worldÿs most common diseases and biggest public-health problem, with an estimated a half-a-billion cases. The hyperglycemic state seen in diabetes mellitus is associated with the development of the diabetes-specific complications. The direct economic cost of diabetes is about 10 percent of the total health care budget, and approximately 90% of the total direct cost is needed for the treatment of the devastating diabetic complications such as neuropathy, retinopathy, nephropathy and vascular diseases. Aldose reductase (AR, ALR2, E.C. 1.1.1.21) enzyme of the polyol metabolic pathway has first found to be implicated in the etiology of the long-term complications of diabetes, as well as many other pathological conditions such as ischemia, abnormal vascular smooth cells' proliferation, cancers, and mood disorders and, therefore, aldose reductase inhibitors (ARIs) have been noted as possible pharmacotherapeutic agents. In the polyol pathway glucose is reduced to sorbitol by the enzyme aldose reductase and then oxidized to fructose by the enzyme sorbitol dehydrogonase. Three main biochemical processes have been proposed to link the activation of the aldose reductase to the diabetic complications: osmotic stress, pseudohypoxia, which causes oxidative stress, and non-enzymatic glycation of proteins.

Although several aldose reductase inhibitors have progressed to the clinical level, only one, namely Epalrestat, is currently on the market, and only in Japan. However, based on recently published articles, we could state that the pharmaceutical research community considers the inhibition of the polyol pathway to be a promising approach to control diabetic complications as well as a number of other pathological conditions.

Aldose reductase is a small monomeric protein composed of 315 amino acid residues. Its primary structure was first determined back in 1992 from rat lens, co-crystallized with citric acid. Its core consists of eight parallel beta strands, the adjacent strands of which are connected by peripheral alpha-helical segments running parallel to the beta sheet. After this initial crystallographic structure determination, several structures of aldose reductase have been resolved with X-ray crystallography. It should be pointed out that all of these contain an organic or organometallic anion, and they differ significantly to each other. For example, crystallization of aldose reductase with the inhibitors sorbinil or tolrestat reveals that their polar head groups both bind near the coenzyme, but that their hydrophobic moieties are at 90° to each other. Sorbinil binds with few changes in the enzyme structure, whereas the binding of tolrestat induces a displacement of Leu300 and Phe122.

A number of studies have been reported aiming to the discovery of novel and potent aldose reductase inhibitors (ARIs) through a three-dimensional-database search followed by design and synthesis. However, only disappointing results have been reported so far, with the best hit being an indole derivative having an in-vitro aldose reductase IC50 value of 0.21 μM. We think that this compound cannot be classified as a novel hit/lead structure, as it is an acetic acid derivative, and also does not conform to either the "rule of five" nor to the "rule of three". These unsuccessful computational SAR results could be either attributed to the diversity of the crystallographic resolved structures of the targeted aldose reductase, or to the recently revealed results of the inability of ten docking programs, and their related scoring functions, to make a useful prediction of the binding affinity of the ligands to eight diverse proteins.

2. OVERVIEW OF RESULTS AND DISCUSSION[9-17]

Our approach in searching the field of aldose reductase inhibitors could be described as more academic oriented. It encompassed the sequential designing and synthesizing of a small library of selected compounds, the biological testing of them, followed by the designing and the synthesis of side chain modified or core modified derivatives.

At this point a reference to a putative aldose reductase inhibitor site model is necessary. This site consists of either two parallel or coplanar lipophilic regions, each one containing a hydrogen-bonding site and a sterically constrained charge-transfer pocket. We designed the first set of compounds based on this schematic pharmacophore of aldose reductase inhibitors proposed by Kador and Sharless. The suggested common features of this pharmacophore are firstly two aromatic regions (a primary and a secondary) that hydrophobically bind with the lipophilic regions on the enzyme. In support with this proposal, it has been shown that half of the van der Waals contacts in the crystallized complex of aldose reductase and the inhibitor zoporlestat are between carbon atoms, attesting to the highly hydrophobic nature of the enzyme-inhibitor interaction. Another structural characteristic of the inhibitors is a carbonyl (or thiocarbonyl) group that could participate in a reversible "charge transfer" interaction. Furthermore, the majority of the aldose reductase inhibitors contain an acidic functionality, mainly an acetic acid or a hydantoin moiety, which has been proposed to contribute directly toward inhibitory potency through interaction with a complementary binding site present on the enzyme. This latter binding site has been assigned to the anion well located within the active site pocket of the enzyme. Although this originally proposed pharmacophore of the aldose reductase inhibitors does not require a direct contribution to enzyme binding of an acidic moiety, it postulates that the presence of a hydroxyl group located about 8-9 Å from the center of the primary aromatic region of the inhibitor can enhance activity. As a central aromatic structure of the prepared compounds we chose the pyrrole ring which favorably combines interesting synthetic chemical reactivity as well as desirable physicochemical properties like electrostatic potential and dipole moment. The first six compounds that were initially prepared and tested for their in-vitro aldose reductase inhibitory activity were a set of six isomeric benzoylpyrrolylacetic acids. The source of the used aldose reductase, for the in-vitro experiments, was the rat lenses. It has been shown that there is an approximately 85% sequence similarity between rat lenses' and human aldose reductase, while the proposed active sites of both enzymes are identical. The performed assay was based on the spectrophotometric monitoring of NADPH oxidation, which has been proven to be a quite reliable method. It was found that the most active isomer was the corresponding 3-benzoylpyrrole-1-acetic acid. Appreciable activity was also showed by the 1-benzoylpyrrole-3-acetic acid and by the 2-benzoylpyrrole-1-acetic acid. It is interesting to note that the calculated distance between the center of the phenyl ring and the carboxylic-hydroxyl in the minimum energy conformation of the active compound (3-benzoylpyrrole-1-acetic acid) was 9.3 Å while the corresponding value for the inactive one (4-benzoylpyrrole-3-acetic acid) was 6.9 Å. Based on this, we suggested that a contributing factor for higher activity is the ability of the compounds to attain an overall more extended low energy conformation. Furthermore, as these compounds bear common structural features with known acidic non-steroidal anti-inflammatories, we also tested them in vivo, at a relatively high dose, for this activity. It was found that the two of the three active aldose reductase inhibitors, (specifically the

3-benzoylpyrrole-1-acetic acid and the 2-benzoylpyrrole-1-acetic acid) did not show any anti-inflammatory activity at the studied dose. This could indicate that these compounds would probably exhibit low gastrointestinal adverse effects. Thus, we selected these two structures as hits for farther activity optimization.

Following the identification of these putative hits as aldose reductase inhibitors, a research effort was undertaken in order to optimize their activity as well as their physicochemical properties, which are considered important for drug likeness. The second series of compounds that were designed, synthesized and tested in vitro for aldose reductase inhibitory could be divided into two groups. The first group comprises of the compounds, which are derivatives of the 3-benzoylpyrrole-1-acetic acid. The changes, which were made in the original structure, were generally selected as to confer variability to the physicochemical properties of the targeted compounds. The biphenyl, naphtyl and benzofuran rings were selected because they have been identified to belong to the class of privileged substructures for protein binding. The second group combines structural features of the 3-benzoylpyrrolylacetic acid as well as that of the 2-benzoylpyrrolylacetic acid. In these compounds there is a gradual addition of methoxy groups, aiming to an increase of the overall hydrogen bond ability of the molecules. It was found that a number of these compounds exhibited considerable potency in the micromolar or submicromolar range. The most active compound was the bis-methoxybenzoylpyrrolylacetic acid derivative, with potency in the nanomolar range. This is favorably compared to the well-established aldose reductase inhibitors, like epalrestat and zopolrestat. It should be noted that important quantative structural-activity relationship features of the potent compounds, in the overall second series, are the presence of substituents with relatively low Hammet sigma values and/or moieties, which increase their overall aromatic area.

The most potent compound, the bis-methoxybenzoylpyrrolylacetic acid derivative was, also, preliminary tested for acute toxicity in an *in-vivo* protocol. It was found that this compound, at a dose of 390mg/kg i.p., did not produce any mortality to experimental rats after a 24 h period. However, in an *in-vitro* (*ex-vivo*) permeability experimental protocol, with the rat intestine (jejunum) in side by side diffusion chambers (cells) this compound was found to be poorly permeable. This is attributed, mainly, to an operating efflux mechanism.

It is of interest to note that the tested arylpyrrolylacetic acid derivative, which does not contain a keto-carbonyl group, exhibited weak inhibitory potency. An inspection of a computational derived low energy conformation of this compound reveals that its two aromatic rings are almost perpendicular to each other. On the contrast, in the case of its aroypyrrorylacetic acid counterpart these two rings are almost in the same plane.

Taken into consideration the overall results of these studies, a third series of compounds (i.e. pyrrolylbenzothiazole and trifluoromethyl bis-benzoylpyrrole derivatives) were designed, synthesized and tested in-vitro for aldose reductase inhibitory activity. The design of the pyrrolylbenzothiazole derivatives was based on the structure of the benzoylpyrrolylacetic acid as well as the putative non-classical bioisosteric relationship between a keto-carbonyl group and the thiazole ring. The design of the trifluoromethyl bisbenzoylpyrrole derivatives was based on the structure of the bismethoxybenzoylpyrrolylacetic acid, where one of its para methoxy groups was replaced with a meta- or para-trifluoromethyl functionality, taking into account a proposed non-classical hydrogen bond acceptor bioisosteric relationship. Furthermore, according to the literature data, we postulated that the replacement of an electron donor on an aromatic moiety with an electron acceptor counterpart would possibly decrease the overall toxicity profile. The in-vitro experimental results showed that the compounds in this third series of aldose reductase inhibitors are active in the submicromolar to the nanomolar range. However, in an *in-vitro* (*ex-vivo*) permeability experimental protocol, with the rat intestine (jejunum) in side by side diffusion chambers (cells) both of these compounds were found to be poorly permeable. This is attributed, mainly, to an operating efflux mechanism.

Literature data indicates that during the last twenty years a considerable number of compounds have been synthesized and have been shown to be effective, *in vitro* and/or *in vivo*, as aldose reductase inhibitors. However, the only aldose reductase inhibitor that is available as a drug is ONO Pharmaceutical's Epalrestat in Japan. The status quo in the field of aldose reductase inhibitors can be attributed to the dearth of new chemotypes that can powerfully block the flux through the polyol pathway. Only two chemical classes of aldose reductase inhibitors have been, so far, evaluated in critical phase three trials. These classes belong to derivatives of either the hydantoin ring or to a carboxylic acid moiety. Hydantoin aldose reductase inhibitors, like sorbinil, are weakly acidic and hence are largely unionized at blood pH, facilitating their efficient tissue penetration and thus being highly potent *in vivo* with a broad spectrum of tissue activity. However, members of this chemical class, including sorbinil, are known to cause skin rash and hypersensitivity or liver toxicity. On the other hand, aldose reductase inhibitors of the carboxylic acid class, like zopolrestat, which have low pKa values, are almost completely in the ionized state at blood pH and are markedly less potent *in vivo* with a narrower spectrum of tissue activity than hydantoins. The question is whether the polyol pathway approaches to diabetic complications, as well as to control other pathologies, merits continued medicinal chemistry attention, and one can discover an aldose reductase inhibitor of distinct chemical structure with sufficient potency and duration of action to effectively block the excess flux at safe doses in the clinic. Several lines of evidence show that there are strong reasons to be optimistic. It is extremely encouraging that minalrestat, a hydantoin-like spiroimide, in an open label short-term clinical study has been reported to strongly suppress both sorbitol and fructose in sural nerve biopsy samples, as well as to improve nerve velocity deficit, at very low doses. Additional support comes from results of aldose reductase knockout mouse studies and from reports that several allelic variants of the human aldose reductase gene are also associated with significant increase in the risk of development of microvascular complications. These developments have lent strong timely support for initiatives aiming to low molecular weight not-hydantoin, non-carboxylic acid structures as possible lead compounds, with in-vitro aldose reductase inhibitory activities lower than 10 micromolar. This inhibitory concentration is generally accepted in the pharmaceutical industry as a first indication that a compound could be a putative hit or lead structure for a possible further pharmacochemical optimization. Thus, in the fourth series of compounds, the main structural changes were focused on the replacement of the acetic acid functionalities with lipophilic and less acidic non-classical bioisostric moieties related to them.

Taken into account that the active hydantoin aldose reductase inhibitors (like sorbinil) have a pKa value close to 8, we focused our attention for novel molecules exhibiting acidity in this range of pKa values. Of course, for the reasons we previously described, we excluded any hydantoin like structures. It has been shown that the 2,6-difluorophenol structures, with a pKa of 7.1, could be successfully incorporated, as a lipophilic isostere of a carboxylic acid. Furthermore, we run computational pKa calculations on a series of structures that could function as putative bioisosteres to a carboxylic acid and could have some common features with the compounds we have, so far, studied as aldose reductase inhibitors. One such structure, we have identified, was a mercaptoethylpyrrolylphenylmethanone with a calculated pKa value of 8.5. Based on the above considerations, a fourth series of compounds was designed, synthesized and tested *in-vitro* as aldose reductase inhibitors. We found that the use of the difluorophenole moiety was quite successful, improving significantly the *in-vitro* inhibitory activity, in comparison with their carboxylic acid counterparts. Furthermore, the first of the new difluorophenole compounds was tested in the *ex-vivo* intestinal experimental protocol and it exhibited satisfactory permeability results. The tested mercaptoethylpyrrolylphenylmethanone derivative has appreciable activity at the concentration of 10 micromolar, which is indicative for being a hit/lead structure. However, it is far less active than its carboxylic acid counterpart. A definite explanation for this result could not put forward at the present and with the limited available SAR data. However, we could suggest that the observed relative inactivity is probably partly due to a combination of its high lipophilic nature as well as its very low aqueous solubility. Work is under way to overcome, via appropriate structural modifications, these adverse physicochemical properties and derive a robust SAR thesis.

Taking into consideration the fact that the replacement of a carboxylic acid moiety with that of the difluorophenol improves the inhibitory activity towards the aldose reductase, we tried to modify few phenylsulfonyl-amino acids and specifically glycine derivatives which have been previously found to be active as aldose reductase inhibitors. Indeed, the IC50 value of the initially synthesized compound is approximatelly 1/3 of the one of the respective glycine derivative, and there is high prospective that there will be farther improvement of the inhibitory activity after the addition of suitable moieties to this main structure. We also tested this compound in the *ex-vivo* intestinal experimental protocol, and it exhibited satisfactory permeability results. It is very interesting to note that there is initial evidence of an operating influx mechanism, and we suggest that this should be taken into consideration for future drug design efforts.

Finally, a number of novel synthetic methodologies were developed during the preparation of the above-mentioned aldose reductase inhibitors, and two of them are briefly described. The first procedure refers to the use of the 4-chloropyridine, as a catalyst, in a Clauson-Kaas type formation of a pyrrole ring. We think that this new methodology will be of general utility in the formation of this heterocyclic ring starting from salts of an aniline derivative. The second procedure refers to the selective, one-pot, formation of an aroylsulfonamide moiety in an aminophenol ring. It includes the sequential addition of chlorotrimethylsilane for the protection of the hydroxyl-group, dimethylaminopyridine as a catalyst for the formation of the sulfonamide and hydrolysis of the silyl-ether at the last step. A very interesting finding was that the addition of a "catalytic" amount of triethylamine by the end of the reaction greatly improves the overall yield. A putative explanation could be that the added amine facilitates the completion of the overall circle of this reaction. The scope and limitations of this methodology is currently investigated in our laboratory.

ACNOWLEDGEMENT

We thank the European Union for the Financial Support, implemented through the General Secretariats of Research and Technology of Greece and Slovenia, as well as for the funding for Graduate Scholarships from a number of Greek, Slovenian and International Organizations.

REFERENCES

1. V. J. Demopoulos, N. Zaher, Ch. Zika, C. Anagnostou, E. Mamadou, P. Alexiou, and I. Nicolaou, Compounds that combine aldose reductase inhibitory activity and ability to prevent the glycation (glucation and/or fructation) of proteins as putative pharmacotherapeutic agents, Drug Design Reviews-Online 2, 293-304 (2005).
2. A. Tsantili-Kakoulidou, I. Nicolaou, D. Vrakas, and V. J. Demopoulos, Modelling of aldose reductase inhibitory activity of pyrrol-1-yl-acetic acid derivatives by means of multivariate statistics, Medicinal Chemistry 1, 321-326 (2005).
3. C. Koukoulitsa, Ch. Zika, G. D. Geromichalos, V. J. Demopoulos, and H. Skaltsa, Evaluation of aldose reductase inhibition and docking studies of some secondary metabolites, isolated from Origanum vulgare L. ssp. hirtum, Bioorg. Med. Chem. 14, 1653-1659 (2006).
4. A. Urzhumtsev, F. TeteFavier, A. Mitschler, J. Barbanton, P. Barth, L. Urzhumtseva, J. F. Biellmann, A. D. Podjarny, and D. Moras, A 'specificity' pocket inferred from the crystal structures of the complexes of aldose reductase with the pharmaceutically important inhibitors tolrestat and sorbinil, Structure 5, 601-612 (1997).
5. B. L. Mylari, S. J. Armento, D. A. Beebe, E. L. Conn, J. B. Coutcher, M. S. Dina, M. T. O'Gorman, M. C. Linhares, W. H. Martin, P. J. Oates, D. A. Tess, G. J. Withbroe, W. J. Zembrowski, A novel series of non-carboxylic acid, non-hydantoin inhibitors of aldose reductase with potent oral activity in diabetic rat models: 6-(5-chloro-3-methylbenzofuran-2-sulfonyl)-2H-pyridazin-3-one and congeners, J. Med. Chem., J. Med. Chem. 48, 6326-6339 (2005).
6. Y. Iwata, M. Arisawa, R. Hamada, Y. Kita, M. Y. Mizutani, N. Tomioka, A. Itai, and S. Miyamoto, Discovery of novel aldose reductase inhibitors using a protein structure-based approach: 3D-database search followed by design and synthesis, J. Med. Chem. 44, 1718-1728 (2001).
7. D. A. Erlanson, R. S. McDowell, T. O'Brien, Fragment-based drug discovery, J. Med. Chem. 47, 3463-3482 (2004).
8. M. H. Lambert MH, Lindvall M, Nevins N, Peishoff CE, Semus SF, Senger S, Tedesco G, Wall ID, Woolven JM, Head MS, A critical assessment of docking programs and scoring functions, Abstracts of Papers of the Papers of the American Chemical Society, 228: 087-COMP Part 1 Aug. 22 2004 (J. Med. Chem. 13 Aug 2005, 10.1021/jm050362n).
9. V. J. Demopoulos, and E. Rekka, Isomeric benzoylpyrroleacetic acids: Some structural aspects for aldose reductase inhibitory and anti-inflammatory activities, J. Pharm. Sci. 84, 79-82 (1995).

10. V. J. Demopoulos, C. Anagnostou, and I. Nicolaou, Validation of a computational procedure for the calculation of the polar surface area (PSA) of organic compounds, Pharmazie 57, 652-653 (2002).
11. N. Zaher, I. Nicolaou, and V. J. Demopoulos, Pyrrolylbenzothiazole derivatives as aldose reductase inhibitors, J. Enzym. Inhib. Med. Ch. 17, 131-135 (2002).
12. C. Anagnostou, I. Nicolaou, and V. J. Demopoulos, Synthesis of [5-(4-pyrrol-1-yl-benzoyl)-1H-pyrrol-2-yl)]-acetic acid and in vitro study of its inhibitory activity on aldose reductase enzyme and on protein glycation, Pharmazie 57, 535-537 (2002).
13. I. Nicolaou, and V. J. Demopoulos, Substituted pyrrol-1-yl-acetic acids which combine aldose reductase enzyme inhibitory activity and ability to prevent the non enzymatic irreversible modification of proteins from monosaccharides, J. Med. Chem. 46, 417-426 (2003).
14. V. J. Demopoulos, I. Nicolaou, and Ch. Zika, A facile synthesis of 1-(6-hydroxyindol-1-yl)-2,2-dimethylpropan-1-one, Chem. Pharm. Bull. 51, 98-99 (2003).
15. I. Nicolaou, Ch. Zika, and V. J. Demopoulos, [1-(3,5-Difluoro-4-hydroxyphenyl)-1H-pyrrol-3-yl]phenylmethanone as bioisostere of a carboxylic acid aldose reductase inhibitor, J. Med. Chem. 47, 2706-2709 (2004).
16. K. Sturm, L. Levstik, V. J. Demopoulos, and A. Kristl, Permeability characteristics of novel aldose reductase inhibitors using rat jejunum in vitro, Eur. J. Pharm. Sci. 28, 128-133 (2006).
17. A. Alexiou, I. Nicolaou, M. Stefek, A. Kristl, and V. J. Demopoulos, Selective, one-pot formation of N-3,5-difluoro-4-hydroxy-phenylobenzosulfonamide. A pharmacodynamic and pharmacokinetic study, 12th Symposium of Pharmacochemistry, 27-28 January 2006, Patras, Greece.

GENETIC EFFECTS OF ELECTROMAGNETIC WAVES

Rouben Aroutiounian, Galina Hovhannisyan, Gennady Gasparian
Department of Genetics and Cytology, Yerevan State University, 375025, Yerevan, Armenia

Abstract: The genetic effects of electromagnetic waves can be detected by different test-systems. The mutagenic effect of ionizing radiation can be developed on the levels of DNA and/or chromosomes. In numerous researches efficiency of micronucleus assay, alkaline single-cell gel electrophoresis, chromosomal aberrations test and FISH-technique and their different combinations for the detection of ionizing radiation-induced genotoxic effects are discussed. Also some molecular-biological approaches developed in the last years are presented.

Key words: electromagnetic waves, mutations, chromosome aberrations, micronuclei, Comet-assay

The effects of radiation exposure on human health include carcinogenesis, genetic diseases affecting future generations and developmental disabilities (Au, 1991). The mutagenic effect of ionizing radiation can be developed on the levels of DNA and/or chromosomes. DNA damage caused by ionizing radiation includes remarkable structural changes to nitrogen bases, forming sites from which a base was removed (apurinic or apyrimidinic sites), breaking hydrogen bonds between two helices, breaking one or two polynucleotide chains and promoting interstrand or intrastrand crosslinking (Ward J.F., 1975). By Maluf (2004) ionizing radiation, as clastogenic agent, induces a considerable number of aneuploidies. Non-ionizing electromagnetic radiations including radio frequency and microwave radiation have great impact in modern society. Increased exposition of radio frequency electromagnetic field (EMF) produced by the appliances used in the telecommunications, industry and medicine may lead to biological effects (Paulraj, Behari, 2006).

Vasili Tsakanov and Helmut Wiedemann (eds.), Brilliant Light in Life and Material Sciences, 251–265.
© *2007 Springer.*

Despite the active research in radiation genetics over the last few decades, there still remains considerable uncertainty in the genetic impact of radiation on human cells, organisms and populations. The main point in that research is the objective data obtained in adequate test-systems. In the present article a panel of genetic markers that could be used in humans and human cells after exposure to electromagnetic waves is reviewed. We will present a part of numerous investigation identifying sensitive biomarkers that can be used as additional complements to physical dosimetry for assessing exposure to radiation. In numerous researches efficiency of micronucleus assay, alkaline single-cell gel electrophoresis, chromosomal aberrations test and FISH-technique and their different combinations for the detection of ionizing radiation-induced genotoxic effects are discussed. Cytogenetic indicators,especially the induction of chromosomal aberrations were used from the early studies of radiation biological effects. They became important tools for the assessment of absorbed radiation doses and their biological effects after occupational exposure or radiation accidents.

The conventional analysis of chromosomal aberrations is usually applied to detect unstable aberrations in the lymphocytes of irradiated personnel (Bochkov, Chebotarev, 1989; Gebhart, Arutyunyan, 1991).

It is shown that the yields of dicentrics (in a metaphase chromosome spread, dicentric chromosomes appear as a single chromosome with two centromeres) and centric rings scored after long-term space flights are considerably higher than those scored prior to the flights. Individual biodosimetry of doses received by cosmonauts who showed a reliable increase in the yields of chromosomal-type aberrations after their first flights were estimated to be from 0.02 to 0.28 Gy (Fedorenko et al., 2001).

However we have to realize, that although cytogenetic analyses for dicentrics and translocations are the most useful techniques for biological dosimetry, these were initially developed for and have been applied to middle and high range dose exposures (Voisin et al., 2004).

Very interesting was the research of superoxide radicals implication in radiation damage by conventional cytogenetic analysis. Clastogenic factors (CFs) - cell substances that induce chromosomal aberrations are exerted via superoxide radicals were first described in the blood of persons irradiated accidentally or for therapeutic reasons. They were found also in A-bomb survivors, where they persisted for many years after the irradiation. The study Emerit et al., 1994 searched for these factors in the plasma of 32 civil workers from Armenia, who had been engaged as "liquidators" around the Chernobyl atomic power station in 1986. The number of aberrations in the test cultures was significantly increased compared to the control cultures.

These liquidators represent a high-risk population. The clastogenic activity was regularly inhibited by superoxide dismutase, indicating the role of superoxide radicals in clastogenesis. (Emerit et al, 1994; 1995) CFs are informative risk factor for the development of cancer, autoimmunological and inflammatory diseases, induced by radiation and their detection give important about the impact of radiation in human health. Fluorescence in situ hybridization (FISH) - in situ hybridization using a fluorescently labeled nucleic acid probe on chromosomes. As classical cytogenetic techniques do not allow to detect all induced types of chromosomal aberrations, the application of FISH technique has greatly increased the possibilities of detecting all types of rearrangements (i.e. reciprocal translocations, insertions, complex rearrangements etc). The relatively rapid disappearance of lymphocytes carrying unstable aberrations limits their use in retrospective dosimetry, years after exposure. Scoring stable aberrations might appear more appropriate in such situations (Leonard et al., 2005) FISH gives the possibility to reveal the majority of such aberrations.

By Tucker (2001) FISH with whole chromosome paints has greatly facilitated the analysis of structural chromosome aberrations and has led to translocations replacing dicentrics as the aberration of choice for many applications. Major challenges remain if we are to go from translocations to an understanding of the health consequences of radiation exposure. Translocations provide a sensitive detection system for low doses, show good specificity for radiation exposure following chronic or low-dose exposure, persist for decades, and are highly relevant to tumorigenesis (Tucker, 2002). An empirical model of the aberrant cell dynamics was utilized for the calculation of mean initial yields of dicentrics and centric rings in groups with different terms and duration of staying in the Chernobyl zone. Corresponding protracted irradiation doses estimated from aberration levels ranged from 79 to 670 mGy (Maznik, Vinnikov, 2005a).

FISH analysis has been successfully applied for qualitative cytogenetic indication of past and chronic radiation exposure to low doses after the Chernobyl accident. In the research of Maznyk, Vinnikov (2005b) cytogenetic analysis using the FISH technique was performed in groups of liquidators, evacuees from 30 km exclusive zone, residents of radioactively contaminated areas and control donors age-matched to exposed persons.

The mean yield of stable chromosome exchanges in liquidators did not correlate with registered radiation doses but had a clear negative dependence on the duration of liquidators' staying in Chernobyl zone, that was in a good agreement with early data based on conventional dicentrics plus rings analysis. The levels of all cytogenetic anomalies,including the unstable ones, in residents of radioactively contaminated areas showed a reasonable positive correlation with levels of 137Cs contamination.

625 Russian Chernobyl cleanup workers and 182 Russian controls in lymphocytes were analyzed for chromosome translocations by FISH. Among the covariates evaluated, some increased (e.g. age, smoking) and others decreased (e.g. date of sample). When adjusted for covariates, exposure at Chernobyl was a statistically significant factor for translocation frequency. The estimated average dose for the cleanup workers based on the average increase in translocations was 9.5 cGy (Jones et al., 2002).

Retrospective dosimetry using FISH detected translocations 10-13 y later of highly irradiated men, mostly reactor crew, from the Chernobyl accident showed good agreement for patients with the lower earlier dose estimates (including dicentric chromosome aberrations in lymphocytes), up to about 3 Gy (Sevan'kaev et al, 2005).

The considerable increase in the incidence of chromosomal aberrations (by G-banding, and molecular cytogenetic methods) in the peripheral lymphocytes was revealed as result of 14-year random cytogenetic monitoring of the Ukraine's population exposed to radiation due to the Chernobyl accident. Even small doses of long-term ionizing radiation could induce specific chromosomal aberrations. There was shown interindividual variability in the chromosomal aberrations under identical radiation conditions (Pilinskaia, 2001).

Another large research of action of chronic irradiation after nuclear testing was done in region of Semipalatinsk. It was revealed that the elevated level of stable translocations analysed by FISH technique in the lymphocytes of residents in the area of the Semipalatinsk nuclear testing site corresponds to a dose of about 180 mSv (Chaizhunusova et al., 2006).

However, not all of the the presented results can't give the exact dosimetric information. By (Leonard et al., 2005) the "...biological dosimetry renders its most useful results when an individual has been exposed to a rather homogeneous high-level radiation over a short time interval. On the other hand, it yielded less satisfactory information even when the most recent techniques were used for situations, where a low level, low dose rate exposure has occurred at some time in the past, for example for persons living in areas contaminated from the Chernobyl accident".

The observed genetic effect depends from the type of irradiation. Patients with cancer were treated with 10 MV X-rays produced at a LINAC accelerator, or high-energy carbon ions produced at the HIMAC accelerator in Chiba. Although carbon ions are 2-3 times more effective than X-rays in tumor sterilization, the effective dose was similar to that of X-ray treatment. However, the frequency of complex-type chromosomal exchanges was much higher for patients treated with carbon ions than X-ray (Durante et al.,1998).

Important research was realized for some professional or medically treated irradiated groups of genetic risk. Three-color FISH technique was applied for monitoring of five different groups of individuals: 30 occupied in radiology, 26 occupied in nuclear medicine or radiation physics, 32 patients with breast cancer, 26 occupied with military waste disposal, all presumably exposed to low doses of radiation or chemical mutagens and a non-exposed control group (N=29). Stable rearrangements mainly characterized the groups of controls, tumor patients, and radiation appliers. Authors state that chromosome painting if included in further attempts of human population monitoring will broaden the basis of argumentation with respect to health risks introduced by mutagen exposure (Verdorfer et al., 2001).

A three-color chromosome painting technique was also used to examine the spontaneous and radiation-induced chromosomal damage in peripheral lymphocytes and lymphoblastoid cells from homozygous and heterozygous patients with ataxia telangiectasia (AT) and Nijmegen breakage syndrome gene (NBS) characterized by genomic instability. X-irradiation was performed using a 6MV linear accelerator at a dose rate of 2.2 Gy/min. The cells from both the homozygous AT and NBS patients showed the highest cytogenetic response. The response of cells from heterozygous carriers was intermediate and could be clearly differentiated from those of the other groups (Neubauer et al., 2002).

Analysis of distribution of chromosomal aberrations in cells shows whether the action is stochastic and is described by Poisson distribution, or it is influenced by other cell processes (interaction of mutagens with cell structures, mixtures of cell subpopulations with different reaction to mutagen etc.). In these cases other models of distribution can fit the obtained results better. We applied geometric and Poisson distributions, that represented the two poles of the general class of negative binomial distribution to our data. The obtained data of distribution patterns, in the higher mentioned groups of patients who are extremely sensitive to radiation permitted to demonstrate specific differences and changes in the distributions of chromosomal aberrations and breaks depending on genotype and levels of described cytogenetic changes (Arutyunyan et al., 2001a). On the base of obtained results we offer the application of comparative analysis of distributions as informative cytogenetic approach to the research of radiosensitivity in the patients with chromosomal instability.

The patterns of distribution of X-ray-induced chromosome aberrations in peripheral lymphocytes of 83 cancer patients who were assigned to or had just undergone radiation therapy were analyzed (Arutyunyan et al., 1998). After in vitro irradiation of the lymphocytes in the both groups of non-exposed and previously irradiated patients great interindividual variation between the patients was revealed,that also is very important feature of their different radiosensitivity.

Another example of application of distributions analysis was presented by Sasaki (2003) "chromosome aberration frequencies were studied in human peripheral blood lymphocytes irradiated in vitro with gamma-rays in a dose range of 0.01-50Gy. Using the dose-yield relationship thus established, a new model of biodosimetry was developed which involved unfolding the chromosome aberration distribution into a mixed Poisson distribution and thence into a dose-distribution profile."

The development of more sophisticated and informative methods of FISH gives more information about cellular processes involved in effects of irradiation. Greulich et al., (2000) found up to 15 different chromosomes involved in rearrangements indicating complex radiation effects. Authors state, that multi-color FISH technique "...will allow researchGreulich ers to rapidly delineate chromosomal breakpoints and facilitate the identification of the genes involved in radiation tumorigenesis."

Chromosome-2, a European Space Agency experiment, continuation of the Chromosome investigation performed on earlier ISS expeditions is realized on the base of University of Duisburg-Essen (Cytogenetic effects...,2005). The lymphocytes collected from crewmembers preflight and postflight are examined using different analytic methods to determine quantity and quality of genetic changes. Except Giemsa staining when the researchers investigate changes in the morphology of the chromosomes also multicolor Fluorescence In Situ Hybridization (mFISH) to score reciprocal translocations and insertions and multicolor Banding Fluorescence In Situ Hybridization (mBAND) of the selected chromosome, to score for inversions (intrachanges) and translocations between homologous chromosomes are applied.

On the base of obtained results the authors concluded that, the complex-type exchanges or intrachanges have limited practical use for biodosimetry at very low doses (Horstmann et al., 2005).

Another modern approaches to fluorescent staining is spectral karyotyping (SKY), when each chromosome can be painted in an individual colour, had revealed that radiation induced chromosomal aberrations may be more complex than expected from conventional and single chromosome painting analyses. Its application permits to develop the knowledge of mechanisms for irradiation associated chromosomal aberrations (Szeles et al., 2006).Controversial results are obtained about the induction of cytogenetic anomalies by low frequency electromagnetic radiation. In their review of investigations conducted to determine the genotoxic potential of exposure to nonionizing radiation (emitted from extremely low frequency electromagnetic fields - EMF during the years 1990-2003 and published in the 63 peer reviewed scientific reports, Vijayalaxmi, Obe (2005) stated that conclusions from 29 studies (46%) did not indicate increased damage, while those from 14 investigations (22%) have suggested an increase in such damage in EMF exposed cells. The observations from 20 other studies (32%) were inconclusive.

In the study of Ahuja et al., (1999) investigations have been carried out at low-level (mT) and low-frequency (50 Hz) lectromagnetic fields in healthy human volunteers. Their peripheral blood samples were exposed to 5 doses of electromagnetic fields (2, 3, 5, 7 and 10mT at 50 Hz) and analysed by comet assay. At each flux density, with one exception, there was a significant increase in the DNA damage from the control value. When compared with a similar study on females carried out by us earlier, the DNA damage level was significantly higher in the females as compared to the males for each flux density.

A micronucleus (MN) may arise from a whole lagging chromosome or an acentric chromosome fragment detaching from a chromosome after breakage (clastogenic event) which do not integrate in the daughter nuclei (Maluf, 2004). In vitro studies allowed detecting increments in MN induction following exposure levels as low as 0.05 Gray (Gy) (Fenech and Morley, 1986). Because of its sensitivity and simplicity the measurement of MN is applied for screening of large populations.

In the study of Muller et al. (2004) 166 children with thyroid tumors (statistically significant increase in the number of children with such tumors after the Chernobyl accident is observed) and 75 without thyroid tumours were analysed for micronucleus formation in peripheral blood lymphocytes. A statistically significant increase in the number of micronuclei was observed for the residents of Gomel compared to other locations, such as Brest, Grodno, and Minsk. It was revealed that the following factors did not significantly affect micronucleus formation: gender, age at the time of the first iodine-131 treatment. The response of the children was characterized by clear individual differences. and the increase/decrease pattern of micronucleus frequencies induced by iodine-131.

The data obtained from MCN investigation usually correlates with the results of chromosomal analysis. Chromosome analysis and MCN assay were performed in villages near the Semipalatinsk nuclear test site area to assess the effects of prolonged radiation. Frequencies of dicentric and ring chromosomes were higher in residents of the contaminated area than those of the non-contaminated area. Frequencies of dicentric chromosomes with fragments and incidences of micronucleus were also higher in the exposed group, than the non-exposed group (Tanaka et al., 2006).

The Comet assay is very sensitive technique for the detection of DNA strand breaks in individual cells. Evaluation of DNA damage in single cells under alkaline conditions developed by Singh et al. (1988) permits efficient determination of single-strand breaks (SSB) and double-strand breaks (DSB), as well as alkali-labile sites when the DNA is incubated at a high pH. The Comet assay has been used extensively in radiobiology, as a sensitive, simple and rapid technique to investigate DNA damage and repair induced by various agents in a variety of mammalian cells.

According to data of Plappert et al., 1995 it is possible to detect a significant increase in DNA damage induced by radiation (gamma-radiation, x-rays) dose as low as 0.05Gy. The obtained data on such sensitivity demonstrate the practicability of the Comet assay for the detection of DNA damage caused by low doses of ionizing radiation. However the sensitivity of assay depends from the type of treatment. By Plappert et al., 1995 "The effect of one acute dose in blood cells is repaired within two h, whereas the effect of chronic (fractionated) irradiation gives a totally different result."

By using the alkaline Comet assay after accidental exposure the highest levels of DNA damage were recorded one day and one week after the radiation incident. The results obtained indicate that the alkaline comet assay is a rapid and sensitive microdosimetric technique and is suitable for in vivo human biomonitoring, especially in cases of incidental exposure to ionizing radiation (Garaj-Vrhovac et al., 2002).

DNA damages and its repair of cultured human cells were studied by exposing to carbon ion beams of HIMAC accelerator. (Nagaoka et al., 1999) Comet assay revealed highly inhomogeneous DNA damages. The damaged fraction of human fibroblast cells was 85% immediately after 4 Gy (100keV/micrometer) irradiation and decreased to 50% after 120 min. incubation indicating a repair of cell DNA. Another experiment was realized on Hiroshima University Radiobiological Research Accelerator (HIRRAC) with irradiation of human peripheral blood lymphocytes by monoenergetic neutrons. The level of DNA damage was computed as tail moment for different doses (0.125-1 Gy). The neutron-irradiated cells exhibited longer comet tails consisting of tiny pieces of broken DNA in contrast to the streaking tails generated by 60Co? The peak biological effectiveness occurred at 0.37 and 0.57 MeV; a further increase or decrease in neutron energy led to a reduced RBE (relative bioefficacy) value. This research confirmed that the Comet assay is a potential tool for use in neutron therapy (Gajendiran et al., 2000).

The Comet assay was used by us to estimate DNA damage induced by UV-C irradiation in leukocytes of Chernobyl accident liquidators, who were engaged during 1986-1987 in the clean-up of the Chernobyl nuclear accident. Blood samples were collected from 12 liquidators in 1997, ten years after exposure. Irradiation doses of liquidators were not higher than 0.25 Gy. Blood plasma of Chernobyl liquidators treated at the Center of Radiation Medicine and Burns of the Armenian Ministry of Health was tested in cooperation with Prof. I. Emerit (Paris, 6 University) for clastogenic activity in blood cultures of healthy donors. Our results show an increased sensitivity of DNA of leukocytes of liquidators to UV-C irradiation compared to leukocytes from controls. There was no statistically significant interindividual differences in the background of DNA damage in leukocytes of each of the investigated groups and between the groups (Arutyunyan et al., 2000, 2001b).

However, there are many publications that Comet-assay permit to register the effect of chronic irradiation. Garaj-Vrhovac and Kopjar (2003) selected the alkaline Comet assay as a biomarker of exposure to ionizing radiation of 50 medical workers. It was found that medical workers showed highly significant increases in levels of DNA damage compared with controls. Differences in comet parameters measured due to smoking and gender were not statistically significant in either exposed or control subjects.

In the research of group of people professionally at risk of exposure to low doses of ionizing radiation there was a significant difference between the control and hazard groups in DNA damage Higher DNA damage was also found for men than for women in the control group. There was no relation of DNA damage to age either in control or hazard group. But it was revealed considerable individual diversity even in the control group (Wojewodzka et al., 1998).

Paulraj and Behari (2006) investigated the effect of low intensity microwave (2.45 and 16.5 GHz, SAR 1.0 and 2.01 W/kg, respectively) radiation on developing rat brain. It was revealed, that the chronic exposure to these radiations cause statistically significant increase in DNA single strand breaks in brain cells of rat.

How the results of Comet assay are correlated with cytogenetic anomalies? We can mention many investigations that apply such combined approach. By Touil et al. (2002) there was a statistically significant interaction between biomarkers assessing the same damage (micronucleus and Comet assays).Multivariate analysis showed that micronucleus frequencies were positively influenced by age and the percentage of (comet) residual tail length was negatively influenced by the interaction between smoking and exposure status. By Kopjar, Garaj-Vrhovac (2005) the Comet assay and chromosome aberrations test are sensitive biomarkers that can be used as additional complements to physical dosimetry for assessing exposure to radiation in nuclear medicine personnel. Scassellati

Sforzolini et al., (2004). suggest that the tested ELF-MF (50 Hz, 5 mT) possess genotoxic (micronucleus test) and co-genotoxic (Comet assay) capabilities.

It is reasonable to combine the mentioned assays to get more information about different genetic effects of radiation.

In the investigation of the role of inherited and acquired factors on individual variation in DNA repair capacity gamma-rays (2Gy)-induced higher frequency aberrations was observed in metabolic genotype GSTM 1-positive individuals irradiated blood compared with GSTM1-null subjects as well as in non-smokers compared with heavy smokers . The observed effect can be "possibly related to the higher expression of enzymes involved in the repair of oxidative DNA damage in heavy smokers and GSTM1-null subjects." (Marcon et al., 2003) Maluf (2004) in his review gives very interesting interpretation of low impact of high levels of radiation obtained by children of irradiated mothers and remarkable increase of chromosomal aberrations in somatic cells nuclear-dockyard workers, that got 20-fold smaller dose. He explain this phenomenon that "because most chromosomal mutations in germinative cells will not be compatible with the gametogenesis and the viability of the embryo or fetus while somatic mutations are not subject to this strong selection".

New molecular-biological approaches were applied in the last years to detect the effect of radiation:Prasanna et al. (2002) presented simple and rapid interphase-based cytological assays that will be applicable to a broad range of radiation exposure scenarios. These assays include analysis of chromosome aberrations (premature chromosome condensation-FISH assay) and mitochondrial DNA mutations assay) using resting human peripheral blood lymphocytes.

In the paper of Grace et al. (2002) was first time presented real-time quantitative reverse transcription-polymerase chain reaction (QRT-PCR) technology using gene expression changes as rapid, sensitive and reproducible tool for early-response accident biodosimetry.

Microarray is a miniature array of different DNA or oligonucleotide sequences on a surface to be used in a hybridization assay (Strachan and Read).

Park et al., (2002) investigated radiation-related expression patterns in cellular culture with nonsense mutation in p53 using cDNA microarray. 384 genes were selected for their ionizing radiation (IR) -specific changes to make 'RadChip'. Radiation-induced responses were clearly separated from the responses to other genotoxic stress. Such approach gives the possibility to identify the IR-specific genes, which can be used as radiation biomarkers to screen radiation exposure as well as probing the mechanism of cellular responses to ionizing radiation. Snyder, Morgan (2004) had analyzed by microarray analysis the radiation-induced genomic instability (RIGI) that manifests in the progeny of cells surviving ionizing radiation (IR), and can be measured using such endpoints as delayed mutation, micronuclei formation, and chromosomal instability. The frequency of RIGI was relatively high, exceeding the gene mutation rate of IR by orders of magnitude. Authors state that such approach can contribute to a greater understanding of the mechanisms of radiation carcinogenesis.

Such test-systems can be complexed with the presented cytogenetic test-systems. Blakely et al. (2003) reports the multiparametric dosimetry system that was developed for medical radiological defense applications. "The system complements the internationally accepted personnel dosimeters and cytogenetic analysis of chromosome aberrations, considered the best means of documenting radiation doses for health records. Our system consists of a portable hematology analyzer, molecular biodosimetry using nucleic acid and antigen-based diagnostic equipment, and a dose assessment management software application."

Biological factors, determining the variation of sensitivity to radiation can be associated with genotype, lifestyle and health of sampled individuals. (Maluf, 2004).There are many proofs of it. For example, the inter-individual variation in DNA repair capacity may be inherited (Cloos J. et al., 1999), determined by the polymorphism of DNA repair genes (Tuimala et al, 2002), or result from alterations in gene expression induced by epigenetic factors such as smoking habits (from Marcon et al., 2003). So, the final formation of DNA and chromosome damage that depends from repair capacity of cells can have many sourses of variability. There exists the opinion that the biological methods can be used for bioindication but not for biodosimetry because of different factors affecting the the levels of radiation-induced damage. Radioadaptive response - one of the most significant factors which can be responsible for incorrect radiation dose evaluation (Mosse, 2002). Only the combination of presented methods can give the objective information about the genetic effects of electromagnetic waves.

REFERENCES

Ahuja, Y.R., Vijayashree, B., Saran, R., Jayashri, E.L., Manoranjani, J.K., Bhargava, S.C., 1999, In vitro effects of low-level, low-frequency electromagnetic fields on DNA damage in human leucocytes by comet assay, Oct, 36(5): 318-22.

Arutyunyan, R., Martus, P., Neubauer, S., Birkenhake, S., Dunst, J., Sauer, R., Gebhart, E., 1998, Intercellular distribution of cytogenetic changes detected by chromosome painting in irradiated blood lymphocytes of cancer patients. Experimental Oncology, 20: 223-228.

Arutyunyan R.M., Hovhannisyan G.G., Ghazanchyan E.G., Oganessian N.M., Nersesyan A.K., 2000, UV-C induced DNA damage in leukocytes of Chernobyl accident clean-up workers. Exp. Oncol., 22, 219-221.

Arutyunyan, R., Neubauer, S.,. Martus, P., Dork, T., Stumm, M., Gebhart, E., 2001a, Intercellular distribution of aberrations detected by means of chromosome painting in cells of patients with cancer prone chromosome instability syndromes. Experimental Oncology, 23: 23-28.

Arutyunyan R.M., Hovhannisyan G.G., Ghazanchyan E.G., Oganessian N.M., Nersesyan A.K., 2001b, DNA damage induced by UV-C irradiation in leukocytes of Chernobyl accident clean-up workers, Central European Journal and Environmental Medicine, 7 (1): 15-21.

Au, W.W., 1991, Monitoring human populations for effects of radiation and chemical exposures using cytogenetic techniques. Occup Med., 4:597-611.

Blakely, W.F., Miller, A.C., Grace, M.B., McLeland, C.B., Luo, L., Muderhwa, J.M., Miner, V.L., Prasanna, P.G., 2003, Radiation biodosimetry: applications for spaceflight. Adv Space Res. 31(6): 1487-93.

Bochkov, N.P., Chebotarev, A.N., 1989, Human heredity and environmental mutagens, Meditsina , Moscow (In Russian).

Chaizhunusova, N., Yang, T.C., Land, C., Luckyanov, N., Wu, H., Apsalikov, K.N., Madieva, M., 2006, Biodosimetry study in Dolon and Chekoman villages in the vicinity of Semipalatinsk nuclear test site. J Radiat Res (Tokyo). 47, A: A165-9.

Cloos, J., Nieuwenhuis, E.J.C., Boomsma, D.I., Kuik, D.J., Van der Sterre, M.L.T., Arwert, F., Snow, G.B., Braakhuis, B.J.M., 1999, Inherited susceptibility to bleomycin-induced chromatid breaks in cultured peripheral blood lymphocytes, J. Natl. Cancer Inst., 91: 1125-1130.

Cytogenetic effects of ionizing radiation in peripheral lymphocytes of ISS crewmembers (http://exploration.nasa.gov/programs/station/ Chromosome-2.html), 2005.

Durante, M., Kawata, T., Nakano, T., Yamada, S., Tsujii, H., 1998, Biodosimetry of heavy ions by interphase chromosome painting Adv Space Res. 22(12): 1653-62.

Emerit, I., Levy, A., Cernjavski, L., Arutyunyan, R., Oganesyan, N., Pogosian, A., Mejlumian, H., Sarkisian, T., Gulkandanian, M., Quastel, M., et al. 1994, Transferable clastogenic activity in plasma from persons exposed as salvage personnel of the Chernobyl reactor, J Cancer Res Clin Oncol., 120(9): 558-61.

Emerit, I., Arutyunyan, R., Oganesian, N., Levy, A., Cernjavsky, L., Sarkisian, T., Pogossian, A., Asrian, K., 1995, Radiation-induced clastogenic factors: anticlastogenic effect of Ginkgo biloba extract. Free Radic Biol Med., Jun, 18(6): 985-91.

Fedorenko, B., Druzhinin, S., Yudaeva, L., Petrov, V., Akatov, Y., Snigiryova, G., Novitskaya, N., Shevchenko, V., Rubanovich, A., 2001, Cytogenetic studies of blood lymphocytes from cosmonauts after long-term space flights on Mir station. Adv Space Res. 27(2): 355-9.

Fenech, M., Morley, A.A., 1986, Cytokinesis - block micronucleus method in human lymphocytes: effect of in vivo ageing and low doses X-irradiation, Mutat Res., 161: 193-8.

Gajendiran, N., Tanaka, K., Kamada, N., 2000, Comet assay to sense neutron "fingerprint", Mut. Res., 452, 2 , 179-187.

Garaj-Vrhovac, V., Kopjar, N., Razem, D., Vekic, B., Miljanic, S., Ranogajec-Komor, M., 2002, Application of the alkaline comet assay in biodosimetry: assessment of in vivo DNA damage in human peripheral leukocytes after a gamma radiation incident, Radiat Prot Dosimetry, 98(4): 407-16.

Garaj-Vrhovac, V., Kopjar, N., 2003, The alkaline Comet assay as biomarker in assessment of DNA damage in medical personnel occupationally exposed to ionizing radiation. Mutagenesis, May; 18(3): 265-71.

Gebhart, E., Arutyunyan, R.M., 1991, Anticlastogens in Mammalian and Human Cells, Springer-Verlag Berlin Heidelberg New York, 125p.

Grace, M.B., McLeland, C.B., Blakely, W.F., 2002, Real-time quantitative RT-PCR assay of GADD45 gene expression changes as a biomarker for radiation biodosimetry. Int J Radiat Biol., 78(11): 1011-21.

Greulich, K.M., Kreja, L., Heinze, B., Rhein, A.P., Weier, H.G., Bruckner, M., Fuchs, P., Molls, M., 2000, Rapid detection of radiation-induced chromosomal aberrations in lymphocytes and hematopoietic progenitor cells by mFISH. Mutat Res., Jul 20; 452(1): 73-81.

Horstmann, M., Durante, M., Johannes, C., Pieper, R., Obe, G., 2005, Space radiation does not induce a significant increase of intrachromosomal exchanges in astronauts' lymphocytes.Radiat Environ Biophys. 44(3): 219-24.

Jones, I.M., Galick, H., Kato, P., Langlois, R.G., Mendelsohn, M.L., Murphy, G.A., Pleshanov, P., Ramsey, M.J., Thomas, C.B., Tucker, J.D., Tureva, L., Vorobtsova, I., Nelson, D.O., 2002, Three somatic genetic biomarkers and covariates in radiation-exposed Russian cleanup workers of the chernobyl nuclear reactor 6-13 years after exposure. Radiat Res., Oct; 158(4): 424-42.

Kopjar, N., Garaj-Vrhovac, V., 2005, Assessment of DNA damage in nuclear medicine personnel-comparative study with the alkaline comet assay and the chromosome aberration test, Int J Hyg Environ Health, 208(3): 179-91.

Leonard, A., Rueff, J., Gerber, G.B., Leonard, E.D., 2005, Usefulness and limits of biological dosimetry based on cytogenetic methods, Radiat Prot Dosimetry, 115(1-4): 448-54.

Maluf, S.W., 2004, Monitoring DNA damage following radiation exposure using cytokinesis-block micronucleus method and alkaline single-cell gel electrophoresis Clinica Chimica Acta 347, 15-24

Marcon, F., Andreoli, C., Rossi, S., Verdina, A., Galati, R., Crebelli, R., 2003, Assessment of individual sensitivity to ionizing radiation and DNA repair efficiency in a healthy population. Mutat Res., Nov 10; 541(1-2): 1-8.

Maznik, N.A., Vinnikov, V.A., 2005a, The retrospective cytogenetic dosimetry using the results of conventional chromosomal analysis in Chernobyl clean-up workers, Radiats. Biol. Radioecol., 45(6): 700-8.

Maznyk, N.A., Vinnikov, V.A., 2005b, Possibilities and limitations of fluorescence in situ hybridization technique in retrospective detection of low dose radiation exposure in post-chernobyl human cohorts. Tsitol. and Genet., Jul-Aug; 39(4): 25-31.

Mosse, I.B., 2002, Modern problems of biodosimetry, Radiats. Biol. Radioecol., 42(6): 661-4.

Muller, W.U., Dietl, S., Wuttke, K., Reiners, C., Biko, J., Demidchik, E., Streffer, C., 2004, Micronucleus formation in lymphocytes of children from the vicinity of Chernobyl after (131)I therapy. Radiat Environ Biophys, May; 43(1) 7-13.

Nagaoka, S., Nakano, T., Endo, S., Onizuka, T., Kagawa, Y., Fujitaka, K., Ohnishi, K., Takahashi, A., Ohnishi, T., 1999, Detection of DNA damages and repair in human culture cells with simulated space radiation. Acta Astronaut., Apr-Jun; 44(7-12): 561-7.

Neubauer, S., Arutyunyan, R., Stumm, M., Dork, T., Bendix, R., Bremer, M., Varon, R., Sauer, R., Gebhart, E., 2002, Radiosensitivity of ataxia telangiectasia and Nijmegen breakage syndrome homozygotes and heterozygotes as determined by three-color FISH chromosome painting, Radiat. Res., 157: 312-21.

Park, W.Y., Hwang, C.I., Im, C.N., Kang, M.J., Woo, J.H., Kim, J.H., Kim, Y.S., Kim, J.H., Kim, H., Kim, K.A., Yu, H.J., Lee, S.J., Lee, Y.S., Seo, J.S., 2002, Identification of radiation-specific responses from gene expression profile. Oncogene, 21(55): 8521-8.

Paulraj, R., Behari, J., 2006, Single strand DNA breaks in rat brain cells exposed to microwave radiation, Mut. Res., 596, 76-80.

Pilinskaia, M.A., Shemetun, A.M., Dybskii, S.S., Dybskaia, E.B., Pedan, L.R., Shemetun, E.V., 2001, The results of 14-year cytogenetic monitoring of priority follow-up groups of Chernobyl accident victims, Vestn. Ross. Akad. Med. Nauk, 10: 80-4.

Plappert, U., Raddatz, K., Roth, S., Fliedner, T.M., 1995, DNA-damage detection in man after radiation exposure--the comet assay--its possible application for human biomonitoring, Stem Cells, 13 Suppl 1: 215-22.

Prasanna, P.G., Hamel, C.J., Escalada, N.D., Duffy, K.L., Blakely, W.F., 2002, Biological dosimetry using human interphase peripheral blood lymphocytes. Mil Med., 167(2): 10-2.

Sasaki, M.S., 2003, Chromosomal biodosimetry by unfolding a mixed Poisson distribution: a generalized model, Int. J. Radiat. Biol., 79(2): 83-97

Scassellati Sforzolini, G., Moretti, M., Villarini, M., Fatigoni, C., Pasquini, R., 2004, Evaluation of genotoxic and/or co-genotoxic effects in cells exposed in vitro to extremely-low frequency electromagnetic fields, Ann Ig., Jan-Apr; 16(1-2): 321-40.

Szeles, A., Joussineau, S., Lewensohn, R., Lagercrantz, S., Larsson, C., 2006, Evaluation of spectral karyotyping (SKY) in biodosimetry for the triage situation following gamma irradiation, Int. J. Radiat. Biol., 82(2): 87-96.

Sevan'kaev, A.V., Lloyd, D.C., Edwards, A.A., Khvostunov, I.K., Mikhailova, G.F., Golub, E.V., Shepel, N.N., Nadejina, N.M., Galstian, I.A., Nugis, V.Y., Barrios, L., Caballin, M.R., Barquinero, J.F., 2005, A cytogenetic follow-up of some highly irradiated victims of the Chernobyl accident. Radiat. Prot. Dosimetry, 113(2): 152-61.

Singh, N.P., McCoy, M.T., Tice, R.R., Schneider, E.L., 1988, A simple technique for quantitation of low levels of DNA damage in individual cells. Exp. Cell Res. 175:184-91.

Snyder, A.R., Morgan, W.F., 2004, Radiation-induced chromosomal instability and gene expression profiling: searching for clues to initiation and perpetuation, Mutat. Res., 568(1): 89-96.

Tanaka, K., Iida, S., Takeichi, N., Chaizhunusova, N.J., Gusev, B.I., Apsalikov, K.N., Inaba,T., Hoshi, M., 2006, Unstable-type chromosome aberrations in lymphocytes from individuals living near Semipalatinsk nuclear test site, J. Radiat. Res. (Tokyo), 47 Suppl A: A159-64

Touil, N., Aka, P.V., Buchet, J.P., Thierens, H., Kirsch-Volders, M., 2002, Assessment of genotoxic effects related to chronic low level exposure to ionizing radiation using biomarkers for DNA damage and repair, Mutagenesis, May; 17(3): 223-32.

Tucker, J.D., 2001, Fish cytogenetics and the future of radiation biodosimetry, Radiat Prot Dosimetry, 97(1): 55-60.

Tucker, J.D., 2002, Sensitivity, specificity, and persistence of chromosome translocations for radiation biodosimetry, Mil Med., 167(2): 8-9.

Tuimala J., Szekely G., Gundy S., Hirvonen A., Norppa H., 2002, Genetic polymorphisms of DNA repair and xenobiotic-metabolizing enzymes: role in mutagen sensitivity, Carcinogenesis, 23: 1003-1008.

Verdorfer, I., Neubauer, S., Letzel, S., Angerer, J., Arutyunyan, R., Martus, P., Wucherer, M., Gebhart, E., 2001, Chromosome painting for cytogenetic monitoring of occupationally exposed and non-exposed groups of human individuals, Mutat Res., 491: 97-109.

Vijayalaxmi, Obe G., 2005, Controversial cytogenetic observations in mammalian somatic cells exposed to extremely low frequency electromagnetic radiation: a review and future research recommendations, Bioelectromagnetics, 26(5): 412-30.

Voisin, P., Roy, L., Benderitter, M., 2004, Why can't we find a better biological indicator of dose? Radiat Prot Dosimetry, 112(4): 465-9.

Ward, J.F., 1975, Molecular mechanisms of radiation-induced damage to nucleic acid. In: Lett JT, Adler H, editors, Advances in radiation biology, New York: Academic Press, 182-239.

Wojewodzka, M., Kruszewski, M., Iwanenko, T., Collins, A.R., Szumiel, I., 1998, Application of the comet assay for monitoring DNA damage in workers exposed to chronic low-dose irradiation. I. Strand breakage, Mutat. Res., 416(1-2): 21-35.

SYNCHROTRON FT-IR SPECTROSCOPY OF HUMAN BONES. THE EFFECT OF AGING

P. Kolovou and J. Anastassopoulou
National Technical University of Athens, Chemical Engineering Department, Radiation Chemistry and Biospectroscopy, Athens, Greece

Abstract: The Synchrotron micro-FT-IR spectroscopy was used to investigate the structural changes of human bones, which are produced upon irradiation. It was observed that after irradiation with a dose of 40 Gy up to 119 Gy the band at 1660 cm^{-1}, which corresponds to the absorption of the Amide I group (-NH-CO-) and indicates that the collagen exists in α-helix, shows considerable changes upon irradiation resulting that the collagen looses its structure from α-helix to random coil. Considerable changes were also observed in the region of the spectra between 900 cm^{-1} and 1200 cm^{-1} where the phosphate groups ($v_3PO_4^{3-}$) of hydroxyapatite absorb. These bands change in intensity and shape. These findings show that the irradiation of human bones leads to damage of the main components of bone tissues.

Key words: Synchrotron, micro-FT-IR, spectroscopy, human bone, irradiation

1. INTRODUCTION

Synchrotron infrared light can be used as a source for micro-FT-IR spectroscopy, which is a quite new technique and is employed to study biological molecules and the complicated biological tissues. Fourier-Transformation Infrared (FT-IR) spectroscopy is a powerful technique, easy-to-use and non-destructive, which gives the characteristic vibrational modes of the functional groups (NH$_2$, -NH-CO-, -COOH, OH, etc.) of biological molecules in cells and tissues[1]. By mapping the spectra of the tissue sections the method shows how the IR approach enables to distinguish the major tissue constituents based on their IR spectral

Vasili Tsakanov and Helmut Wiedemann (eds.), Brilliant Light in Life and Material Sciences, 267–272.
© 2007 Springer.

alterations. On the other hand, free radicals have been shown to be involved in bone resorption[2]. It is also known that ionizing radiation is used for cancer therapy and in the treatment of pain due to cancer. However, in the pathway of the treatment of cancer, ionizing radiation affects also the surrounding healthy tissues. The purpose of this study is to determine in vitro the effect of free radicals produced by the ionizing radiation on the biological molecules, their structure and more generally the physicochemical properties of bone tissue, which are induced during radiotherapy. By analyzing the infrared spectra and the changes that take place upon irradiation it is possible to investigate the new changes in the structure and conformation of major molecules that are involved in the tissues.

2.   MATERIAL AND METHODS

Cortical and cancellous bone biopsies were obtained intra-operatively from head of 25-75 year patients undergoing osteoarthritic surgery. The histological evaluation showed normal bone, with no evidence of metabolic or inflammatory disease or malignancy (Goldner Trichrome staining).

The samples were prepared in slices in order to preserve the natural characteristics of the bone. The bone was cut into slices of 5 mm perpendicular to the longitudinal axis of the head for Synchrotron micro-FT-IR measurements (Fig. 1).

Figure 1. Sheme of bone slice perpendicular to the longitudinal axis.

The histological evaluation of representative sections of the biopsies showed no evidence of any metabolic disease, osteopenia, or bone cancer. The bone sections were immersed successively in hydrogen peroxide solution (H_2O_2) and in acetone, according to a modification of the method described[1]. Synchrotron micro-FT-IR spectrometer was used equipped with a microscope, a liquid nitrogen cooled detector and a KBr beam splitter, at a frequency resolution of 4cm^{-1}, in the transmission mode. The light source was from MIRAGE- port SA5- SuperACO at LURE, beamline LURE, Centre Universitaire Paris-Sud- BP 34 F-91898 Orsay Cédex; France.

Bones were irradiated with 40, 79 and 119 Gy using a Gammacell γ-Co(60) source at room temperature.

3.   RESULTS AND DISCUSSION

The FT-IR spectra in the region 1800-700 cm^{-1} of the examined non-irradiated bone section (0 Gy) are shown in Fig. 2.

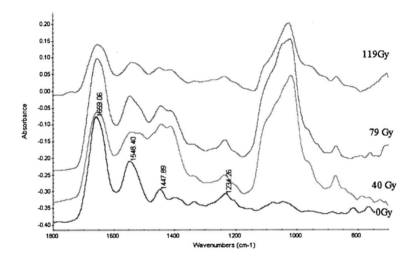

Figure 2. FT-IR spectra of non-irradiated and after irradiation with a dose of 40, 79, 119 Gy.

The intense band near 1660 cm^{-1} corresponds to the absorption of the Amide I group (-NH-CO-). This Amide I band is a combined vibration primarily of C=O stretching vibration and the N-H deformation modes. This band indicates that the secondary structure of the collagen exists in α-helix[3,4]. The peak at 1548 cm^{-1} is due to the Amide II vibration in combination with C-N stretching mode. These two bands give useful information that is correlated with the structural properties of collagen and non-collagenous proteins[3,4]. The band situated at 1234 cm^{-1} has been assigned to Amide III.

The presence of this band is associated with the tertiary structure of proteins and corresponds to β-sheets. This band increases upon irradiation (Fig. 2, 40 Gy). The non-ionized carboxyl group v_sCOOH of the proteins gives a band at 1400 cm^{-1} together with the absorption of the $v_3CO_3^{2-}$ mode which is assigned to AB carbonate of hydroxyapatite of the bone. The asymmetric absorptions of the phosphate groups ($v_3PO_4^{3-}$) of hydroxyapatite are located at 1080 cm^{-1} and 968 cm^{-1}. The bands at 1151 cm^{-1} and 1145 cm^{-1} are assigned to the absorption of $v_3HPO_4^{2-}$, which correspond to non-stoichiometric hydroxyapatite.

After irradiation, there is a dramatic increasing of the bands phosphate in the region 1190-850 cm^{-1} (Fig. 2, 40 Gy and Table), where the phosphate groups of hydroxyapatite absorb.

Table 1. Comparison of some characteristic infrared bands of non-irradiated and irradiated human bones together with their assignments.

Non-irradiated, cm^{-1}	40 Gy, cm^{-1}	119Gy, cm^{-1}	Assignments[1-4,7]
1735	1740	1738	C=O non-ionized
	1670		β-turn
		1663	β-turn, random coil
1650			Amide I, α-helix
1635	1631	1633	Collagen random coil
1593		1588	AmideII
1548	1552	1548	AmideII, α-helix
		1510	$v_{as}COO^-$
1461			$\delta_{as}CH_3$, $v_3CO_3^{2-}$ aragonite
1448	1444	1450	δCH_3
1430			C-OH of HA
	1410		$v_3CO_3^{2-}$ calcite
1400		1400	$vCOO^{-2}$
1379			CH$_3$ umbrella of lipids
		1278	AmideIII, α-helix
1234	1237	1235	Amide III, β-sheet
1151		1160	$v_3HPO_4^{2-}$
1145			$v_3HPO_4^{2-}$
	1113	1115	v_1 CO$_3^{2-}$ calcite
1080		1060	$v_3PO_4^{3-}$
		1016	
968	960	953	$v_3PO_4^{3-}$
	874	869	$v_2CO_3^{2-}$ calcite
		792	
		760	

The intensity increasing of the bands suggests that upon irradiation the phosphate salts predominate as compared to the non-irradiated spectra (Fig. 2, 0 Gy). The lipids are being damaged, which is in agreement with literature, where there was observed oxidative damage of lipids after irradiation of bone marrow of male Wistar rats[5,6]. The new band, which appears at 1670 cm^{-1} after irradiation, suggests the presence of prime amines with short aliphatic chain. This oxidation damage process leads to chain scission that lowers chain length of collagenous and non-collagenous proteins. This band shows also that the proteins change their structure from α-helix to β-turn.

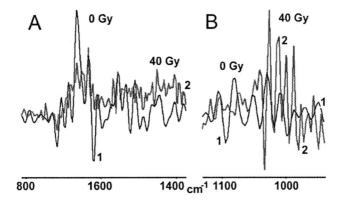

Figure 3. Deconvoluted spectra of non-irradiated and irradiated with a dose of 40 Gy human bones in the region 1800-1500 cm⁻¹ (A) and 1200-800 cm⁻¹ (B).

The band at 1735 cm⁻¹ due to non-ionized stretching vibration of vC=O carbonyl group increases in intensity and shits to higher wavenumbers upon irradiation. It is known that lipid peroxides, which are produced from radiolysis, generate unsaturated aldeydes, ketones and hydrocarbons[8]. This carbonyl band has been proposed to be used to characterize the apoptotic cells[9].

In Fig. 3 is shown the deconvoluted spectra of non-irradiated human bones (1) and after irradiation with a dose of 40 Gy (2). Considerable changes are observed in the region where the Amide I and II absorb there, as well as in the region of the phosphate groups. The deconvoluted spectra show a finer spectrum with more bands due most likely to the alterations of the bone upon irradiation. It was observed from the deconvoluted spectra that the band at 1735 cm⁻¹ increases with the irradiation dose and this increasing was analogous up to 60 Gy. After this dose the system becomes more complicated because more products from radiolysis are produced, which are shown in the spectra. More work is needed in order to analyze the samples and the spectra.

REFERENCES

1. M. Petra, J. Anastassopoulou, T. Theologis, T. Theophanides, Synchrotron micro-FT-IR spectroscopic evaluation of normal paediatric human bone, J. Mol. Str., 733 (2005) 101–110, and references therein.
2. Samar Basu Karl Michaëlsson, Helena Olofsson,Sara Johansson, and Håkan Melhus Biochem Biophys Res Comm Association between Oxidative Stress and Bone Mineral Density 288, 275–279 (2001).
3. T. Theophanides, "Interaction of Metal Ions with Nucleic Acids", in *Infrared and Raman Spectroscopy Applied to Biological Molecules*, (ed. T. Theophanides), D. Reidel Publishing Co, Dodrecht, Holland, p. 185 (1979).

MICRO-FT-IR SPECTROSCOPIC STUDIES OF BREAST TISSUES

J. Anastassopoulou[1], P. Arapantoni[2], E. Boukaki[1], S. Konstadoudakis[2], T. Theophanides[1], C. Valavanis[2], C. Conti[3], P. Ferraris[3], G. Giorgini[3], S. Sabbatini[3] and G. Tosi[3]

[1]*National Technical University of Athens, Chemical Engineering Department, Radiation Chemistry and Biospectroscopy, Zografou Campus, 15780 Zografou, Athens, Greece.* [2]*Metaxa Hospital, Piraeus, Greece,* [3]*Dipartimento di Scienze dei Materiali e della Terra, Universitö Politecnica delle Marche, via Brecce Bianche, 60131 Ancona, Italy*

Abstract: Micro-FT-IR spectroscopy was used to study breast cancer tissues and, in particular osteosarcoma tissue. By analysing the spectra, we have found characteristic bands in the infrared regions, where the main components of these signature bands are located. In the region between 1680-1660 cm-1 are found the characteristic bands of Amide I and II of proteins. The bands, which correspond to the vibrations of the phosphate groups, are found in the region near 1140-900 cm-1. These characteristic bands have been monitored as a function of the degree of cancer progression. The results have been obtained with chemometric methods, such as cluster analysis, principal component analysis and custom analysis in order to distinguish the neoplastic zones from the normal zones.

Key words: breast cancer, osteosercoma. Micro-FT-IR, infrared, spectroscopy

1.   INTRODUCTION

Breast cancer affects a large number of women between 40 and 55 years of age. It is estimated that 1 to 14 women in Greece (1 to 7 in USA) will develop breast cancer sometime during their life. An early detection of this disease could expand the survival of women. In 1929 M. Cutler had the idea that in breast cancers the cancer tissue should have different optical

Vasili Tsakanov and Helmut Wiedemann (eds.), Brilliant Light in Life and Material Sciences, 273–278.
© 2007 *Springer.*

properties from the normal tissue and should give shadow images of the breast[1]. This diaphanography gave a "shadowgram" without significant changes for clinical trials. Mammography is the only method now that is used for detection of breast cancer and for screening of population. It is estimated that in screening mammography 5-30% of breast cancers have a false-negative mammogram because the result depends on fat and density of the breast. The Magnetic Resonance Imaging (MRI) technique gives 0.1% negative false result and is not used by women due to claustrophobia[2]. Positron emission tomography (PET) has better detection results but has a resolution of 2 mm and a large number of women are excluded for several reasons. For the early diagnosis and therapy of breast cancer it is crucial to develop a non-destructive bio-analytical technique to obtain images of the breast, which must be independent of breast shape and mass density in order to detect lesions which are difficult to scan with the existing techniques. Breast tissue is too complex with high heterogeneity. It contains different proportions of components, such as glandular epithelium, fibrous and adipose tissues, carcinomatous cells and necrotic areas to be directly determined. Micro-FT-IR spectroscopy promises a lot in this field, since it provides information about the structure and composition of biological materials[3,4]. The infrared spectra also do preserve histo-morphological aspects. In this work we used micro-infrared spectroscopy, which enables us to detect and quantify the changes of the molecular structure of biological molecules based on the measurements of functional spectral differences between healthy and cancer tissues.

2.   MATERIALS AND METHODS

Human breast cancer tissues were obtained from the tumor bank of Anticancer Metaxa Hospital, Histopathology Department, Piraeus, Greece. All the surgical specimens of the human breast tumor were placed in 10% neutral buffered-formalin for a maximum of 24 hours. The next day the fixation of the specimens was succeeded by placing it in xylene, alcohol and then embedded in paraffin. The whole procedure was automated. The tissues were enclosed with paraffin in a block and two microtome sections of 5 μm were cut. The first was fixed in a glass with hematoxilin and eosin (H & E) for histological examination and to identify the interesting regions. The second slice was used for infrared analysis.

Spectral data were achieved with Perkin-Elmer Spectrum One FT-IR and with Spotlight FT-IR Imaging System 300 spectrometers equipped with a Perkin Elmer Auto-image microscope. The spectral resolution was 2 and

4 cm-1. The spatial resolution was 6.25 × 6.25 and 30 × 30 mm. Each spectrum was the result of 256 scans. Background scans were obtained from a region of no sample and rationed against the simple spectrum. The thin sections were deposited on a steel support and reflectance spectra were collected. Specific areas of interest were identified by means of the microscope television camera. Baseline (polynomial line fit) was performed in all cases while Second Derivative, Fourier Self Deconvolution and Curve Fitting (Gaussian character) procedures were used to determine the absorbance ratio between bands of interest. All spectra were scaled for equal intensity in the Amide II band. Attribution of the bands was done according to literature data[5]. For data handling the following software packages were used: Spectrum v.303 (Perkin-Elmer), Grams AI (Galactic Corp.) and Pirouette 3.11 (Imfometrix Corp) for multivariate analysis.

3. RESULTS AND DISCUSSION

In Fig. 1A and B are shown the micro FT-IR spectra of a human breast osteosarcoma across the horizontal line 2 of the tissue, in the region 4000 cm-1 to 700 cm-1.

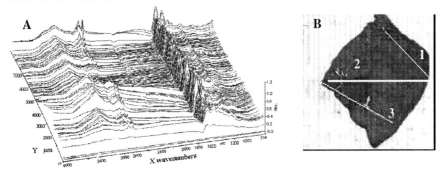

Figure 1. A: micro-FT-IR infrared spectra collected across the line 2 in the region 4000 - 700 cm-1, B: a 5μm thickness of breast tissue section.

Considerable intensity and shift changes are shown in the spectra through out the line. Important changes are observed in the intense band at 3307 cm-1, where the N-H stretching vibration of the Amide I of the proteins and nucleic acids absorb[6-8]. This band usually is used to characterise the normal and carcinomatous tissues[5]. The bands in the region 2950-2850 cm^{-1} are assigned to stretching vibrations of methyl and methylene groups of the lipids, proteins and nucleic acids[6,9]. These bands change also with the normality of the tissue due to dilution[10].

Fig. 2 shows some characteristic micro-FT-IR mapping spectra from the breast tissue. The spectra 1 and 2 correspond to neoplastic zone and the spectra 3 and 4 correspond to normal zone of the osteosarcoma tissue. In Fig. 3 are shown the principal component (PCA) imaging and custom analysis.

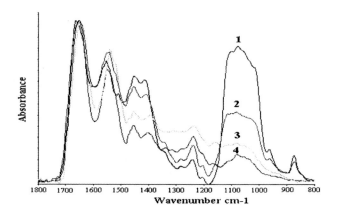

Figure 2. Micro-FT-IR mapping infrared spectra of human breast osteosarcoma. Curves 1,2 represent absorptions in the neoplastic zone and 3,4 in the normal zone of the breast tissue.

Considerable changes are shown in both intensities and shape of the bands in the region at 900 to 1150 cm^{-1}, where the phosphate groups of the proteins and nucleic acids absorb[11].

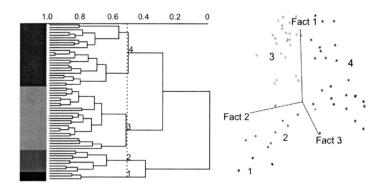

Figure 3. Chemometric analysis of human breast osteosarcoma

The intensity of the phosphate bands increases from normal to neoplastic tissue. This band could serve as criterion to distinguish normal tissue from osteosarcoma. The intense band at 1651 cm^{-1} is assigned to

vC=O vibration coupled to Amide I (N-H) groups of proteins, suggesting that the proteins have a secondary structure of α-helix[12-13]. This band is weaker in normal tissue. The peak at 1549 cm[-1], which is assigned to Amide II, is also lowered in normal tissue. The second derivative and deconvolution spectra in the regions of phosphates and Amide I, II are more complicated by showing more bands in the case of osteosarcoma tissue (Figure not shown).

4. CONCLUSIONS

From the mapping micro-FT- IR spectra it is concluded that the cancer tissue spectra are markedly different from those of the normal tissue. The IR spectra vary gradually in the region of the absorption bands of Amide I and II as well as in the phosphate region with the degree of the progression to cancer.

REFERENCE

1. B. Chance et al, Breast cancer detection based on incremental biochemical and physiological properties of breast cancers. Acad. Radiol., 12 (2005) 925-933.
2. Breast MRI Web Site.
3. M. Petra, J. Anastassopoulou, T. Theologis, T. Theophanides, Synchrotron micro-FT-IR spectroscopic evaluation of normal paediatric human bone, J. Mol. Str., 733 (2005) 101–110, and references therein.
4. L.G. Carr, Resolution limits for infrared micro-spectroscopy explored with synchrotron radiation. Rev. Sci. Instrum. 72 (2001) 1.
5. B. Schrader, Ed., Infrared and Raman Spectroscopy, Weinheim, VCH Verlagsgesellschaft mbH, 1995.
6. T. Theophanides and J. Anastassopoulou, Spectroscopy-Symmetry, National Technical University of Athens Press, Athens, 1998.
7. R. Eckel, et al, Characteristic infrared spectroscopic patterns in the protein bands of human breast cancer tissue, Vibrat. Spectr. 27 (2001) 165-173.
8. M. Meurens, J. Wallon, J. Tong, H. Noël and J. Haot, Breast Cancer Detection by Fourier Transform Infrared Spectrometry, Vibrat. Spect. 10 (1996) 341-346.
9. T Theophanides, Fourier Transform Infrared Spectroscopy, D. Reidel Publishing Co. Dodrecht, 1984.
10. J Anastassopoulou and T Theophanides, Raman studies of model vesicle systems, J. Appl. Spectrosc. 44 (1990) 523-524.
11. T. Theophanides, in: Infrared and Raman Spectroscopy Applied to Biological Molecules, (ed. T. Theophanides), D. Reidel Publishing Co, Dodrecht, Holland, p. 185 (1979).

12. T. Theophanides, in: Proceedings of 6th International Conference on Raman Spectroscopy in Biochemistry Section, (Eds. E.D. Schmid et al), Heyden, London, Vo 1, p. 115 (1978).

13. M. Jackson, L.-P. Choo, P.H. Watson, W.C. Halliday, H.H. Mantch, Beware of connective tissue proteins: assignment and implications of collagen, Bioph. Biocim. Acta, 1270 (1995) 1-6.

PRIMARY MECHANISM OF EM INTERACTION WITH THE LIVE TISSUES

Viktor Musakhanyan
Inst

Abstract: There is a prevailing opinion that the theoretical explanation of electromagnetic (EM) fields influence on live organisms is impossible to explain theoretically and even the play between parameters of waves and tissues is unknown to us. The explanation of mechanism of this influence is vitally important owing to the development of new types of electronic devices operating in different frequency ranges and due to the still continuing controversy about their adverse health effect. It is shown that the application of newly developed procedure of shutting-on of the interaction of charged particles with electromagnetic fields allows explaining their influence on live tissue by origination of macroscopic polarization currents due to the joint action of electric and magnetic components of electromagnetic waves. The currents originate in the case of resonance between the proper frequency of the medium and of frequency of external electromagnetic fields. Thus, the experiments to measure these polarization currents can provide information about dangerous frequency ranges and these ranges, with maximal polarization currents, should be excluded during construction of electronic devices.

Key words: EM waves, magnetic component, macroscopic polarization currents

1. INTRODUCTION

The conventional mechanism of interaction of external electromagnetic wave with matter, including live structures, has not been developed until now (see, for example, reviews [1,2] and the References cited there).

Below, a theoretical model is proposed, based on taking into consideration the *magnetic component* of the electromagnetic wave

Vasili Tsakanov and Helmut Wiedemann (eds.), Brilliant Light in Life and Material Sciences, 279–282.
© 2007 *Springer.*

interacting with the *bound electrons* in matter. This component is usually neglected in literature since one is considered to be small as compared to the electric component of electromagnetic field.

2. THREE DIMENSIONAL OSCILLATORS IN AN EXTERNAL ELECTROMAGNETIC WAVE FIELD

The system of 3-D oscillators with the proper frequency ω_0 in a field of a plane electromagnetic wave with frequency ω_0 and wave vector $\vec{k}$ is considered. The equation of motion for such an oscillator with the charge q and mass m can be written as the non-relativistic motion equation of Lorenz[3].

$$\ddot{\vec{x}} + \gamma \cdot \dot{\vec{x}} + \omega_0^2 \cdot \vec{x} = \frac{q}{m} \cdot \left(\vec{E}_0 + \frac{1}{c}\left[\dot{\vec{x}} \cdot \vec{B}_0 \right] \right) \cdot \cos\left(\omega t - kz + \varphi_0 \right)$$

Where φ_0 is an arbitrary initial phase.

Equation (1) bears a resemblance to the well-known Drude-Lorentz model[4] and was written also in [5].

After averaging over initial phases, we obtain the polarizing current

$$j = q \cdot n \cdot c \frac{q^2 \cdot E_o \cdot B_o}{4 \cdot m^2 \cdot c^2 \cdot \omega} \cdot \left[\frac{\cos(\omega_0 - \omega)t - \cos\omega_0 t}{2\omega_0 - \omega} - \frac{\cos(\omega_0 + \omega)t - \cos\omega_0 t}{2\omega_0 + \omega} \right] \quad (2)$$

where n is the density of oscillators.

3. THE PROCESSES WHERE THE POLARIZING CURRENT CAN BE REVEALED

3.1 Live tissues

The estimations for therapeutic devices of sub-millimeters range with the medium frequency $v \approx 50 GHZ$, averaged power $P = 10^{-8} \div 10^{-10} W$ and cross-section $2 \times 3 mm^2$ show that the macroscopic currents density j can reach value $10^3 A/m^2$ for moderate density of resonant oscillators $n \approx 10^{15} 1/m^3$ in microscopic volumes of live tissues.

The influence of power lines on human organism may be considered. Close to these lines, the values of parameter of intensity ξ may even exceed this value for the laser radiation by many orders of magnitude, that is, the macroscopic currents may also cause dangerous effects.

Cell phone influence on live tissues should also be considered experimentally by measuring the polarization currents to obtain the information about dangerous frequency ranges with maximal polarization currents.

3.2   The light-induced "magnetization"

In a simple experimental situation, when laser pulse is propagating in Rb vapors, the pick-up coil shows the induced electromotive force [6], and some models have been proposed to explain such a phenomenon[7], however, without quantitative success.

Our estimations show that for real values of laser parameters and densities of oscillators the current densities described by Eq. (2) can also reach the macroscopic values. So, for field strengths $E_0 \approx 10^7 V/m, \omega \approx 10^{15} \sec^{-1}, n \approx 10^{21} 1/m^3$, we have the amplitude of current density $j \approx 10 A/m^2$.

Similar polarization currents have been revealed also in optical waveguides [8].

3.3   Getmantsev's phenomenon

The origination of combinational frequencies during irradiation of ionosphere by vertical SW beam with power $P \approx 150 kW$ at the frequency $v = 5,75 MHz$ has been registered at the distance of $180 km$ using synchronous detector, where, according to the theory of SW propagation the amplitude of the wave should vanish. The amplitude of irradiation was modulated within the range of frequencies from $1,2 kHz$ to $7,2 kHz$. The registered field amplitude [9] at the modulation frequencies E_{reg} was equal to $0,02 \mu V/m$.

To calculate the polarized currents in this case there is a need to know the densities of resonant oscillators, however, even the small value of $n \approx 10^{10} 1/m^3$ will result in current densities up to $0.6 A/m^2$ at the modulation frequency. This time $E_0(t)$ is a slowly varying function and calculation of field strength amplitude results in $E_{reg} \approx 10 \mu V/m$. Evidently, concurrent processes also take place and more investigations are needed to describe this process exactly.

4.   SUMMARY

The concept of origination of polarizing currents due to cooperative action of electric and magnetic components of the EM wave may be applied to various fields of physics, where interaction of electromagnetic field with matter, including live tissues, takes place.

REFERENCES

1. N.G. Ptitsina, G. Villoresy, L.I. Dorman, N. Iucci, and M.I. Tyasto, Natural and man-made low-frequency magnetic fields as a potential health hazard, Physics Uspekhi, 41, 687-709 (1998).
2. V.N. Binhi and A.V. Savin, The effect of weak magnetic fields on biological systems: physical aspects, UFN, 173 (3), 265-300 (2003).
3. V.V. Musakhanyan and A.M. Khanbekyan, On the inphase polarizing currents in the system of oscillators interacting with electromagnetic wave, VANT - Nuclear-Physical Investigations, 6(18), 32-34 (1990).
4. P. Grosse, Freie Elektronen in Festkorpern (Springer-Verlag, Berlin, Heidelberg, New-York, 1979).
5. L.M. Ball and J. Dutt, Harmonics on the scattering of light by bound electrons, Nuovo Cimento, 35(3), 805-810 (1965).
6. R.E. Movsesyan and A.M. Khanbekyan, Light-induced magnetization of Rb - vapor, Proceedings of the Armenian NAS, Physics, 22(1), 53-55 (1987).
7. A.V. Andreev, V.I. Emel'yanov, and Yu.A. Il'insky, Cooperative Phenomena in Optics, (Nauka, Moscow, 1988).
8. B. K. Nayar, Optical Fibres with Organic Crystalline Cores. Nonlinear Opt.: Mater, and Devices. Proc. Int. Sch. Mater. Sci. and Technol., Erice, July 1-14, 1985. -(Berlin, 1986) pp. 142-153.
9. G. Getmantsev, N.A. Zuykova, D.S. Kotik, et al., Combination frequencies in the interaction between high-power short-wave radiation and ionospheric plasma, JETP Lett., 20(1), 101-102 (1974).

DYE SENSITIZING AND PROTECTIVE PROPERTIES DURING ELECTROMAGNETIC IRRADIATION OF LIVING SYSTEMS

Hasmik Avetisyan, Yegishe Elbakyan, Vladimir Hovhannisyan
Yerevan Physics Institute, 2 Alikhanyan Brothers. str., Yerevan 0036, Armenia

Abstract: The protective properties of some dyes using Drosophila melanogaster line are investigated. It is shown that some photochemically inactive dyes can partly or completely suppress photodynamic effect in flies at visible light, UV ionizing (254 nm) and solar radiation irradiations. Energy and charge transfer from photoactivated biomolecules to dye is considered as the possible mechanism of inhibition of light induced biological effect.

Key words: Sensitization, photoprotection, PDT, Drosophila, ionization

1. INTRODUCTION

Photodynamic therapy (PDT) is successfully used in the treatment of a wide variety of localized malignancies. Unfortunately, deeper tumors cannot be treated by PDT method, because light, used for photodynamic treatment, isn't able to penetrate into the tissue more than 1 cm. In this context new approach to enhance the effectiveness and expansion of PDT is necessary. One of them may be the combination of PDT with radiotherapy[1-3].

The common mechanism is that both modalities produce free radicals, but unlike γ-irradiation, which generates free radicals by developing secondary electrons at the site of the absorbers, such as nucleic acids in the cell, PDT generates them through the absorption of light by photosensitizers.

Vasili Tsakanov and Helmut Wiedemann (eds.), Brilliant Light in Life and Material Sciences, 283–287.
© 2007 *Springer.*

Combined treatments of PDT and radiotherapy could be more useful since they allow the reduction of the ionizing radiation dose to obtain the same effect as one obtainable by radiotherapy alone.

Several groups have investigated the combination effect of PDT and γ-radiation and have concluded that optical and ionizing radiations produce not just additive but in many cases synergistic effect.

Moreover, it is shown that the same PDT-active porphyrins can act as radio sensitizer (e.g., it is shown, that 5 Gy radiation combined with hematoporphyrin exerts the same effect as a 15 Gy irradiation by itself), as well as radioprotector[4].

The present work is the first part of research devoted to electromagnetic irradiation (visible and UV light) effect suppression by some dyes using *Drosophila melanogaster* line. Energy transfer and charge transfer from biomolecules to the dyes is explored as a possible mechanisms of dye-protection.

2. MATERIALS AND METHODS

The experiments were carried out on *Drosophila melanogaster D-32* in laboratory conditions. Drosophila cultures were bred in test tubes with sugar-barm environment. The test-tubes were stored in thermostat at temperature 22-25°C. Both photodynamically active sensitizers: chlorin e_6 (CH_e_6, Institute of molecular and nuclear physics, Minsk, Belorus), Zn, Al and Mg phthalocyanines (Fluka), methylene blue, ("Reactant", Lvov) (group A) and inactive compounds- brilliant green, pyronin, fluorescein, trypaflavine ("Reactant") (group B) were used as dyes.

Before starting experiments the flies were translated into empty test tubes (8-10 individuals in each test tube) and after 1,5-2 hours was given the forage with addition of one or two types of dyes. Dyes of 10^{-5}-10^{-4} M concentration were added to the food for drosophila 24 hours before irradiation. The irradiation was carried out at 25° C.

Filtered sunlight (400-800 nm), narrow band halogen lamps (540-570, 580-610, 650-700 nm), light-emitting diodes -LEDs (590, 630, 655, 670 nm) and UV (254 nm) radiation (filtred mercury lamp) were used as photostimulators depending on spectrum of dye absorption. Intensity of irradiation was less than I = 45 mW/cm^2 and the exposition t was from 10 minutes to 1.5 hours.

3. RESULTS AND DISCUSSION

In the control test-tubes (without any dye) at all irradiation doses the percentage of survived insects was approximately 100%, except of UV irradiation. In the darkness with the use of all dyes observations also indicated no damage to the insects. When using dyes of group B the insects were not practically affected, while with the dyes of group A findings showed dose-dependent insect mortality. Combination of dyes from A group with some dyes from B group leads to the partial disappearance of photodynamic effect (Fig. 1). A protective action of these dyes also is detected at UV ionising (254 nm) radiation (Fig. 2).

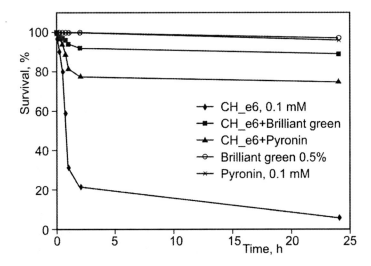

Figure 1. Drosophila survival at using of Chl.e6 with combination of inactive dyes depending on time after the start of sunlight irradiation (I = 30 mW/cm2, irradiation time t = 40 min.).

Estimations show that due to weak absorption (less than 1 %) the optical shielding by dye molecules may be neglected. Suppression of photodynamic action to flies can be explained by the toxic photoproducts (singlet oxygen, etc.) neutralization by the inactive dyes. However, more simple physical mechanism, namely quenching of excited states (singlet and /or triplet) of photodynamic dyes by inactive chromophores via mechanism of inductive resonance transfer of energy is represented by more probable in this case. In the first experiments with a solution of albumin+chlorin e_6 complex the quenching of the dye fluorescence was observed at addition of brilliant green, pyronin, fluorescein, and was not observed at addition of trypaflavine.

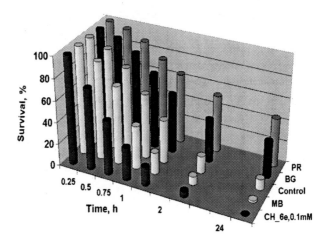

Figure 2. Photosensitized drosophila survival value during UV (254 nm) irradiation (I = 20 mW/cm2, t(ir) = 25 min).

Recently, some attempts have been made to protect insects against photodynamic action, a typical oxidative process, by the administration of antioxidizing agents, such as carotenes, tocopherol and ascorbic acid[5]. In all cases, no appreciable protection was obtained, even though these compounds are powerful inhibitors of photooxidative reactions in vitro[6]. In this study it is shown that the second generation photodynamic dyes can be used for the non-polluting control of insects population. The sunlight, which intensity in summertime reaches 70 mW/cm^2 can be used as a source of photodynamic activation. It allows to provide the lethal doze for insects nearly in 10 min. At the same time, it is possible to control the density of insects in the given district by adding photodynamic process quenchers in a forage. Photoinsecticides represent a possible alternative to traditional chemical insecticides which have a number of lacks, such as: toxicity for plants, animals and partially for people, mutagenicity, chemical resistance that results in environmental contamination, etc.,[6-7]. Photodynamic compounds are synthesized or selected so that the dark toxicity would be at very low level and there would be no mutagen effect Experimental model of drosophila allows to investigate photosensitization impact when using combination of dyes, as well as to study photodynamic effect on reproductive functions of insects. The carried out studies can help to increase the efficiency and safety of photodynamic therapy of people and application of photodynamic method in science, medicine, agriculture and the food-processing industry.

REFERENCES

1. Madsen S.J, Sun Ch.H, Bruce J.et al, Effects of combined photodynamic therapy and ionizing radiation on human glioma spheroids, Photochem.&Photobiol 76(4), 411–416(2002).
2. Luksiene Z., Experimental evidence on possibility to radiosensitize aggressive tumors by porphyrins, Medicina 40(9), 868-874(2004).
3. Pogue B., O'Hara J.A., Demidenko E. et al, Photodynamic therapy with verteporfin in the radiation-induced fibrosarcoma tumor causes enhanced radiation sensitivity. Cancer Research 63(1), 1025–1033(2003).
4. Schwartz S., Keprios M., Modelevsky G. et al, Modification of radiosensitivity by porphyrins. Studies of tumours and other systems, in:.Diagnosis and therapy of porphyrias and lead intoxication, edited by Doss M (Springer-Verlag, 1978), pp. 227–37.
5. Robinson, J.R., Beatson, E.P. Effect of selected antioxidants on the phototoxicity of erythrosin B toward house fly larvae, Pestic. Biochem. Physiol. 24, 375–383 (1985).
6. Jori, G., Photosensitizers: approaches to enhance the selectivity efficiency of photodynamic action, J. Photochem. Photobiol., B: Biol. 36, 87–93 (1996).
7. Ben Amor T. and Jori G. Photodynamic insecticides: an environmentally friendly approach to control pest populations, Photodynamics News 6, 6–8 (2001).

CRYOGLOBULINS IN ACUTE ISCHEMIC STROKE

L.A. Manukyan, V.A. Ayvazyan, A.S. Boyajyan
Laboratory of Macromolecular Complexes, Institute of Molecular Biology NAS RA, 7 Hasratyan St., 375014, Yerevan, Armenia

Abstract: Cryoglobulins (Cgs) are pathogenic immune complexes, non specific markers of the inflammatory and autoimmune responses. In this study we for the first time, revealed Cgs in the blood of ischemic stroke patients and analyze their composition.

Key words: complement proteins; cryoglobulins; immunoglobulins; ischemic stroke; LP-X

1.   INTRODUCTION

Cryoglobulins (Cgs) are immune complexes, where both antigen and antibody are presented by immunoglobulins (Igs) that reversibly precipitate at low temperatures. The presence of Cgs in circulation is considered as a nonspecific marker of the activation state of the immune system, inflammatory response and autoimmune sensitization[1-6]. According to Brouet et al. Cgs can be classified into three types[7]. Cgs may lead to immune complex vasculitis[8,9], may activate complement[1,10] and stimulate production of tumor necrosis factor-α (TNF-α)[11]. These findings generate interest in studying Cgs in ischemic stroke (IS), as vasculitis, undesirable complement activation and TNF-α expression contribute to the pathology of IS by damaging tissue and promoting inflammation[12-19].

Vasili Tsakanov and Helmut Wiedemann (eds.), Brilliant Light in Life and Material Sciences, 289–293.
© 2007 *Springer.*

2.   MATERIALS AND METHODS

Blood serum samples of 60 patients with acute IS and 40 age and sex matched healthy volunteers (control group) were examined.

Cgs precipitation and purification procedures were based upon earlier developed approaches and recommendations[1,20,21]. Total concentration of protein in Cgs was determined according to the method of Lowry et al[22]. For characterization of Ig-composition of Cgs, we used immunoblotting procedure described by Musset et al,[21] with slight modification. To analyze the presence of low-density lipoproteins (LDL), including β-lipoprotein and abnormal lipoprotein (LP-X), in the serum or in Cgs we performed electrophoresis in Bacto agar geles followed by polyanion precipitation in situ in the gels after electrophoresis according to Seidel et al[23] with slight modification. To identify the presence of the complement C1q and C3 proteins and their split products in Cgs, sodium SDS-PAGE and subsequent Western blotting were carried out according to Bio-Rad Laboratories, Inc., (USA) instruction manuals[24,25].

3.   RESULTS AND DISCUSSION

The results obtained demonstrated elevated levels (about 1.8-fold; $p < 0.0001$) of Cgs in the blood of IS patients (0.16 ± 0.075 mg/ml of serum) comparing to controls (0.088 ± 0.045 mg/ml of serum). According to the results obtained Cgs presenting in the blood of IS patients belong to type III Cgs, accounting for the classification developed by Brouet et al[7]. With regard to the presence of complement proteins, both C1q and C3 (and their split products) were found in all Cgs isolated from the serum of IS patients, whereas in case of Cgs isolated from the serum of the healthy subjects nothing was detected (Figure 1).

Results of the experiments on identification of the presence of low density lipoproteins in Cgs demonstrated that all Cgs isolated from the serum of IS patients contained both LP-X and β-lipoprotein, whereas in case of the healthy subjects only the latest component was detected.

We propose that elevated levels of type III Cgs in the blood of IS patients are direct consequence of stroke and reflect development of both inflammatory and autoimmune reactions occurring after stroke onset[14,26,27]. The presence of the complement C1q and C3 proteins and their split products in Cgs of IS patients provide further evidence for this suggestion.

Concerning LP-X, the present study, for the first time, demonstrated the presence of LP-X in the blood of IS patients in both a free form and a

Cg-bound form (complexed to Igs and complement proteins). Currently it is difficult to identify the mechanism responsible for formation of LP-X in IS. and a further study on the origin of LP-X in IS should be developed.

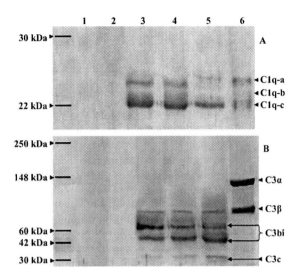

Figure 1. Typical western blot patterns of C1q (A) and C3 (B) detection in Cgs isolated from the blood of stroke patients. 1-2 – Cgs isolated from the blood of controls; 3-5 – Cgs isolated from the blood of stroke patients; 6 – C1q standard (A) and C3 standard (B).

We suggest that there may exist at least two pathomechanisms of the involvement of Cgs in IS progression. First of all, cryoglobulinemia leads to systemic vasculitis due to direct obstruction of the small blood-vessels, arteries and veins, by Cgs. Additionally, Cgs may also induce or enhance the inflammatory response through activation of the complement cascade and cytokine production.

REFERENCES

1. U. Kallemuchikkal, P.D. Gorevic. Evaluation of cryoglobulins. Archives of Pathology and Laboratory Medicine, 123(2), 119–125 (1998)
2. A. Dispenzieri. Symptomatic cryoglobulinemia. Current Treatment Options in Oncology, 1(2), 105-118 (2000)
3. K. Mohammed, H.U. Rehman. Cryoglobulinaemia. Acta Med. Austriaca, 30(3), 65-68 (2003)
4. P. Hilgard et al. Cryoglobulin-associated uptake of hepatitis C virus into human hepatocytes. Hepatogastroenterology, 52(65), 1534-1540 (2005)
5. C. Ferri et al. Cryoglobulins. J. Clin. Pathol., 55(1), 4-13 (2002)

6. M.M. Newkirk. Rheumatoid factors: host resistance or autoimmunity? Clinical Immunology, 104(1), 1-13 (2002)

7. J.O. Brouet. et al. Biologic and clinical significance of cryoglobulins. A report of 86 cases. Am. J. Med., 57(5), 775-788 (1974)

8. C. Ferri, M.T. Mascia. Cryoglobulinemic vasculitis. Curr Opin Rheumatol., 18(1), 54-63 (2006)

9. B. Hellmich. et al. Early diagnosis of vasculitides. Z. Rheumatol., 64(8), 538-546 (2005)

10. S.M. Weiner et al. Occurrence of C-reactive protein in cryoglobulins. Clin. Exp. Immunol., 125(2), 316-322 (2001)

11. L. Mathsson et al. Cryoglobulin-induced cytokine production via FcgammaRIIa: inverse effects of complement blockade on the production of TNF-alpha and IL-10. Implications for the growth of malignant B-cell clones. Br. J. Haematol., 129(6), 830-838 (2005)

12. P. Vilela, A. Goulao. Ischemic stroke: carotid and vertebral artery disease. Eur. Radiol., 15(3), 427-433 (2005)

13. U.S. Vasthare et al. Complement depletion improves neurological function in cerebral ischemia. Brain Research Bulletin, 45(4), 413-419 (1998)

14. A.L. D'Ambrosio et al. The role of the complement cascade in ischemia/reperfusion injury: implications for neuroprotection. Molecular Medicine, 7(6), 367-382 (2001)

15. M.D. Ginsberg. Adventures in the pathophysiology of brain ischemia: penumbra, gene expression, neuroprotection: the 2002 Thomas Willis Lecture. Stroke, 34(1), 214-223 (2003)

16. A. Boyajyan et al. Involvement of alternative and classical pathways of complement activation in the pathogenesis of ischemic stroke. Clinical Biochemistry, 38(9), 857-858 (2005)

17. A.S. Boyajyan et al. Dynamics of complement activation in acute ischemic stroke. Immunology (Moscow), 25(4), 221-224 (2004)

18. K. Saito et al. Early increases in TNF-alpha, IL-6 and IL-1 beta levels following transient cerebral ischemia in gerbil brain. Neurosci. Lett., 206(2-3), 149-152 (1996)

19. L. Zheng et al. Induction of apoptosis in mature T cells by tumor necrosis factor. Nature 377(6547), 348-351 (1995)

20. J.A. Hardin. Cryoprecipitogogue from normal serum: mechanism for cryoprecipitation of immune complexes. Proc. Natl. Acad. Sci. USA, 78(7), 4562-4565 (1981)

21. L. Musset et al. Characterization of cryoglobulins by immunoblotting. Clin.Chem., 38, 798-802 (1992)

22. O.H. Lowry et al. Protein measurement with the Folin phenol reagent. J. Biol. Chem., 193(1), 265-275 (1951)

23. D. Siedel et al. Improved techniques for assessment of plasma lipoprotein patterns. I. Precipitation in gels after electrophoresis with polyanionic compounds. Clin.Chem., 19(7), 737-739 (1973)

24. Mini-Protean 3 Cell. Instruction Mannual (catalog number 165-3301), Bio-Rad Laboratories, Inc., 1-23

25. Mini Trans-Blot Electrophoretic Transfer Cell. Instruction Mannual (catalog number 170-3930), Bio-Rad Laboratories, Inc., 1-24

26. G. Stoll. Inflammatory cytokines in the nervous system: multifunctional mediators in autoimmunity and cerebral ischemia. Inflammatory cytokines in the nervous system: multifunctional mediators in autoimmunity and cerebral ischemia. Rev. Neurol. (Paris), 158(10 Pt 1), 887-891 (2002)

27. K.J. Becker et al. Sensitization to brain antigens after stroke is augmented by lipopolysaccharide. J. Cereb. Blood Flow Metab., 25(12), 1634-1644 (2005)

LEVEL AND CHEMICAL COMPOSITION OF CRYOGLOBULINS IN SCHIZOPHRENIA

Aren Khoyetsyan[1], Anna Boyajyan[1], Maya Melkumova[2]
[1]Laboratory of Macromolecular Complexes, Institute of Molecular Biology NAS RA, 7 Hasratyan St., 375014, Yerevan, Armenia, [2]Nubarashen Republic Psychiatric Hospital MH RA, Nubarashen-Avan, 375017, Yerevan, RA

Abstract: The blood samples of 40 schizophrenic patients were tested for the presence of cryoglobulins (Cgs) and composition of Cgs was examined. The elevated levels of type III Cgs, containing complement components, were detected in all study subjects.

Key words: complement, cryoglobulins, immune system dysfunction, schizophrenia.

1. INTRODUCTION

Schizophrenia is a severe form of mental illness with neurodevelopmental etiology. The pathogenesis of schizophrenia still remains obscure. According to data available, immune system dysfunction, including alterations in mechanisms of both specific and non-specific immune response and development of autoimmune processes, is involved in the pathogenesis of this disease[1, 2].

Cryoglobulinaemia (monoclonal - type I, mixed - type II or III) is a nonspecific marker of the activation state of the immune system, inflammatory and autoimmune processes[3]. Cgs are abnormal immune complexes, which reversibly precipitate at temperatures below 37°C, and where both antigen and antibody are presented by immunoglobulins. Cgs may bind complement components and activate complement[4,5]. On the other hand, considerable evidence[6,7], including our results[8,9], suggests about the involvement of complement system dysfunction in the autoimmune

Vasili Tsakanov and Helmut Wiedemann (eds.), Brilliant Light in Life and Material Sciences, 295–298.
© 2007 *Springer.*

pathomechanisms and aberrant apoptosis contributing to the development of schizophrenia. In the present study we have investigated the Cgs levels and composition in the blood of schizophrenic patients.

2.   MATERIALS AND METHODS

Forty patients with chronic paranoid form of schizophrenia (ICD-10) and age and sex matched thirty healthy volunteers were involved in this study. The patients were hospitalized at Nubarashen Republic Psychiatric Hospital of the Ministry of Health of Armenia. Blood samples were obtained by venipuncure at 9:00-10:00 a.m. After 1h of coagulation serum was separated by centrifugation. Cgs were isolated by exposure of blood serum samples to precipitation at low temperature followed by extensive washings of Cg-enriched pellets[4,10]. Total concentration of protein in Cgs was determined according to the method of Lowry et al[11]. For characterization of Ig-composition of Cgs, we used immunoblotting procedure described by Musset et al.[12] with slight modification. To identify the presence of the complement C1q and C3 proteins and their split products in Cgs, sodium SDS-PAGE and subsequent Western blotting were carried out according to Bio-Rad Laboratories, Inc., (USA) instruction manuals (catalog numbers 165-3301 and 170-3930, 2002). For data analyses ordinal descriptive statistics and Mann-Whitney U-test were applied, using Sigma Plot (Sigma, USA) software.

3.   RESULTS AND DISCUSSION

According to the data obtained, the increased levels (more than 1.8-fold; $p < 0.00001$) of Cgs were detected in the blood of schizophrenia affected subjects as compared to healthy controls. Notably, the levels of Cgs in female patients were higher (1.33 times, $p<0.0368$) than in male. This difference was not detected in case of healthy volunteers.

Identification of Ig-composition of Cgs indicated that Cgs isolated from the blood of schizophrenia-affected patients belong to type III Cgs (polyclonal IgG, IgA and IgM, see Figure 1).

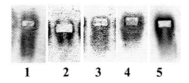

Figure 1. Typical patterns of Ig content in Cgs isolated from the blood of schizophrenic patients (1- Ig γ-chain, 2- Ig μ-chain, 3- Ig α-chain, 4- Ig λ-chain, and 5- Ig κ-chain).

With regard to the presence of complement proteins, both C1q and C3 (and their split products, see figure 2a and 2b) were found in all Cgs isolated from the serum of schizophrenic patients, whereas in case of Cgs isolated from the serum of the healthy subjects nothing was detected.

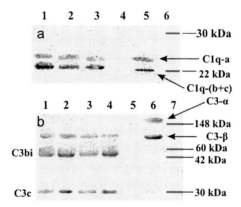

Figure 2. Typical patterns of C1q (a) and C3 (b) detection in Cgs isolated from blood of schizophrenic patients: a. 1-3 – Cgs isolated from the blood of schizophrenic patients; 4 - Cgs isolated from the blood of healthy subjects, 5 – C1q standard, 6 – marker proteins; b. 1-4 – Cgs isolated from the blood of schizophrenic patients; 5 - Cgs isolated from the blood of healthy subjects, 6 – C1q standard, 7 – marker proteins.

The results of the present study indicate that pathogenesis of schizophrenia is characterized by the elevated level of type III Cgs, containing C1q and C3 complement components. As regards to gender differences, our observation probably reflects different manifestation of disease in male and female patients as well as sex chromosome aberrations detected in schizophrenia[13,14].

REFERENCES

1. N. Muller, M. Riedel, M. Ackenheil, and M. J. Schwarz, Cellular and humoral immune system in schizophrenia: A conceptual re-evalution, World Biol Psychatry 1, 173-179 (2000)
2. M. Rothermandt, V. Arolt, A.T. Bayer, Review of immunological and immunophatological findings in schizophrenia, Brain, Behavior and Immunity 15, 319-339 (2001)
3. C. Ferri, A. Zignego, S. Pileri, Cryoglobulins, J Clin Pathol 55, 4-13 (2002)
4. U. Kallemuchikkal, P.D. Gorevic, Evaluation of cryoglobulins, Archives of Pathology and Laboratory Medicine, 123(2), 119–125 (1998)
5. S.M. Weiner, V. Prasauskas, D. Lebrecht, S. Weber, H.H. Peter, P. Vaith, Occurrence of C-reactive protein in cryoglobulins, Clin. Exp. Immunol., 125(2), 316-322 (2001)

PART 4

Material and Environmental Sciences

RECENT ADVANCES IN IMAGING WITH SPECTROSCOPIC ANALYSIS AT ELETTRA

L. Aballe, A. Barinov, M. Bertolo, L. Gregoratti, B. Kaulich, A. Locatelli, T.O. Mentes, L. Quaroni, S. La Rosa, and M. Kiskinova
Sincrotrone Trieste, Area Science Park, 34012 Trieste, Italy

Abstract: The paper is a brief overview of the efforts and achievements at Elettra in development of instrumentation for imaging and spectroscopic analysis using different spectral range of the synchrotron radiation. Selected results will illustrate the potential of these techniques for multi-disciplinary applications.

Key words: X-ray and IR microscopy, synchrotron radiation, chemical and magnetic imaging, contrast mechanisms, XPS, XANES.

1.   INTRODUCTION

Imaging with spectroscopic analysis, based on the interaction of matter with electromagnetic radiation, has undergone a very fast progress with the construction of the third generation synchrotron sources, which have provided very bright, polarized and tunable radiation with wavelengths from μm to Å range. The contrast mechanisms in microscopes built at the synchrotron laboratories are based on photon absorption, phase shift and emission of photons or electrons. Chemically-specific information is provided by the following well developed spectroscopic methods: (i) X-ray Absorption Spectroscopy (XAS) and in particular X-ray Absorption Near-Edge Spectroscopy (XANES), which monitors the transmitted photons or the total emitted photons or electrons while varying the energy of the incident photons[1]; (ii) X-ray Fluorescence Spectroscopy (XRFS), which detects and energy filters the photons emitted as a result of secondary de-excitation processes occurring in the atoms interacting with

Vasili Tsakanov and Helmut Wiedemann (eds.), Brilliant Light in Life and Material Sciences, 301–316.
© 2007 *Springer.*

monochromatized X-rays[2]; (iii) X-ray Photoelectron Spectroscopy (XPS), which detects and energy filters the electrons emitted from atomic core and valence levels as a result of interaction with monochromatic photons, or electrons emitted as a result of de-excitation Auger processes[3]; and (iv) Infrared (vibrational) spectroscopy (IRS)[4]. Due to the higher penetrating power of the X-rays as compared to electrons, the photon-in/photon-out spectroscopic mode is mostly bulk sensitive and can be used for characterization of both conductive and non-conductive materials. In the photon-in/electron-out spectroscopic mode the probing depth is determined by the kinetic energy of the emitted electrons and using soft X-rays usually does not exceed 100 Å. Thus XPS and XANES based on the electron yield are surface sensitive and very appropriate for investigations of surface and interfacial phenomena.

High spatial resolution and image formation in microscopy are achieved using two different approaches, which classify the instruments as scanning and full-field imaging[5-9]. In the scanning approach X-ray photon optics or an aperture (IR) demagnify the incident photon beam to a small spot onto the sample, and an image is acquired by detecting the photon or electron signal while rastering the sample. The microprobe in the scanning X-ray microscopes is formed using reflective or diffractive optical elements[10,11]. The scanning approach offers maximum flexibility for use of different photon and electron detectors, so that in principle scanning X-ray microscopes can work as Scanning Transmission X-ray Microscope (STXM) or Scanning PhotoEmission Microscope (SPEM). The IR microscopes also can be used in transmission and reflective mode.

The full-field imaging microscopes obtain a magnified image of the irradiated sample area by projection of the transmitted photons or emitted electrons using appropriate photon or electron tailoring optics. The instruments based on projection of transmitted photons, known as Transmission X-ray Microscopes (TXM), work similarly to a visible light microscope: a condenser lens illuminates the specimen, and a second objective lens behind the specimen generates a magnified image onto a spatially resolving detector[5,6]. The instruments based on projection of emitted electrons, called X-ray PhotoEmission Electron Microscopes (XPEEM), use suitable electron optical imaging system for magnification and projection of the emitted electrons[7,12,13].

The potential and the applications of the microscopes are determined by their operational principle and the wavelength dependence of the interaction of matter with the electromagnetic radiation. Because of its static design TXM and XPEEM can more easily achieve high spatial resolution. Another advantage of the full-field imaging is the short acquisition time (down to a msec range at present), which makes these microscopes very suitable for dynamic studies and 3D imaging. The disadvantage of TXM-XPEEM, compared to STXM-SPEM is that the spectroscopic analysis, based on XANES or XPS, requires recording of subsequent images at photon energies across a selected absorption edge or at electron energies within a selected energy window.

The complementary capabilities of different microscopy approaches in terms of imaging, spectroscopy, spatial and time resolution are strongly requested by the multi-disciplinary research programs at the synchrotron facilities and have motivated continuous investments in development of instrumentation for imaging with spectroscopic analysis. In this article we present some of the recent achievements in application of imaging with spectroscopic analysis in material and life science research using the different microscopes built and operated at Synchrotron Laboratory Elettra in Trieste.

2.   SCANNING PHOTOEMISSION MICROSCOPY (SPEM)

2.1   Instruments

The two SPEM instruments operated at Elettra are designed and constructed in-house and are operated at dedicated beamlines using different photon energy ranges, below 100 eV[14] and above 350 eV[15]. They are complementary because the low photon energy one is very well suited for probing valence band and shallow electron levels, whereas the second one has achieved excellency probing deeper core electron levels. The different photon energy imposes the choice of the focusing optics, Schwarzschild objective and high resolution Fresnel zone plate (ZP) lenses, respectively. The main advantage of the SPEMs is that the lateral and spectral resolutions are independent, i.e. the lateral resolution, being determined only by the focusing optics, is the same in imaging and spectroscopy modes. Thus SPEMs use the full power of XPS with submicron spatial resolution (0.1-0.5 μm) and probing depth up to a few tens Å. The information provided by core or valence level imaging and spectromicroscopy can be summarized as follows. Most often the research is focused on the correlation between the image contrast and the local concentration and/or chemical state of the sample constituents. In this case the emission from a specific electronic level of the elements is mapped. The core level energy shifts and changes in the valence spectra due to interactions with other atoms or molecules are fingerprints of the actual chemical state and electronic structure. Shifts in the energy position of the XPS spectra can also be induced by band-bending and local charging providing additional important characterization abilities.

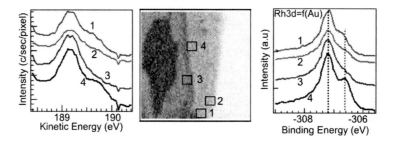

Figure 1. (center) An image illustrating the attenuation of the total Rh 3d5/2 core level emission by the deposited Au patch which has undergone spreading creating areas with different Au coverage. In the darkest area the Au film thickness is more than 4 layers and the Rh 3d5/2 emission is very low. (left) Reconstructed Rh 3d5/2 spectra from the image, corresponding to the areas indicated with squares of the same color. (right) Rh 3d5/2 spectra taken in spectroscopy mode from microspots within the squares.

The potential and efficiency of the SPEM instruments have been significantly improved by implementation of multichannel electron detectors allowing simultaneous collection of images equal to the number of the channels[16]. Since each channel corresponds to a specific kinetic energy, defined by the selected energy window, the multi-channel imaging opens an opportunity not only for mapping of different electronic and chemical states with a single sample scan, but also reconstruction of the spectrum corresponding to the covered energy window, so-called spectro-imaging. Fig. 1 illustrates spectro-imaging using the 48-channel detector developed and fabricated at Elettra. The example is an laterally heterogeneous Au/Rh(110) interface, formed by depositing Au patch through a mask on the Rh(110) surface followed by cycles heating and exposure to different gas ambient[17]. This leads to expanding of the patch due to diffusion of Au atoms away from the patch and development of regions with different Au coverage. The contrast in the Rh $3d_{5/2}$ core level image summing all channels corresponds to the total Rh $3d_{5/2}$ electron emission, which is attenuated with increasing the thickness of the Au layer. The brightest region corresponds to an Au-free Rh surface. From the Rh $3d_{5/2}$ map we reconstructed the Rh $3d_{5/2}$ spectra from the indicated areas, which are shown in left panel. The lineshape of these spectra illustrates how the Rh $3d_{5/2}$ core levels change as a function of the Au coverage. For comparison the right panel shows the Rh $3d_{5/2}$ spectra measured in a conventional spectroscopy mode from a microspot within the indicated areas. The comparison illustrates the good energy resolution achieved using the spectro-imaging mode.

The use of XPS as a spectroscopic method limits the SPEM applications to studies of specimen in solid state and in ultra-high vacuum environment (pressures lower than 10^{-6} mbar), but on the other hand XPS is the best analytical method for probing the composition and electronic structure at morphologically complex surfaces and interfaces. Interfacial processes, including also wear, corrosion and mass transport phenomena, lie at the heart of key technologies, such as industrial and environmental catalysis, fabrication of electronic and magnetic devices, chemical sensors, biocompatible implants etc.

2.2   Characterization of degradation processes in electronic devices

One of the most recent SPEM studies was focused on understanding the degradation mechanisms leading to appearance of dark spots in Organic Light Emission Devices (OLEDs), which are the major reason for their short lifetime[18]. OLEDs were fabricated and operated under ultra-high vacuum (UHV), exposed to air before UHV operation or exposed and operated in air. SPEM was used to map *in-situ* the changes in morphology and chemical composition of the Al cathode surface and of the damaged areas after the OLEDs have undergone degradation during operation. The comparison confirms that the dark spots are fractures in the Al anode and they occur independently on the exposing and operation conditions.

Fig. 2 illustrates that the fractures appear in the Al 2p image as a dark hole because there is no Al 2p emission in the places where the Al film is disrupted. A natural result of disruption of the Al film is exposing the organic layers and InSn oxide (ITO) anode below. Indeed, the central part of the fracture appears very bright in the In 3d map, whereas it has a low intensity in the C 1s map. This indicates that the organic layer has also been locally damaged, exposing the ITO surface. A common and distinct feature of the In maps is the presence of In-containing species deposited on the Al surface around the fractures, the amount of these species decreasing gradually moving away from the hole, as illustrated by the In 3d concentration profile. Since the deposited film screens the Al 2p emission these parts appear darker in the Al 2p map. This is direct evidence that In-containing volatile species are generated from local degradation of the ITO anode depositing on the Al anode around the hole. Similar contrast, observed in the C 1s maps indicates co-deposition of organic species from the locally decomposed organic layers. The In, Al and C maps contain the same topographic features, resulting from the disruption of the Al film causing local enhancement or shadowing effects[19]. More detailed information about the chemical composition within the fractured area and the surrounding cathode surface were provided by the XPS spectra in Fig. 3. The spectra, measured in the dark holes are very similar to the ITO spectra, except the additional new Si 2p line, originating from the ITO glass support, indicating the reduced thickness of the ITO due to material loss in the damaged area. The very small C 1s peak in the spectra confirms that locally the organic layer has also been removed. Comparing the XPS spectra from the cathode areas in the vicinity and away from the fracture confirm the deposition of the ITO constituents, In, small amount of Sn, and C-containing species from the organic layers. The conclusion was that the lateral surface imprefections of the ITO films should be one of the major reasons for electric spikes causing the cathode disruption and device failures.

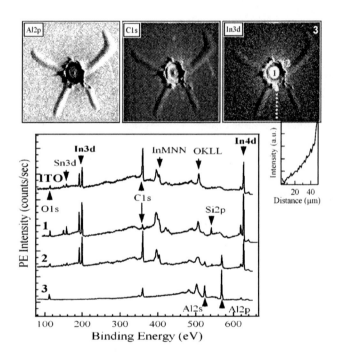

Figure 2. (Top) Al 2p, C 1s and In 3d maps in a damaged region of the Al cathode, (a 'dark' spot in the OLEDs). The In concentration profile is shown in the panel below the In map (along the arrow in the In map). (Bottom) XPS spectra taken inside the hole (1), on the Al cathode in the vicinity (2) and far away (3) of the hole. The top spectrum corresponding to the anode (ITO) is shown for the sake of comparison with spectrum 1.

2.3   Characterization of nano-structured and low dimensional materials

One of the best examples for SPEM achievements in characterization of nano-structured materials are the results obtained with C and MoS_x-based nanotubes. The low dimensionality and high anisotropy the nanotubes determine a large variety of mechanical and electronic properties, which makes them an excellent material for nano-technological applications.

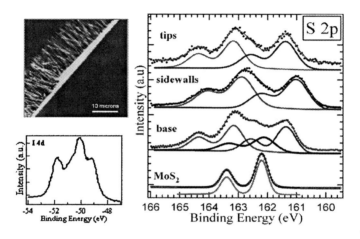

Figure 3. Left (top) Mo 3d image of aligned MoSx nanotubes grown on a Si substrate; (bottom) I 4d spectrum measured on the nanotube bundles. Right S 2p spectra and fitting components taken showing different bonding configurations of S at different parts along the nanotube bundles, The S 2p spectrum of the MoS_2 compound is shown as well.

The high spatial resolution of SPEM allowed us to localize the nanotube bundles (~ 10 microns long and up to 0.5 micron wide) and select the areas for systematic spectroscopic measurements along the axes of the nanotubes[20-22]. An important finding in these studies is the position dependence of the photoemission VB and core level spectra of aligned nanotubes, reflecting the different structural organization along the wall and the tips of the nanotubes Fig. 3 shows a typical cross sectional Mo 3d image of bundles of aligned MoS_x-based nanotube bunches, produced by a catalyzed transport reaction with iodine as a transport agent. The S 2p, Mo 3d, and the valence band spectra taken at the tips and sidewalls and the base (interface with Si) appear different and all differ significantly from the corresponding spectra taken on a reference MoS_2 crystal[22]. Fig. 4 illustrates the significant difference between the S 2p spectra from the tips, base and sidewalls of the nanotubes. An important finding of the spectroscopic analysis was the presence of I in the structure of the MoS_x nanotubes. The 'multicomponent' S 2p and I 4d spectrum indicates coexistence of different S and I bonding configurations within the structure of the nanotube bundles, which are under evaluation.

Another field where the low-energy SPEM has demonstrated its exclusive potential is in imaging the lateral variations in the density of electronic state at the Fermi edge of doped manganese oxides exhibiting colossal magneto-resistance[23]. It has been predicted existence of different phases in these materials, which may affect significantly their properties but the length scales and origin of these phases are very disputable[24] Imaging the evolution of the emission at the Fermi edge during the heating-cooling cycles of $La_{1-x-y}Pt_yCa_xMnO_3$ systems, which are undergoing metallic to insulator phase transitions at low temperatures we observed the development of unexpectedly large (in μm range) insulating domains inside the metallic phase[25]. Fig. 4 displays a series of images taken as the sample is cycled through a range of temperatures. Instead of being in the expected homogeneously metallic phase at low temperature, the sample shows the existence of rather large (several microns across) insulating patches sprinkled around inside the metallic phase, while the compound becomes homogeneously insulating at elevated temperatures. As the temperature is lowered again, the insulating patches reappear in the metallic phase, interestingly roughly retaining their shapes and positions, thereby suggesting that these patches dynamically re-form themselves with an approximate memory. These results are expected to provide an insight into the mechanism of such phenomenon known as electronic phase separation.

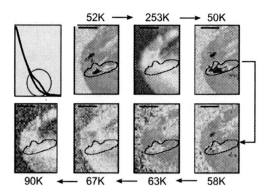

Figure 4. Images of the electron emission at the Fermi edge (left-top spectrum) which reveal spatial variations in the electronic structure of $La_{1-x-y}Pt_yCa_xMnO_3$ sample across the phase transition temperatures. The region where insulating patches re-appear is indicated with the black line.

3.   X-RAY PHOTOEMISSION ELECTRON MICROSCOPE (XPEEM)

3.1   Instrument

The full-field imaging microscope, XPEEM, operated at Elettra[26] has an energy filter which adds XPS to the standard XANES used in the simpler PEEM instruments, which monitor only the total emitted electron signal. This opens multiple spectroscopic capabilities, including angular resolved photoelectron spectroscopy and photoelectron diffraction from a microspot[27]. The energy filter also improves the resolution because it reduces the lens chromatic aberrations[7]. In XPEEM the lateral resolution and spectral resolution are 'inversely' correlated because of the instrument transmission, so that the analyzer energy resolution should be relaxed when very high lateral resolution is required. The XPEEM image obtained in the energy filtered imaging mode contains similar information as the SPEM image. In order to make spectroscopic analysis of a micro-spot the kinetic energy of the imaging photoelectrons is scanned while acquiring a set of images. The instrument also works in *area selective spectroscopy mode* providing spectroscopic information from an area of ~ 1 μm^2 with superior spectral resolution. This mode uses modified settings of projector lens and analyzer allowing the full dispersive plane of the analyser to be monitored on the screen. An important advantage of XPEEM at Elettra is that the instrument has a magnetic sector field and an additional electron source for performing mirror electron microscopy (MEM) and low energy electron microscopy (LEEM) with a lateral resolution better than 8 nm. This multifunctional instrument allows complete spectroscopic and structural characterization of the specimen with comparable spatial resolution for correlating chemical, magnetic and structural properties of the materials.

3.2   Studies of magnetic materials

The dichroic effects for differently polarized light observed in XANES mode, namely X-ray Magnetic Circular Dichroism (XMCD) or X-ray Magnetic Linear Dichroism (XMLD) combined with elemental specificity at length scales of a few tens nm fit excellent to studies of magnetic memory elements with fast shrinking dimensions. Their magnetic state and switching properties depend on the relative energies of the various domain configurations, which can be verified in the XPEEM experiments. Imaging of domain structures of nano-structured magnetic materials collecting stacks of images while scanning the photon energy has allowed theoretical predictions to be verified. Fig. 5 shows the magnetization configurations for a ring-shaped Co structures predicted by the theory and confirmed by XPEEM measurements[28].

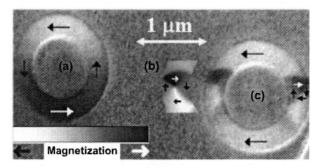

Figure 5. Magnetization configurations for ring shaped Co micro-structures grown on Si measured at the across the Co 2p edges in XMCD mode (a) and (c). (b) is a theoretical simulations of the onion state (c).

Figure 6. Mn 3p XMCD images of the domain structure of ferromagnetic (FM) MnAs film on GaAs during the phase transition to the paramagnetic (PM) state with increasing the temperature above 40° C. Field-of-view 10 µm.

Another example highlighting the imaging potential of XPEEM in following dynamic processes is the study of phase transitions in epitaxial MnAs films grown on GaAs[29]. This system combines ferromagnetic and conventional semiconductor materials into novel devices that may find applications for spin injection into semiconductors. It undergoes a phase transition from a ferromagnetic hexagonal α-phase to a paramagnetic orthorhombic β-phase above 40°C, the two phases coexisting within a temperature range of nearly 30°C. The XMCD images in Fig.6. show the evolution of the magnetic structure with increasing temperature. In the early phase of the transition the paramagnetic phase appears as short stripes, growing together into continuous broadening features, separating the ferromagnetic phase into continuously thinning stripes. When the width of the ferromagnetic stripes falls below certain dimensions, which depends on the thickness of the MnAs film, domains with opposite magnetization nucleate and grow inside. With further decreasing the width the ferromagnetic stripes break into individual pieces. Simultaneously the magnetization in the domains decreases until all magnetic contrast is lost (not shown here) and the conversion into the paramagnetic phase is completed. The LEEM images taken during the phase transition temperatures confirmed that the observed magnetic transitions are intimately connected with the structural transition.

4. SCANNING AND FULL-FIELD IMAGING TRANSMISSION MICROSCOPES

4.1 TwinMic instrument

In the frame of a RT&D European project, with the contribution of three European large scale facilities (ELETTRA, ESRF and SLS) and five other European institutions we used a novel approach integrating the scanning and full-field imaging transmission microscopes into a single instrument with fast switching between the two microscopy modes, called TwinMic[30]. Preserving the performance of each of the microscopes it has opened an opportunity to perform fast imaging for dynamic studies combined with micro-tomography and micro-spectroscopy (μ-XANES, μ-XRFS).

The outline of this microscope is shown in Fig. 7. According to Zeitler's theorem of the reciprocity of detector response functions in scanning microscopes and source intensity distributions for full-field imaging similar imaging performance can be achieved in both microscope types, but these conditions can be inconvenient in complementary instruments[31]. The similarity and complementarities of the two types of microscopes prompted us the idea to integrate them into a single instrument.

The optical set-up illustrated in Fig. 7 includes (1) the different focusing elements tailoring properly the incoming beam for illumination the sample in STXM or TXM modes; (2) a sample stage which can be scanned for the STXM mode; (3) a projection ZP optics behind the sample that can be inserted (TXM) or retracted (STXM); (4) appropriate detectors for the STXM and TXM modes. The most important issue is the possibility to tailor the coherence of the illumination, because the scanning type microscopy requires coherent illumination for diffraction limited imaging, whereas the performance the full-field imaging microscopy benefits from partial or incoherent illumination32. In the TwinMic, a conventional ZP[33] forms the microprobe in scanning microscopy mode, while in the full-field imaging the a novel X-ray diffractive optical element was developed, which allows tailoring the X-ray beam to illuminate homogeneously the object field[34,35].

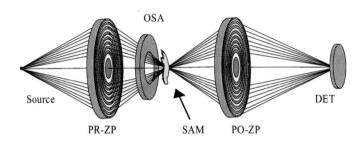

Figure 7. Schematics of the optical set-up of the twin microscope with the pre-specimen ZPs (PR-ZP), the aperture (OSA), the specimen (SAM), the post-ZP (PO-ZP) and the detectors (DET).

For STXM mode we use a set of photodiodes, multi-segment detector and fast read-out CCD camera[36], whereas for TXM we use a high resolution CCD camera. The detecting systems allow using of different contrast mechanisms to extract important image information in addition to spectroscopic information, e.g. absorption, dark-field imaging differential phase contrast, differential interference contrast techniques[37] etc.

The Twinmic station offers the possibility to implement versatile sample environments like, e.g. a liquid cell, a chemical reaction cell, a cryo stage and a tomography stage to be operated either in air, helium atmosphere or vacuum. The end-station is completely computer controlled for data acquisition and processing.

4.2   First applications of twinmic

Although this new instrument is still under commissioning it has attracted a lot of interest for life science research. Fig. 8 shows an example from marine biology where the morphology of the organic biosilicate skeleton of a planktonic diatom *"Coscinodiscus sp"* has been analysed[32]. Higher penetration power of X-rays compared to charged particles has allowed revealing a double hexagonal structure of the pillbox-like biosilicate skeleton of the diatoms. The example demonstrates also the potential of using a configured electron-multiplied CCD detector to acquire simultaneously brightfield and differential phase contrast images[36].

Figure 8. X-ray transmission micrographs of planktonic diatoms *"Coscinodiscus sp."* acquired with the TwinMic X-ray microscope in scanning mode. Using configured CCD camera allows acquiring simultaneously a brightfield image (left) and the differential phase contrast image (right). Photon energy - 718 eV, the dwell time per pixel was 80 ms.

5. SYNCHROTRON-BASED INFRARED SPECTROMICROSCOPY (IRSM)

The broadband characteristics, high collimation, polarization and pulsed structure of infrared radiation from synchrotron light sources, have allowed a new class of experiments with significant multidisciplinary impact[4]. For chemistry and biology using exclusively the mid-infrared (3-20 μm) range, IRSM has reached diffraction limited resolution[38]. Synchrotron-based IR beamlines use commercial FTIR spectrometers and microscopes working in both transmission and reflection mode. The microprobe is determined simply by an aperture and the maps of selected IR spectral features were formed by scanning the specimen. Although the IRSM has modest spatial resolution the great advantage of this chemically sensitive technique is that IR is non-destructive radiation allowing studies of living objects as well.

Fig. 9 illustrates the possibility to image the composition of sub-cellular structures in an intact cell. The contrast in the IR images is generated by mapping the absorption bands characteristic of amide (A) and methyl/methylene (B) groups throughout the sample, fingerprints of the local protein and lipids contents. In the light microscope image of the same cell the nucleus is clearly visible in the lower right of the cell, together with some other organelles. Comparison of the IR and light microscope images clearly demonstrates the differences in the distribution and local concentrations of proteins and lipids within the cell which can be correlated with the cell structure.

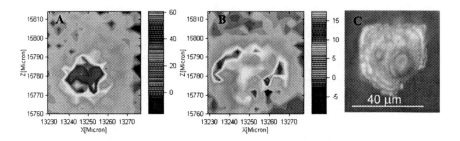

Figure 9. IR images of amide band absorption (A) and methyl/methylene band absorption (B) of a formalin fixed cell measured in reflection mode. (C) The corresponding visible light image.

The complementary of IRSM with the X-ray spectromicroscopy has been recognized and the potential of the combined studies has already been demonstrated[38]. In particular, in life science research the IRSM of the organic matter, which undergoes irreversible changes under X-rays, combined with elemental mapping with XAS or FS analysis of the same sample is becoming a routine approach at facilities hosting different microscopes.

6.   OUTLOOK

Some of the near-future upgrades aim at improving the lateral resolution, spectral resolution and data acquisition velocity of X-ray microscope. The progress is directly linked to further development of efficient, high resolution X-ray focusing optics for SPEM, STXM and TXM, in order to work routinely with laterally resolution ≤ 50 nm. Implementation of aberration corrections systems for XPEEM is expected to approach the theoretical resolution limit of 2 nm. Improving the transmission of the electron energy analyzers for SPEM and implementation of angle-resolved photoemission microscopy for band mapping and the efficiency of the detectors in order to work with better spectra resolution are the two projects that are under commissioning. Another important development is implementation of exchangeable sample holders for complementary IRSM and STXM-TXM characterization of biological and environmental samples. The construction of ultra-fast position sensitive detectors for time-resolved measurements using the pulsed character of synchrotron radiation and future X-ray laser sources is to be considered as well.

REFERENCES

1. J. Stöhr, NEXAFS Spectroscopy (Springer Verlag, Berlin 1992).
2. R.O. Müller, Spectrochemical Analysis by X-ray Fluorescence (Plenum Press, New York 1972).
3. K. Siegbahn, J. Electr. Spectr. Rel. Phenom. Vol. 51 (1990) 11.
4. Proceedings of SPIE Volume 3775, Editor(s): G. L. Carr and P. Dumas 1999.
5. J. Kirz, C. Jacobsen, M. Howells, Soft-X-ray Microscopes and Their Biological Applications, Q. Rev. Biophys. Vol. 28 (1995), p. 33.
6. B. Niemann, D. Rudolph, and G. Schmahl, Optics Comm. Vol. 12 (1974), p. 160.
7. E. Bauer, Rep. Prog. Phys. Vol. 57 (1994), 895.
8. J. Electr. Spectr. Rel. Phenom. 84 (1997), Ed. H. Ade, special issue on Spectromicroscopy.
9. S. Günther, B. Kaulich, L. Gregoratti and M. Kiskinova, Progr. Surf. Sci. Vol. 70 (2002).
10. D. Attwood, Soft X-rays and Extreme Ultraviolet Radiation: Principle and applications (Cambridge University Press 1999).
11. E. Spiller, Soft X-ray Optics (SPIE, Bellingham 1995).
12. O. H. Griffith, W. Engel, Ultramicroscopy Vol. 36 (1991), p. 1.
13. O. H. Griffith, G. F. Rempfer, Advances in Optical and Electron Microscopy, (Academic Press 10, New York 1987), p. 269.
14. F. Barbo, M. Bertolo, A. Bianco, G. Cautero, S. Fontana, T.K. Johal, S. La Rosa, G. Margaritondo, K. Kaznacheyev, Rev. Sci. Instrum.71, pp. 5-10, 2000; http://www.elettra.trieste.it/experiments/beamlines/spectro/index.html
15. L. Casalis L, L. Gregoratti L, M. Kiskinova et al, Surf. Int. Analysis 25 (1997) 374; http://www.elettra.trieste.it/experiments/ beamlines/esca/index.html
16. L. Gregoratti, A. Barinov, E. Benfatto, G. Cautero, C. Fava, P. Lacovig, D. Lonza, M. Kiskinova, R. Tommasini, S. Mähl, and W. Heichler, Rev. Sci. Instr. 75 (2004) 64.
17. A. Barinov et al, unpublished data.
18. P. Melpignano, A. Baron-Toaldo, V. Biondo, S. Priante, R. Zamboni, M. Murgia, S.Caria, L. Gregoratti, A. Barinov, M. Kiskinova, Appl. Phys. Lett. 85 (2005), 41105.
19. S. Günther, A. Kolmakov, J. Kovac, M. Kiskinova, Ultamicroscopy 75 (1998) 35.
20. S. Suzuki , Y. Watanabe, T. Ogino, S. Heun, L. Gregoratti, A. Barinov, B. Kaulich, M. Kiskinova, W. Zhu, C. Bower, O. Zhou, Phys. Rev. B 66 (2002) 035414.
21. A. Barinov et al, in preparation
22. J. Kovač, A. Zalar, M. Remskar, A. Mrzel, D. Mihailović, L. Gregoratti, M. Kiskinova, Elettra Highlights 2003.
23. M. Uehara, S. Mori, C.H. Chen, and S.-W. Cheong, Nature 420 (2002) 797.
24. A. Moreo, S. Yunoki, E. Dagotto, Science 283 (1999) 2034.
25. D.D. Sarma, D. Topwal, U. Manju, S.R. Krishnakumar, M. Bertolo, S. La Rosa, G. Cautero, T.Y. Koo, P.A. Sharma, S.-W. Cheomg, A. Fujimori, Phys. Rev. Lett.
26. A. Locatelli, A. Bianco, D. Cocco, S. Cherifi, S. Heun, M. Marsi, M. Pasqualetto, and E. Bauer, Phys. IV 104 (2003) 99 – 102; http://www.elettra.trieste.it/experiments/ beamlines/nano/index.html.
27. Th. Schmidt, S. Heun, J. Slezak, J. Diaz et al, Surf. Rev. Lett. Vol. 5 (1998), p. 1287.
28. M.Klaui , J. A. Vaz, C. Bland, T. L. Monchesky, J. Unguris, E. Bauer, S. Cherifi, S. Heun, A. Locatelli, L. J. Heyderman, and Z. Cui C., Phys. Rev. B 68 (2003) , 134426.

29. L. Daeweritz, C. Herrmann, J. Mohanty, T. Hesjedal, and K. H. Ploog; E. Bauer, A. Locatelli, S. Cherifi, R. Belkhou, A. Pavloska, S. Heun; J. Vac. Sci. Technol. B 23, 1759 (2005).

30. B. Kaulich et al., Synchr. Rad. News 16 (2004), pp. 22-25.

31. E. Zeitler, M.G.R. Thomson, Optik 31 (1970), pp. 258-280 and 359-366.

32. B. Kaulich et al., IPAP Conf. Series 7 (2006), pp. 22-25.

33. E. Di Fabrizio et al. Synchrot. Radiat. News 12 (1999), pp. 37-39.

34. U. Vogt et al., Optics Letters 31 (2006), pp. 1465-1467.

35. D. Cojoc et al, Microelectronic Eng. 83, (2006), pp. 1360-1363.

36. G.R. Morrison et al., IPAP Conf. Series 7 (2006), pp. 337-339.

37. T. Wilhein et al., Appl. Phys. Lett. 78 (2001), pp. 2082-2084.

38. L. Miller and P. Dumas, Vibr. Spectr. 32, 3 (2003).

ADVANCED MATERIALS RESEARCH WITH 3RD GENERATION SYNCHROTRON LIGHT

P. Soukiassian, M. D'angelo, H. Enriquez and V.Yu. Aristov
Commissariat à l'Energie Atomique, Saclay, Laboratoire SIMA associé à l'Université de Paris-Sud/Orsay, DSM-DRECAM-SPCSI, Bâtiment 462, 91191 Gif sur Yvette Cedex, France

Abstract: H and D surface nanochemistry on an advanced wide band gap semiconductor, silicon carbide is investigated by synchrotron radiation-based core level and valence band photoemission, infrared absorption and scanning tunneling spectroscopy, showing the 1st example of H/D-induced semiconductor surface metallization, that also occurs on a pre-oxidized surface. These results are compared to recent state-of-the-art *ab-initio* total energy calculations. Most interestingly, an amazing isotopic behavior is observed with a smaller charge transfer from D atoms suggesting the role of dynamical effects. Such findings are especially exciting in semiconductor physics and in interface with biology.

Key words: Nanochemistry, hydrogen, deuterium, silicon carbide, surfaces, interfaces, metallization, passivation, synchrotron radiation, core level and valence band photoemission, infrared absorption, and scanning tunneling spectroscopies.

1.   INTRODUCTION

The last 2 decades marked the arrival of two state-of-the-art experimental techniques that brought a revolution in physics and materials science. First, the discovery in the early 80's of scanning tunneling and atomic force microscopy (STM, AFM) has allowed to probe real space surface structures leading to much deeper understanding and to birth and tremendous development of nanoscience and nanotechnology. At about the same time, synchrotron radiation, primarily known for many years for its

Vasili Tsakanov and Helmut Wiedemann (eds.), Brilliant Light in Life and Material Sciences, 317–328.
© 2007 *Springer.*

detrimental effects in high energy physics machines, was made available to users of the atomic and condensed matter physics community, initially in the "parasitic mode" with the first generation machines, and then with fully dedicated 2nd generation storage rings as e.g. Tantalus or Aladdin (Wisconsin), SuperACO (Orsay), SPEAR (Stanford), DESY (Hamburg), NSLS (Brookhaven) and many others. The early/mid 90's have witnessed 3rd generation synchrotron sources with ALS (Berkeley), NSRRC (Hsinchu) or Elettra (Trieste). Since then, many advanced storage rings are now in operation world-wide (ESRF-Grenoble, APS-Argonne, Spring 8-Himeji, PLS-Pohang, SLS-Villigen, BESSY II-Berlin, MAX II-Lund, CLS-Saskatoon), under construction or commissioning (Diamond-UK, Soleil-France, Boomerang-Australia, ALBA-Spain), or in project (e.g. CANDLE in Armenia). These facilities have driven very important progress in various fields helping to develop novel materials having advanced characteristics. Synchrotron "white" light most important features include i) continuous spectrum/large energy range from infrared to hard-X-rays, ii) polarized light, iii) very low emittance/high brilliance up to 10^{22} photons/mm^2/s/mrad^2 at 0.1% bandwidth. All these characteristics scale well above those of conventional plasma or X-ray sources. They are especially useful in surface and interface science with the even higher brilliance achieved at 3rd generation synchrotron facilities. This resulted in features as very high energy and spatial resolutions combined with fast data acquisition that are so important for systems sometimes having limited lifetimes. It gave access to information not accessible previously and deep understanding into complex systems, leading to new discoveries[1].

Silicon carbide (SiC) is not a new material since it is older than the solar system. But it is definitively an advanced semiconductor (IV-IV compound), especially suitable for high power, high temperature, high voltage and high frequency electronic devices and sensors[2]. SiC has also other interesting capabilities including exceptional mechanical properties, biocompatibility and resistance to radiation damages[2]. SiC average factors of merit scale well above those of conventional semiconductors by up to 3 orders of magnitude[2,3]. Unlike other wide band gap semiconductors, ultra high quality wafers have been successfully grown[4]. Furthermore, promising nano-objects could be self-assembled on SiC surfaces[3,5]. It is only rather recently that SiC surfaces could be controlled and understood at the atomic level, thanks to such advanced experimental techniques as STM, synchrotron radiation-based diffraction, and state-of-the-art theoretical calculations[3].

In this article, we review the amazing nanochemistry at SiC surfaces that lead to the discovery of the 1st example of H-induced semiconductor surface metallization with a surprising isotopic effect taking place when deuterium is used instead. These studies are based on core level and valence band photoemission spectroscopy using synchrotron radiation, and additional infrared absorption and scanning tunneling spectroscopy experiments.

2. EXPERIMENTAL

The photoemission experiments are conducted at the Elettra storage ring (Trieste) on the VUV beam line. Atomic H exposures are performed at a 300°C sample temperature using research grade molecular H_2 dissociated by a heated tungsten filament. Details about high quality 3C-SiC(100) 3x2 surface preparation and experiments can be found elsewhere[3].

3. HYDROGEN-INDUCED 3C-SIC(100) 3X2 SURFACE METALLIZATION

The 3C-SiC(100) surface has more than 10 different reconstructions ranging from Si-rich to C-rich, all having very different properties and atomic structures[3]. Atom-resolved STM, grazing incidence X-ray diffraction (GIXRD) at ESRF-Grenoble, photoelectron diffraction (PED) at ALS-Berkeley, and *ab-initio* total energy calculations have been used to determine the complex atomic structure of the 3x2 reconstruction (Si-rich)[3]. The latter includes 3 Si atomic layer (1/3+2/3+1 Si ML) lying on the 1st C plane[3]. The 1st Si plane includes rows of asymmetric dimers all tilted along the same direction. The 2nd Si plane has rows of dimers having alternating lengths. The 3rd Si plane, located just above the 1st C plane also has Si dimer rows.

Atom-resolved STM shows that the clean surface has asymmetric Si dimers[3]. The surface exposed to atomic H at a surface temperature of 300°C have its dangling bonds terminated by an H atom, leading to the top Si dimers becoming symmetric with, contrary to previous beliefs, the surface keeping its 3x2 array as for the clean surface[3,6]. We use scanning tunneling spectroscopy (STS) that could provide insights about the local electronic properties around the Fermi level. As can be seen in Fig. 1 (left) from the I-V characteristics for the clean and H-covered surfaces, H exposure results in band gap closing[6]. Such a behavior clearly suggests surface metallization upon hydrogen atom interaction. Amazingly, the H-induced metallization is not affected by oxygen with no band gap opening upon O_2 exposures[7,8].

The surface metallization could be traced looking at a possible Fermi level build-up using VB-UPS. H interaction on the 3C-SiC(100) 3x2 surface results in the quenching of the S electronic surface state, which indicates H atoms interacting with the top-most surface dangling bonds, in excellent agreement with the above STM results[6]. Most interestingly, the hydrogen leads to an electronic state developing at the Fermi level with a clear Fermi level edge clearly indicating surface metallization as can be seen in Fig. 1 (right), in excellent agreement with the STS measurements[6].

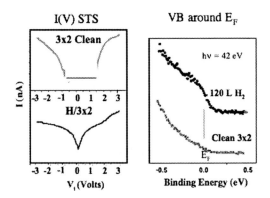

Figure 1. Metallization of the H/3C-SiC(100) 3x2 surface evidenced by STS through Band Gap closing in I[nA](V[volts]) (left) and Fermi level build-up in valence band photoemission at hn = 42 eV (right) for clean and H covered 3x2 surfaces[6].

Additional insights on H atoms interaction with the 3x2 surface could be found using infrared absorption spectroscopy - IRAS (Agere Systems) with absorbance spectra shown in Fig. 2. The broadband electronic absorption extending over the whole frequency range (1000 - 6000 cm^{-1}) is seen in p-polarization only. The vibrational band at 2100 cm^{-1} is associated with H on the Si back-surface of the multiple internal reflection plate, while the two modes centered at 2118 cm^{-1} and 2140 cm^{-1} are assigned to the top layer H-Si-Si-H dimer, and to the Si 3rd layer H-Si structure. This indicates that the H atoms break the Si-dimers located in the 3rd plane below surface[6].

The nature of the H atom interaction at the 3C-SiC(100) 3x2 surface is understood in depth using synchrotron radiation based-high energy resolution core level photoemission spectroscopy at the Si 2p core level which provides very valuable and deep insights about charge transfers and Si bonding configurations[3].

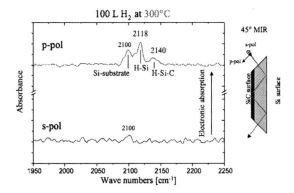

Figure 2. Multiple internal reflection infrared absorption spectroscopy (MIR-IRAS) providing the absorbance spectra vs wave numbers in p- (top) and s- (bottom) polarization[6]. It shows the Si-H stretching vibrational modes on 100 L of H2 ~ 5 L H/3C-SiC(100) 3x2 surface. A schematic shows the MIR-IRAS set-up on the right.

Fig. 3 shows Si 2p core level spectra in surface sensitive mode (hv = 150 eV) for clean and H-exposed 3x2 surfaces. The clean surface spectrum is decomposed into one bulk and 3 surface & sub-surface shifted components that are specific of the 3x2 reconstruction[3,7]. S1 is related to the up-atom of the top asymmetric dimer. S2 accounts for the down atom of the top asymmetric Si-dimer and to the atoms located in the second Si plane. S3 is related to the 3rd Si atomic plane including the Si-Si dimers located in this atomic plane and to the other Si atoms in the 2nd plane that are not related to S2[3,7]. The effect of H atoms on the surface Si atoms could be traced in Fig. 4 which shows that the S1 surface component is shifted by 450 meV to lower binding energy as a result of charge transfer into the top Si-dimer leading to become symmetric.

The other interesting feature is the appearance of a reacted shifted component R shifted by 470 meV to higher binding energy from the S3 sub-surface component. This indicates H strong interaction below the surface, breaking Si-dimers in the sub-surface region. Therefore, the 2 Si atoms belonging to the latter have now very different electronic status[3,7]. Indeed, such a finding could easily be correlated to the above MIR-IRAS results, with specific vibrational mode indicating H atom terminating a Si atom that is bonded to a C atom[6]. From the 3C-SiC(100) 3x2 surface structure[3], one can therefore easily deduce that this could occur only in the 3rd Si plane that is located just above the first C plane[3,6].

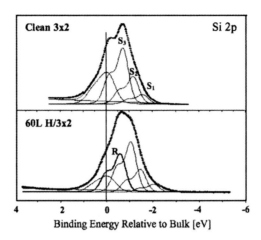

Figure 3. H/3C-SiC(100) 3x2: Si 2p core level peak decomposition for clean and H covered 3x2 surfaces. S1, S2, S3 are the surface and sub-surface shifted components and R the reactive component[3,7]. The photon energy is 150 eV.

So, the H atoms asymmetric attack on the 3rd plane Si-Si dimers leads to dimer breaking, leaving two Si dangling bonds that are expected to become terminated by an H atom just like for the top surface dangling bonds. However, due to steric limitation (the SiC lattice parameter is 20% smaller when compared to silicon[3]), only one Si dangling bond can be terminated by a H atom, leaving the other one empty[6]. Such an asymmetric attack results in H creating dangling bond defects. Subsequently, this situation results in having a very different electronic status for the two Si atoms initially forming a dimer, leading to an additional charge transfer that significantly contributes to the observed surface metallization[6,9]. Indeed, one has to keep in mind that C has a higher electron affinity compared to Si and therefore a charge transfer to the C plane is likely. In addition, as evidenced above by the Si 2p surface and sub-surface components (S1, S2 and S3) that are shifted to lower binding energy by rather large values (≈ 0.5 eV) similar to those observed for alkali metal/semiconductors interfaces[10], H atoms interaction with the topmost Si-dimers are also contributing to the surface metallization[9].

A side view of the clean 3C-SiC(100) 3x2 surface atomic model is displayed in Fig. 4 (left). The atomic structure has been determined using real-space atom-resolved STM and synchrotron radiation-based diffraction (GIXRD, PED) experimental techniques, and theoretical *ab-initio* total energy first principle calculations[3]. Fig. 4 (right) shows the H atoms interaction in the surface and sub-surface regions with the different steps described above, including topmost Si dimer becoming symmetric, dangling bond defect formation, with the facing Si dangling bond terminated by a H atom[6,9]. Therefore, one can easily imagine that, at room temperature, the H atoms located in the 3rd atomic plane would not necessarily remain permanently "attached" to the same Si dangling bond and instead, hop from one dangling bond to another one, leading to a dynamical situation possibly contributing also to the metallization process.

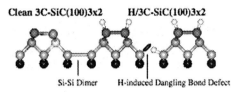

Figure 4. Atomic side view models of clean (left) and H covered 3C-SiC(100) 3x2 (right) surfaces[6].

4.   PREDICTIONS FROM AB-INITIO TOTAL ENERGY THEORETICAL CALCULATIONS

In 2005, the interaction of hydrogen atoms with the 3C-SiC(100) 3x2 surface has attracted a lot of interest from theory. Indeed, this system has been investigated by four different groups all using state of-the-art *ab-initio* total energy local density functional calculations. All of them also conclude that the surface is metallized by hydrogen interaction[9]. While H atoms decorating the top-most surface dangling bonds is also favored, in excellent agreement with real-space atom-resolved STM, VB-PES and MIR-IRAS experiments[5], a different geometry is proposed for H atoms interaction with the 3rd plane Si-Si dimer, favoring for the H atom a bridge-bond adsorption site between two Si atoms belonging to the same dimer[10] as depicted below in the Fig. 5 model.

Figure 5. Atomic side view model of the H/3C-SiC(100)3x2 surface as proposed by ab-initio total energy calculations[11] with a bridge bond position for the H atom located in the 3rd atomic plane.

However, such a bridge bond position for H atoms is not consistent with the experimental results in particular with the 2140 cm^{-1} vibration frequency observed in infrared absorption spectroscopy specific of a H-Si-C bond, not of H atoms in bridge bond positions between 2 Si atoms[5]. Such an assignment is further supported by MIR-IRAS experiments of H interaction on the 3C-SiC(100) c(4x2) surface reconstruction[6]. A bridge bond position would give a much lower frequency (between 1100 cm^{-1} to 1400 cm^{-1} as calculated[11]) that is not observed experimentally. In addition, in case of H atoms in bridge bond positions, no reactive component would be observed at the Si 2p core level (Fig. 3) since the two Si atoms bonded to the same H would have the same electronic status[9]. The difference may come from the fact that these calculations are "0 K" total energy calculations while experimentally, H exposures are made at a surface at 573 K, with data recorded at room temperature which is a very different situation. Therefore, it is rather difficult to compare predictions based on calculations not taking into account the entropy and performed for a "frozen" system. Such "static" calculations cannot address complex dynamical processes occurring here during H atoms interaction and safely derive information about the later.

5.   ISOTOPIC EFFECT IN DEUTERIUM-INDUCED 3C-SIC(100) 3X2 SURFACE METALLIZATION

Deuterium (D) has the same electronic properties than hydrogen, with a single "s" valence electron. So in principle, it should exhibit the same behavior. However, it was shown to behave differently e.g. in electrical defect passivation at SiO_2/Si interfaces with surprisingly enough, a higher efficiency for deuterium compared to hydrogen. It is therefore challenging to explore if D atoms would also metallize the 3C-SiC(100) 3 × 2 surface as H atoms do and if so, some difference in the behavior would take place. As can be seen from the valence band spectra (Fig. 6), a clear Fermi step built-up upon 100L atomic D exposure, indicating surface metallization[13].

Also, as for H, D interaction quenches the SS electronic surface state at 1.3 eV below Fermi level since D atoms terminate topmost surface Si dangling bonds as H atoms do[6,13].

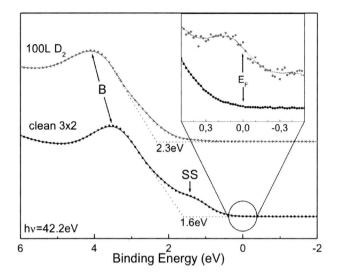

Figure 6. Valence band spectra for the clean 3C-SiC(100) 3x2 surface and exposed to 100 L of D2[13] dissociated by a heated tungsten filament. The inset shows the region corresponding to the Fermi edge. The photon energy is 42.2 eV.

Also of interest is the D-induced surface metallization that leads to band bending as can be seen from a 0.6 eV rigid shifts of valence band maximum and bulk component B (Fig. 6). Similar rigid shifts (≈ 0.7 eV) are observed at Si 2p & C 1s core level bulk components[13]. Interestingly, these values are comparable to the band bending observed at metal-semiconductor interfaces, especially with alkali metals that are also group I elements[10]. They are significantly smaller than those for H/3C-SiC(100) 3x2 that reach 0.84 eV, stressing the significant role of isotopic effects[3,7,13].

In order to understand better the origin of such a metallization especially the possible role of isotopic effects, we look at Si 2p and C 2s CL-PES to probe fine charge transfers occurring upon D interaction. As can be seen in Fig. 7a at Si 2p core levels, the S1, S2 and S3 surface components are shifted to lower biding energy by 0.24, 0.26 and 0.18 eV respectively upon 100 L D exposure[13]. As for H interaction, the reactive R component indicates an asymmetric attack on the 3rd Si plane dimers leading to D-induced dangling bond defects. Turning to the C 1s core level (Fig. 7b), one can see that D induces a new component shifted by 0.34 eV to lower binding energy. It indicates that the C plane is attracting a larger charge transfer, in agreement with its higher electron affinity at 2.55 compared to Si at 1.9.

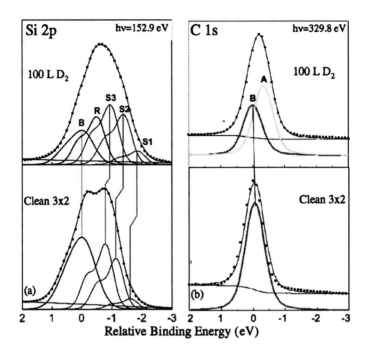

Figure 7. Si 2p (a) and C 1s (b) core level spectra for clean 3C-SiC(100)3x2 and atomic D exposed showing Si 2p B bulk, S1, S2 and S3 surface, sub-surface shifted and R reacted components, and C 1s B bulk & A chemically-shifted components[13].

Another feature of strong interest is the significantly smaller Si 2p surface core level shifts upon D interaction (0.25 eV) compared to H (0.45 eV)[10,18], indicating a smaller D charge transfer to the surface. Since D and H have exactly the same electronic properties, such an isotopic effect is likely to result from the mass difference between the two atoms, suggesting that some dynamics may be at the origin of the smaller charge transfer from heavier D atoms. One can easily imagine that, at 300 K, the D or H atoms would hop in the 3rd Si plane from one dangling bond to the other one and vice-versa, with a higher hoping probability for lighter H atoms. In turn, it suggests that the metallization is likely to be more stable with D than with H.

6. H/3C-SIC(100) 3X2 PRE-OXIDIZED SURFACE

Finally, notice another very interesting & unique property of H-induced (also likely with D) metallization that is not removed by oxygen interaction[7,8]. Furthermore, such H-induced metallization could also be achieved on pre-oxidized surface[12]. It suggests H and O interactions not occurring at same sites, achieving on same surface, both metallization and passivation.

7.   CONCLUSIONS

Atomic H/D interaction on clean & pre-oxidized 3C-SiC(100) 3x2 surface is studied by synchrotron radiation-based VB-PES & CL-PES, STS, and MIR-IRAS. H/D are found to drive an unprecedented semiconductor surface metallization, triggered through asymmetric attack at sub-surface Si-dimers. This results from competition between H/D terminating top surface dangling bonds and steric hindrance generated below surface. Most interestingly, H metallization also occurs on pre-oxidized 3C-SiC(100) 3x2 surface while an amazing isotopic effect takes place with smaller charge transfer for D, likely due to a lower mobility. These findings directly impact defect passivation at semiconductor interfaces and provides means to develop electrical contacts on chemically passive SiC, particularly exciting for interfacing with biology.

REFERENCES

1 Giorgio Margaritondo, Introduction to Synchrotron Radiation, Oxford, New York (1988); J. Phys IV (France) 132 (2006) 23-29; and references therein.

2 SiC: Review of Fundamental Questions and Applications to Current Device Technology, W.J. Choyke, H. Matsunami & G. Pensl editors, Akad. Verlag, Berlin (1998) Vol. I & II.

3 P. Soukiassian and H. Enriquez, J. Phys. Cond. Mat. 16 (2004) 1611; & references therein.

4 D. Nakamura et al., Nature 430 (2004) 1009; R. Madar, Nature 430 (2004) 974.

5 P. Soukiassian, F. Semond, A. Mayne and G. Dujardin, Phys. Rev. Lett. 79 (1997) 2498.

6 V. Derycke, P. Soukiassian, F. Amy, Y.J. Chabal, M. D'angelo, H. Enriquez and M. Silly, Nature Mat. 2 (2003) 253.

7 D'angelo, H. Enriquez, M. Silly, V. Derycke, V.Yu. Aristov, P. Soukiassian, C. Ottaviani, M. Pedio and P. Perfetti, Mat. Sci. For. 457 (2004) 399.

8 F. Amy and Y.J. Chabal, J. Chem. Phys. 119 (2003) 6201.

9 P. Soukiassian, Appl. Phys. A 82 (2006) 421-430; and references therein.

10 V. Aristov, G. Le Lay, P. Soukiassian, K. Hricovini, J.E. Bonnet, J. Oswald and O. Olsson, Europhys. Lett. 26 (1994) 359; W. Chen, A. Kahn, P. Soukiassian, P.S. Mangat, J. Gaines, C. Ponzoni and D. Olego, J. Vac. Sci. Technol. B 12 (1994) 2639.

11 R. di Felice et al., Phys. Rev. Lett. 94 (2005) 116103; F. de Brito Mota et al., J. Phys. Cond. Mat. 17 (2005) 4739; ibid 18 (2006) 7505; Hao Chang et al., Phys. Rev. Lett. 95 (2005) 196803; Xiagyang Peng et al., Phys. Rev. B 72 (2005) 013793.

12 M. Silly, C. Radtke, H. Enriquez, P. Soukiassian, S. Gardonio, P. Moras and P. Perfetti, Appl. Phys. Lett. 85 (2004) 4893.
13 J. Roy, V.Yu. Aristov, C. Radtke, P. Jaffrennou, H. Enriquez, P. Soukiassian, P. Moras, C. Spezzani, C. Crotti and P. Perfetti, Appl. Phys. Lett. 89 (2006) 042114.

DYNAMICS OF DIMERS AND ADATOMS AT SILICON AND GERMANIUM SURFACES

G. Le Lay[1*],V. Yu Aristov[2], F. Ronci[3], S. Colonna[3] and A. Cricenti[3]

[1]*CRMCN-CNRS, Campus de Luminy, Marseille Cedex 20 and Université de Provence, Marseille, France, [2]Institute of Solid State Physics, Russian Academy of Sciences, Chernogolovka, Moscow Distr., Russia, [3]Istituto di Struttura della Materia del CNR, via Fosso del Cavaliere 100, I-00133, Rome, Italy. *lelay@crmcn.univ-mrs.fr*

Abstract: Asymmetric dimers at the bare silicon (or germanium) (100) surfaces undergo a rapid flip-flop motion above an order-disorder phase transition temperature. We discuss, here, an intriguing controversial issue, namely the possibility of a kind of re-entrant dynamical behavior at very low temperatures. We further show that metal adatoms (typically, tin adatoms) adsorbed on the (111) surfaces behave also dynamically. The toolkit to address these questions is a synergetic combination of local probes and synchrotron radiation techniques.

Key words: surface, dynamics, dimers, adatoms, silicon, germanium

1. INTRODUCTION

The emergence of scanning proximal probes in the 80s in parallel with the development of synchrotron radiation light sources and advances in computer science have given a very strong impetus to surface studies and especially give birth to what is now called nanosciences. In this respect, the silicon (100) surface of paramount practical importance is certainly the most studied object. Yet, as we will see in section 2, open questions still remain, especially concerning its ground state structure and, eventually, the dynamical behavior of its constitutive moieties. Recently, metal adsorbed (111) surfaces of elemental semiconductors have also shown surprising dynamical features. We will address this question through typical examples

329

Vasili Tsakanov and Helmut Wiedemann (eds.), Brilliant Light in Life and Material Sciences, 329–337.
© 2007 *Springer.*

in section 3. In all cases the synergetic use of local probes and advanced synchrotron radiation spectroscopic and structural methods, combined with detailed simulations, have revealed these astonishing behaviors and eventually give the clue for their understanding.

2. GROUND STATE STRUCTURE AND DYNAMICS AT SI,GE(100) SURFACES

Until recently the atomic arrangements at the bare Si,Ge(100) surfaces were considered as well understood with the formation of asymmetric (buckled) dimers to lower the surface energy by reducing the number of dangling bonds. The c(4x2) reconstruction, due to an "antiferromagnetic"-like ordering of these buckled dimers, is seen in scanning tunneling microscopy (STM) images as an out-of-phase zig-zag arrangement. This superstructure is nearly degenerated with the p(2x2) reconstruction, displaying, instead, an in-phase zig-zag configuration (see Fig. 1).

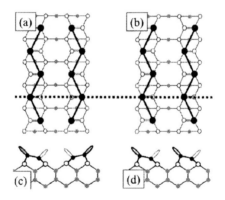

Figure 1. Top and side views of the structural models of the c(4x2) (a-c) and p(2x2) (b-d) reconstructions of the Si(100) surface (adapted from Fig. 6 of Ref. 8).

Past an order-disorder phase transition temperature, about 150 K for Si(100), straight dimer rows, where the dimers appear symmetric, build up the 2x1 reconstruction, typically observed at room temperature (RT). This is due to the thermally activated flip-flop motion of the buckled dimers, which interchange rapidly their two possible degenerate orientations. With asymmetric dimers mapped to an Ising spin model, the c(4x2) to 2x1 phase transition corresponds to an anti-ferromagnetic to paramagnetic transition. Since years 2000, diverse experiments carried out at very low temperatures

(LT) on the Si(100) surface, i.e., below about 40 K, have been indicative of, possibly, a novel phase transition to a new 2x1 phase, where the dimers may appear again symmetric in STM imaging (see Fig. 2 and, for a recent review, see[1] and references therein).

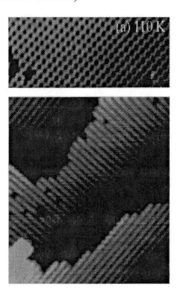

Figure 2. a) STM topographs of the Si(100) c(4x2) structure at 110 K and b) of the p(2x1) one at 5 K (adapted from Fig. 2 of Ref. 8).

In low energy electron diffraction (LEED) observations, carried out down to 24 K, the quarter order spots gradually fade, which, eventually, could corroborate a transition toward a LT 2x1 new phase. It could correspond to a kind of re-entrant dynamical flipping of the asymmetric dimers, or, instead, could indicate the presence of static, symmetric dimers, a possibility, eventually supported by correlated calculations on cluster models[2]. At variance, down to 30 K, the high-resolution (HR) synchrotron radiation (SR) photoelectron spectroscopy (PES) Si 2p core level (CL) spectra preserve their identity. They especially comprise clear signatures of the up and down atoms within each buckled dimers, with essentially the same intensities as for the frozen-in c(4x2) phase at 100-140 K (see Fig. 3). Since, PES is generally considered as a non-invasive probe, the author concludes that the c(4x2) structure is the ground state one[3].

The origin of symmetric images in STM is assigned to extrinsic effects on the buckled dimers by tip-surface interactions and/or by the tunneling current. Similarly, a current-injection effect is considered to perturb the actual structure for LEED experiments, below about 40 K[4]. To avoid such

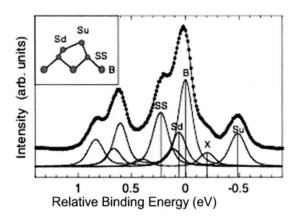

Figure 3. High resolution Si 2p core level spectrum from the Si(100) surface taken at 30 K. (adapted from Fig. 9 of Ref. 3).

current effects, non-contact atomic force microscopy (NC-AFM) studies have been recently performed.

Experiments carried out at 5 K, for weak tip-surface interactions conditions, reveal the characteristic zig-zag arrangement of the c(4x2) structure (see Figure 1a of Ref. 5). *Ab initio* density functional theory (DFT) simulations with a sharp silicon tip terminated by a single dangling bond at its apex indicate that at small frequency shifts, that is, for weak tip-surface interactions, the surface reconstruction should be determined easily[6]. Hence, the static c(4x2) reconstruction is considered as the ground state. Instead, at closer approach to the surface at large frequency shifts in the strong interaction regime (typically: $\Delta f = -40$ Hz), the tip flips a surface dimer when positioned close to its lower atom through the formation of a chemical bond (note that once a chemical bond is formed, either with the lower or upper atom, further approach of the tip induces the dimer into a symmetric configuration). Thus, at very low temperatures (T ≤70 K), the imaging process is dominated by tip induced dimer flip events, which results in a permanent deformation of the surface, giving an "apparent p(2x1) *symmetric* phase" (see left panel of Fig. 5 of Ref. 6). As a matter of fact, this is not really the case for the imaging conditions of figure 1c of ref. 5 ($\Delta f = -30$ Hz): perusal at this topographic image reveals a systematic asymmetry, which we have quantified by intensity measurements along and perpendicular to the dimer rows, as shown in Fig. 4. Actually, a "ferromagnetic"-like p(2x1) arrangement is observed in which all dimers in each row are buckled in the same way, with a small, but significant, asymmetry (the intensity difference between the lower and the upper atoms being of the order of 10%). This p(2x1) *slightly asymmetric* phase cannot be produced by tip induced flips. Instead, it corresponds rather well with the simulated image, calculated at $\Delta f = 32$ Hz, in conditions where the tip does not come "so close that a bond is formed with the tip Si apex atom and the dimer is induced to flip", for the asymmetric (i.e., "ferromagnetic"-like) static p(2x1) phase (see Fig. 3 of Ref. 6). At intermediate frequency shifts ($\Delta f = -22$ Hz, see Fig. 1b of Ref. 5) a surface phase with bright lines along dimers rows and flicker noise is observed. The authors of ref. 5 emphasize that "appearance of the flicker noise strongly suggests that surface dimers are frequently flipped by the tip-surface interaction". We believe that this may indicate that, actually, the surface converts between the c(4x2) and the new, slightly asymmetric, p(2x1) phases. Yet, at this stage, according to the detailed calculations, no chemical bond has been formed and, thence, no tip-induced event should occur. In any case, in conditions of frequency shifts below 32 Hz, the NC-AFM topographs should reflect real atomic structures. As a consequence, we believe that, presently, the question of the ground state geometry, whether c(4x2) or just slightly asymmetric p(2x1), remains open. We hope to solve this issue in a near future by HR-SR Si 2p core-level spectroscopy measurements carried out at 5K: eventually the CL line shapes for these two structures should be significantly different, allowing for a clear assignment. Finally, here, we want to stress that dimer dynamics at semiconductor surfaces is not restricted to Si(100) and Ge(100) single crystals; it has also been evidenced at other dimerized surfaces like, e.g., β-SiC(100)[7], α-Sn(100) and Ga-terminated GaAs(100) surfaces (see[8] and references therein).

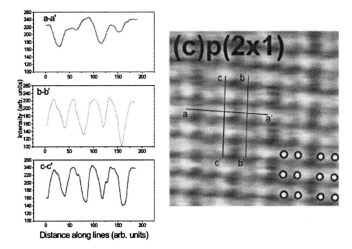

Figure 4. Intensity profiles along indicated segments of the NC-AMF 2x1 image of the Si(100) surface acquired at 5 K (adapted from fig. 1c of ref. 5).

3.   DYNAMICS AT METAL-ADSORBED SI,GE(111) SURFACES

Few years ago this dynamics concept has been extended to adsorbate covered surfaces, namely, to the Sn or Pb induced √3x√3 reconstructions (in short √3α) at 1/3 monolayer (ML) metal coverage on Ge(111) and Si(111) (see Fig. 5)[9-11]

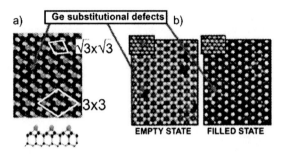

Figure 5. Geometrical model and STM images of the Sn/Ge(111) surfaces at LT and at RT (insert) (from Fig. 10 of Ref. 9).

STM at RT temperature shows nearly perfect surfaces where all adatoms, which occupy so-called threefold-hollow (T_4) sites, appear identical (see[12] and references therein). Yet, surprisingly, the Sn 4d shallow CL has a two-component line-shape, reflecting tin in two different environments, structural and/or electronic. That is, the Sn 4d CL in the √3α reconstructions possesses a high binding energy (HBE) and a low binding energy (LBE) component. This two-component scenario is valid for Sn or Pb on Ge or Si(111), yet, for Sn, the Sn 4d (HBE:LBE) intensity ratios are inversed[8], being 1:2 for Sn on Ge(111) and 2:1 for Sn/Si(111)[9]. Below – 50°C, a reversible √3α ⇔ 3x3 symmetry breaking phase transition (where the tin adatoms appear in STM images in two different states in the LT 3x3 phase), was discovered ten years ago at the Sn/Ge(111) surface[13]. The origin of this intriguing phase transition has been a matter of intense discussion. Yet, it was the analysis of the HR synchrotron radiation Sn 4d CL spectra from both phases that gave the clue. Basically, these CLs remain identical (BE positions of the two components and their intensity ratios), besides a slight narrowing at LT due to reduced phonon broadening (see Fig. 6). We proposed a dynamical model, later supported both theoretically and through different experiments by several groups (see[8] and references therein).

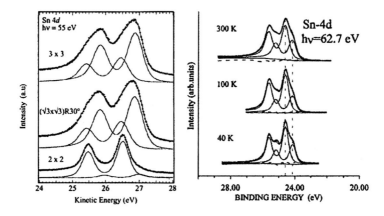

Figure 6. HR-SR Sn 4d core levels of Sn/Ge(111) at LT (3x3 phase) and RT (Ц3a phase and low coverage 2x2 phase) and Sn/Si(111) Ц3a phase at the indicated temperatures.

This order-disorder scenario bears strong analogy to the behavior of dimers at Si,Ge(100) surfaces. Namely, from the static, corrugated (two height levels), LT 3x3 structure, above the phase transition temperature, the Sn adatoms would oscillate rapidly between the two non-equivalent vertical positions, displaying, for slow probes like STM, a time-averaged, apparently flat, √3α reconstruction. This oscillating behavior has just been directly observed by detecting stepped variations between two levels upon recording tunneling current traces at different temperatures during LT STM measurements (see Fig. 7)[14].

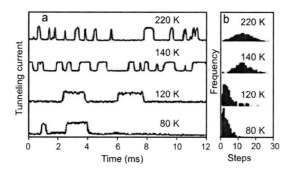

Figure 7. STM tunneling current traces and step histograms at diverse temperatures for Sn/Ge(111).

6400 current traces were recorded on top of the Sn adatoms (in spectroscopy mode with the feedback loop off) over a 80x80 grid on every 10x10 nm^2 STM image. The location of the oscillating Sn adatoms on each STM image has been identified upon plotting σ-maps showing the value of the standard deviations of all current traces; the clear correlation between left and right sides of Fig. 8 demonstrates that at 140 K, all the Sn adatoms effectively oscillate as expected[14].

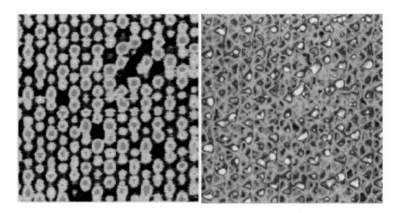

Figure 8. Left: STM image obtained at 140 K for Sn/Ge(111). Right: corresponding s-map.

Although such a behavior has not yet been identified for Sn/Si(111) the symmetric two-components Sn 4d CLs is a clear indication that a parallel behavior exist also in this case, although the transition temperature must be much lower.

4. SUMMARY AND CONCLUSION

To summarize, we have addressed the question of the ground state structure of the most utilized Si(100) surface. Evaluation of recent results shows that a re-entrant dynamical flipping of the asymmetric dimers seems unlikely at very low temperature. Instead, either the static c(4x2) reconstruction or a novel, just slightly asymmetric, 2x1 phase appear possible. If really a non-intrusive spectroscopic tool, high-resolution synchrotron radiation measurements of the Si 2p core-level at 5 K may give the final answer. Besides, we have further shown that surprising dynamical effects also take place, past a certain phase transition temperature, at some metal adsorbed semiconductor surfaces, typically, for the Sn/Ge(111) system.

REFERENCES

1. T. Uda et al., Prog. Surf. Sci. 76, 147 (2004)
2. P. Bokes, I. Stich and L. Mitas Chem. Phys. Lett. 362, 559 (2002)
3. Yoshinobu, Prog. Surf. Sci. 77, 37 (2004)
4. S. Mizuno, T. Shirasawa, Y. Shiraishi and H. Tochihara Phys. Rev. B 69, 241306(R)
5. Yan Jun Li et al., Phys. Rev. Lett. 96, 106104 (2006)
6. L. Kantorovich and C. Hobbs, Phys. Rev. B 73, 245420 (2006)
7. V. Yu. Aristov, L. Doulliard, O. Fauchoux and P. Soukiassian Phys. Rev. Lett. 79, 3700 (1997)
8. M.E. Dávila, J. Avila, M.C. Asensio and G. Le Lay, Surf. Rev. Lett. 10, 981 (2003)
9. M. Göthelid, M. Björkqvist, T.M. Grehk, G. Le Lay and U.O. Karlsson Phys. Rev. B 52, R14352 (1995)
10. G. Le Lay et al., Appl. Surf. Sci. 123/124, 440 (1998)
11. J. Avila et al., Phys. Rev. Lett. 82, 442 (1999)
12. M.E. Dávila, J. Avila, M.C. Asensio and G. Le Lay, Surface Sci., in press
13. J.M. Carpinelli, H.H. Weitering, E.W. Plummer and R. Stumpf, Nature 381, 398 (1996)
14. F. Ronci, S. Colonna, S.D. Thorpe, A. Cricenti and G. Le Lay, Phys. Rev. Lett. 95, 156101 (2005)

IMPURITY STATES IN QUANTUM WELL WIRES AND QUANTUM DOTS WITH COATING

A.A. Kirakosyan[1], A.Kh. Manaselyan[1], M.M. Aghasyan[2]
[1]Yerevan State University,Yerevan, Armenia [2]CANDLE, Yerevan, Armenia

1.   INTRODUCTION

Many researchers are interested in quazi-one-dimensional (Q1D) and quazi-zero-dimensional (Q0D) heterostructures because of the scientific aspects of the phenomena and the extraordinary possibilities of numerous applications[1-3]. The properties of these kind of structures are due to their geometrical sizes and forms, and their component characteristics. As a consequence, the electron gas topology for nanoheterostructures becomes as a new degree of freedom[4-7].

The impurity states in low-dimensional semiconductor heterostructures have been a subject of extensive investigation in basic and applied research[8-10]. Thus, an understanding of the nature of impurity states in mentioned systems is important, since the presence of impurities can dramatically alter the properties and performance of a quantum device.

In designing the semiconductor heterostructures, it is important to take into account the difference of its dielectric constants (DC). As it is known in most semiconductors the Coulomb interaction of charge carriers with the impurity centre is reduced, and the energy of binding states has characteristic values of about a few meV[9].

In low-dimensional systems the interaction between charged particles increases with the decreasing of characteristic dimensions of the system, since the field of charges in the surrounding medium begins to play a

Vasili Tsakanov and Helmut Wiedemann (eds.), Brilliant Light in Life and Material Sciences, 339–348.

marked role. If the DC of the environment is less than the DC of the system, the interaction becomes stronger than in the uniform system. In the size-quantized semiconductor layer the mismatch of the DCs is taken into account in the works[11-14].

Many papers are devoted to the study of impurity states in quantum well wires (QWR) and the quantum dots (QD), but in most of them the DC mismatch inside and outside the QWR or the QD is neglected[4,5,7,15-25]. The effect of DC mismatch *GaAs* rectangular quantum wire surrounded by $Ga_{1-x}Al_xAs$ on shallow donor impurity states in the case of both finite and infinite potential barrier is studied in Ref.[26].

In this paper we report the calculation of the binding energy of a hydrogenic impurity in the circular semiconductor wire and spherical quantum dot with the coating in the environment surrounding the system. The calculations are done with regard to DC mismatches of wire (dot), coating and surrounding medium and its effect on binding energy of the impurity centre located on the wire axis (in the centre of the sphere) is considered. The effect of the magnetic field on the binding energy with presence of DC mismatch is studied as well.

2.   IMPURITY CENTRE POTENTIAL IN QWR

Consider a system consisting of the semiconducting wire of radius R_1, with the DC χ_1, having the coating of radius R_2 and DC χ_2, immersed in the infinite environment with the DC χ_3. Solving the Poisson equation for the above-mentioned nonhomogeneous system taking into account standard continuity conditions on the interfaces of "wire-coating" ($r = R_1$) and "coating-environment" ($r = R_2$) we get the expression for the potential of impurity centre as:

$$\varphi(r,z) = \frac{2e}{\pi\chi_1} \int_0^\infty dt \cos(tz) \begin{cases} K_0(tr) + N_1 I_0(tr), & r < R_1, \\ N_2\left[K_0(tr) + N_3 I_0(tr)\right], R_1 \le r \le R_2, \\ N_4 K_0(tr), & r > R_2, \end{cases} \quad (1)$$

where $I_m(tr)$ and $K_m(tr)$ are the modified Bessel functions of the second and third kinds of m th order, respectively,

$$N_1 = \frac{\gamma_1 K_1(tR_1)A_3 + K_0(tR_1)A_2}{\gamma_1 I_1(tR_1)A_3 - I_0(tR_1)A_2}, N_2 = \frac{\gamma_1 A_1}{tR_1[\gamma_1 I_1(tR_1)A_3 - I_0(tR_1)A_2]},$$

$$N_3 = \frac{1}{A_1}(\gamma_2 - 1)K_1(tR_2)K_0(tR_2), N_4 = \frac{\gamma_1\gamma_2}{t^2 R_1 R_2[\gamma_1 I_1(tR_1)A_3 - I_0(tR_1)A_2]},$$

$$(2)$$

$$A_1 = \gamma_2 I_1(tR_2)K_0(tR_2) + I_0(tR_2)K_1(tR_2),$$
$$A_2 = (\gamma_2 - 1)K_1(tR_2)K_0(tR_2)I_1(tR_1) - K_1(tR_1)A_1,$$
$$A_3 = (\gamma_2 - 1)K_1(tR_2)K_0(tR_2)I_0(tR_1) + K_0(tR_1)A_1,$$

$$(3)$$

$$\gamma_1 = \frac{\chi_1}{\chi_2}, \ \gamma_2 = \frac{\chi_2}{\chi_3}$$

In the case of a homogeneous system ($\chi_1 = \chi_2 = \chi_3$), using Eq. (2)-(3) and equation $K_0(x)I_1(x) + K_1(x)I_0(x) = 1/x$[27], we get the well-known expression for the Coloumb potential in a dielectrically homogeneous medium. Note that to the three different cases: $\chi_1 = \chi_2 \neq \chi_3$, $\chi_1 \neq \chi_2 = \chi_3$, $\chi_1 \neq \chi_2 \neq \chi_3$ at $R_2 \to \infty$, corresponds the same physical situation: wire in the infinite environment with a different DC. In the above considered cases, from (1)-(3) follow the known expression for the potential of point charge in the wire ($r \leq R$) and surrounding environment ($R \leq r < \infty$) (see e.g., Ref.[28]).

3. IMPURITY CENTRE POTENTIAL IN QD

Solving the Poisson equation in the sphere, coating and surrounding medium, and using the boundary conditions at the surfaces "sphere-coating" ($r = R_1$) and "coating-surrounding medium" ($r = R_2$), we derive the expression for the potential energy of an electron in the considered system:

When $\chi_1 = \chi_2 = \chi_3$, from Eq. (4) follows the potential energy expression for an electron in the field of a Coulomb centre in a dielectrically homogeneous medium. The terms proportional to $(1 - \chi_1/\chi_2)$ and $(1 - \chi_2/\chi_3)$ take into account the difference of DC of the coating layer and surrounding medium. It is necessary to note that $U(r)$ undergoes the greatest change at $\chi_3 = 1$, i.e., when QD is in vacuum.

$$U(r) = \begin{cases} -\dfrac{e^2}{\chi_1 r} + \dfrac{e^2}{\chi_1 R_1}\left(1 - \dfrac{\chi_1}{\chi_2}\right) + \dfrac{e^2}{\chi_2 R_2}\left(1 - \dfrac{\chi_2}{\chi_3}\right), & r < R_1, \\[3mm] -\dfrac{e^2}{\chi_2 r} + \dfrac{e^2}{\chi_2 R_2}\left(1 - \dfrac{\chi_2}{\chi_3}\right), & R_1 \le r \le R_2, \\[3mm] -\dfrac{e^2}{\chi_3 r}, & r > R_2. \end{cases} \tag{4}$$

4.   DISCUSSION OF RESULTS

In numerical calculations carried out for the *GaAs* wire and dot coated by $Ga_{1-x}Al_xAs$ following values of parameters have been used[29]: $E_R = 5.2\text{meV}$, $a_B = 104\text{Å}$, $m_1 = 0.067m_0$, $m_2 = (0.067 + 0.083x)m_0$ (m_0 is the free electron mass) and $V_0 = 1.247xQ_e\,\text{eV}$ ($Q_e = 0.6$ is the conduction-band discontinuity fraction) for the concentration x within the limits $0 \le x \le 0.45$. The calculations are made in the frame of staircase infinitely deep (SIW) potential well model[30], using the expressions obtained in[30–33] for the ground state wave functions and the binding energy both in QWR and QD and in the absent and with present of a homogeneous magnetic field. In the calculations we also neglect the role of the Ã-Õ mixing, which in the $GaAs$ - $Ga_{1-x}Al_xAs$ systems begins to play a decisive role for the values $R < 50\text{Å}$ and $x > 0.5$[34].

In Fig.1 the impurity binding energy dependences on the wire (A) and the dot (B) radius for various values of alloy concentration x and DC χ_3 are presented for $R_2 = a_B$. From the comparison of curves it follows that as the alloy concentration x increases the maxima of the curves shift to the wire axis (A) and to the dot centre (B). The increase of binding energy is conditioned by the decrease of the electron localization region as a consequence of increasing potential barrier height at the border of the wire or the dot and the coating, and the strengthening of the system inhomogeneity as a consequence of coating and surrounding environment DC changes.

Curves 1 ($x = 0,1$) and 3 ($x = 0,3$) correspond to the model calculations for a fully uniform system with the DC $\chi = 13,18$.

From the comparison of curves 2 and 1 it follows that at $\chi_3 = 10,06$ (*AlAs*) the change of alloy concentration from 0,1 to 0,3 (the DC of coating decreases by about 5%) increases the binding energy by 27% in the wire and by 36.9% in the dot.

Because of small linear dimensions of the system ($R_1 \leq R_2 \sim a_B$) the impurity centre field is concentrated out of the wire and the dot, essentially in the surrounding environment, so the DC changes of environment have a considerable effect on binding energy. If the system is in a vacuum (curves 1, 2 (a)), then $\Delta\chi_3 = 9,06$ and relative change of binding energy at $x = 0.3$ equals 2.8, and at $x = 0.1$ is about 2.24. The relative change of the binding energy for the dot is 3.57 at $x = 0.3$ and is about 4.5 at $x = 0.1$. Thus the comparison of binding energy values for the wire and the dot with the same parameters shows a considerable strengthening of the role of size-quantization in the 0D system. Note that the decrease of the relative change of binding energy is conditioned by the electron removal to the wire axis, at the same time, away from the environment border. As the wire and the dot radius increases the effect of the DC mismatch on binding energy decreases, and the minima at $R_1 \approx 0.9a_B$, caused by the effective mass mismatch are smoothed.

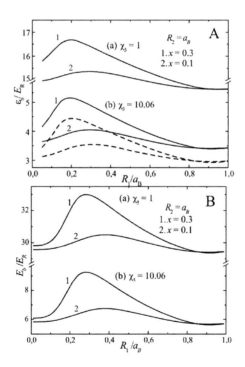

Figure 1. The impurity centre binding energy dependence on wire (A) and dot (B) radius ($R_2 = a_B, \chi_1 = 13.18$): (a) surrounding environment is vacuum and (b) surrounding environment is AlAs.

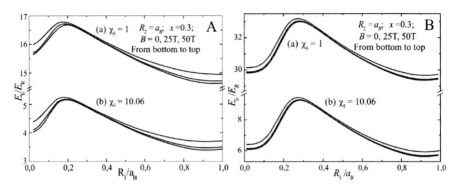

Figure 2. The impurity centre binding energy dependence on the wire (A) and dot (B) radius for various values of magnetic field at $x = 0.3$, $R_2 = a_B$: (a) surrounding environment is vacuum and (b) surrounding environment is AlAs.

The dependence of binding energy ε_b on coating radius R_2 is examined too. For a fixed value of R_1, $\varepsilon_b(R_1, R_2)$ has a maximum at $R_2 = R_1$ and then falls abruptly, tending to the value at $R_2 \to \infty$. The curves corresponding to large values of x, and bigger value of χ_2 and χ_3 (at fixed x) decrease relatively slowly.

In Fig. 2 the impurity centre binding energy dependences on the wire (A) and the dot (B) radius are presented for various values of magnetic field at $x = 0.3$, $R_2 = a_B$, $m_2 \neq m_1$. With the increase of B the binding energy increases quickly within the region $R_1 \leq 0.1a_B$ and $R_1 \geq 0.6a_B$ and slowly in the region $0.2a_B \leq R_1 \leq 0.5a_B$. Such behaviour of binding energy is the consequence of the fact that at $0.2a_B \leq R_1 \leq 0.5a_B$ the size-quantization prevails over the magnetic one. As was noted above, the minima at $R_1 \approx 0.9a_B$, caused by the effective mass mismatch in the wire or the dot and coating, are smoothed because of the systems inhomogeneity. With the increase of B the depth of minimum becomes smaller and tends to zero because the strong field localizes electron in the central region. As one can see, this minima completely vanish for the heterostructure *GaAs - Ga$_{0.7}$Al$_{0.3}$As* -vacuum at $B = 25$ T, and for *GaAs - Ga$_{0.7}$Al$_{0.3}$As - AlAs* at $B > 50$ T, since the impurity field in the main is in the surrounding medium because of the small sizes of the system. The curves corresponding to the large values of B and χ_2, χ_3 (for fixed x) decrease relatively quickly.

For the fixed value of R_1 the $\varepsilon_b(R_1, R_2)$ has a maximum at $R_2 = R_1$ and then falls abruptly, tending to the limit as $R_2 \to \infty$.

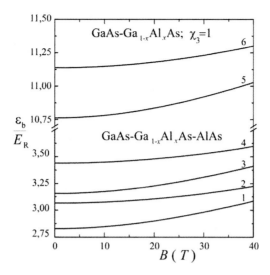

Figure 3. The impurity center binding energy dependence on magnetic field for various parameters of problem: $R_1 = 0.75a_B$, $R_2 = 1.5a_B$, $\chi_1 = 13.18$: $x = 0.1$: 1. $\chi_2 = \chi_3 = 13.18$, 3. $\chi_2 = 12.268$, $\chi_3 = 10.06$, 5. $\chi_2 = 12.268$, $\chi_3 = 1$: $x = 0.3$: 2. $\chi_2 = \chi_3 = 13.18$, 4. $\chi_2 = 12.268$, $\chi_3 = 10.06$, 6. $\chi_2 = 12.268$, $\chi_3 = 1$:

In Fig. 3 the impurity centre binding energy dependence on magnetic field is presented for various parameters values of the problem. From a comparison of curves 1 ($x = 0,1$) and 2 ($x = 0,3$), to which corresponds the model calculation $\chi_2 = \chi_3 = 13,18$, and for 3 and 4, 5 and 6, it follows that the ε_b rise velocity depending on B decrease with the alloy concentration rise, owing to increase of the potential barrier on the border between wire and the coating. Although with the increase of B the electron localization radius in the wire axis region decreases, at the fixed x, with the decreasing of χ_2 (curves 1 and 3, 2 and 4) and χ_3 (curves 3 and 5, 4 and 6) the rise velocity ε_b, increases depending on B. This is the consequence of field concentration in the main in surrounding medium.

In Fig. 4 the dependence of the impurity centre binding energy on the sphere radius is presented for various values of magnetic fields at $x = 0.3$, $R_2 = 2a_B$. According to calculations, the location of the maximum (for the given B and χ_3) is shifted to the region of large R_1, which is caused by the increase of the linear size of the localization region, conditioned by the removal of the confining barrier. The most important, however, is the effect of dielectric environment, which, although moderates because of the increased coating radius R_2, but still considerably affects the binding energy.

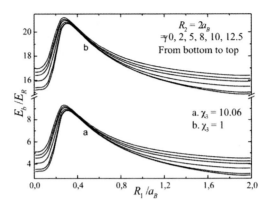

Figure 4. The impurity centre binding energy dependence on sphere radius R_1 for various values of magnetic field at $x = 0.3$, $R_2 = a_B$ ($B[\text{in T}] \approx 6\gamma$).

In Fig. 5 the dependence of the impurity centre binding energy on the magnetic field is presented for a fixed coating radius $R_2 = 2a_B$ and various parameter values of the problem. From a comparison of the curves it follows that, with the increase of sphere radius R_1, the dependence of the binding energy on the magnetic field is strengthened. Indeed, according to calculations for "weak" magnetic fields, the dependence of binding energy on γ can be presented by the expression $E_b(\gamma) = A + \gamma^2 C$, where the coefficients A and C depend on the values of R_1 and x. At fixed $x = 0.3$, and for $R_1 = 0.5a_B$, the magnetic field is "weak" when $\gamma \ll \gamma_0 = (A/C)^{1/2} = 23.6$. With the increase of R_1 the parameter γ_0 decreases, taking the value $\gamma_0 = 9.2$ for $R_1 = a_B$, and $\gamma_0 = 4.7$ for $R_1 = 1.5a_B$. Such behaviour of the binding energy is the consequence of the fact that, as was mentioned before, for $0.2a_B \leq R_1 \leq 0.7a_B$ the size-quantization prevails over the magnetic one.

5.   CONCLUSION

According to the obtained results, the dielectric constant mismatch of the wire or sphere, coating, and surrounding environment appreciably affects the binding energy of the impurity centre. This effect increases with the rise of system inhomogeneity, caused both by the increase of the alloy concentration and the decrease of the DC of the coating and environment.

The presence of a magnetic field leads to a rise of binding energy, at that effect increases with both decreasing alloy concentration and decreasing DC of the coating and environment.

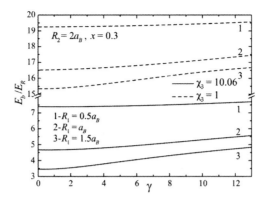

Figure 5. The impurity centre binding energy dependence on magnetic field for fixed coating radius $R_2 = a_B$, $x = 0.3$ and for various values of sphere radius R_1.

The obtained results show that the mentioned effects in the QD's quantitatively exceed the same effects in the QWW's, hence it is necessary to take them into account, especially for small radii of the QD's and relatively large values of alloy concentration.

REFERENCES

1. Optical Properties of Semiconductor Quantum Dots, ed. by U. Woggon, Springer-Verlag, Heidelberg, 1997.
2. N.V. Tkach, I.V. Pronishin, A.M. Makhanec, Fiz. Tv. Tela (russian), **40**, 557 (1998).
3. R.R.L. De Carvalho, J.R. Filho, G.A. Farias, and V.N. Fieire, Superlattices and Microstructures, **25**, 221, (1999).
4. C.L. Foden, M.L. Leadbeater, J.H. Burroughes, and M. Pepper. J. Phys. Cond. Mat., **6**, L127 (1994).
5. C.L. Foden, M.L. Leadbeater, and M. Pepper. Phys. Rev. B, **52**, R8646 (1995).
6. Jeongnim Kim, Lin-Wang Wang, and A. Zunger. Phys. Rev. B, **56**, R15541 (1997).
7. J.W. Brown and H.N. Spector. J. Appl. Phys., **59**, 1179 (1986).
8. G. Bastard, Wave Mechanics Applied to Semiconductor Heterostructures, Les editions de Physique, Cedex, France, 1988.
9. P. Harrison, Quantum Wells, Wires, and Dots. Theoretical and Computational Physics. John Wiley & Sons, New York, 1999.
10. A. Corella-Madueno, R.A. Rosas, J.L. Marin and R. Riera, Phys. Low-Dimens. Semicond. Struct. **5/6**, 75 (1999).
11. N.S. Rytova, Vestnik MGU, **3**, 30 (1967).
12. L.V. Keldish, Pisma v ZHETF, **29**, 716 (1979).
13. S. Fraizzoli, F. Bassani and R. Buczko. Phys. Rev. B, **41**, 5096, (1990).
14. J. Cen, K.K. Bajaj, Phys. Rev. B, **48**, 8061 (1993).
15. F.A.P. Osorio, M.H. Degani, and O. Hipolito. Phys. Rev. B, **37**, 1402 (1988).
16. N. Porras-Montenegro, S.T. Peres-Merchancano, Phys. Rev. B **46**, 9780 (1992).
17. D.S. Chuu, C.M. Hsiao and W.N. Mei, Phys. Rev. B **46**, 3898 (1992).

18. N. Porras-Montenegro, S.T. Peres-Merchancano, and A. Latge, J. Appl. Phys., **74**, 7624 (1993).
19. J.L. Zhu, Xi Chen, Phys. Rev. B **50**, 4497 (1994).
20. F.J. Ribeiro and A. Latge, Phys. Rev. B **50**, 4913 (1994).
21. Z. Xiao, J. Zhu, F. He, Superlatt. Micristruct. **19**, 137 (1996).
22. C. Bose, C.K. Sarkar, Physica B **253**, 238 (1998).
23. C. Bose, J. Appl. Phys. **83**, 3089 (1998).
24. C.A. Duque, A. Montes, N. Porras-Montenegro and L.E. Oliveira, Semicond. Sci. Technol. **14**, 496 (1996).
25. C.Y. Hsieh, Chin. Journ. Phys. **38**, 478 (2000).
26. Z.Y. Deng, S.W.Gu, Phys. Rev. B **48** 8083 (1993).
27. Handbook of Mathematical Functions With Formulas, Graphs and Mathematical Tables, Natl. Bur. Stand. Appl. Math. Series No. 55, edited by M. Abramowitz and I.A. Stegun, (U.S. GPO, Washington D.C., 1964).
28. D.D. Ivanenko, A.A. Sokolov, Clasical Field Theory, Moscow & Leningrad, 1951.
29. S. Adachi. J. Appl. Phys., **58**, R1 (1985).
30. M.M. Aghasyan and A.A. Kirakosyan, J. Contemp. Phys. (Armenian Ac. Sci), **34**, 17 (1999).
31. M.M. Aghasyan, A.A. Kirakosyan, Physica E **8**, 281 (2000).
32. A.Kh. Manaselyan, A.A. Kirakosyan, J. Contemp. Phys. (Armenian Academy of Sciences) **38**, 15 (2003).
33. A.Kh. Manaselyan, A.A. Kiarakosyan, Physica E, **22**, 825 (2004).
34. S. Pescetelli, A.Di Carlo, and P.Lugli. Phys. Rev. B, **56**, R1668 (1997).

VUV SPECTROSCOPY OF WIDE BAND-GAP CRYSTALS

V.N. Makhov

P.N. Lebedev Physical Institute, Russian Academy of Sciences, Leninskii Prospect 53, 119991 Moscow, Russia; Institute of Physics, University of Tartu, Riia 142, 51014 Tartu, Estonia

Abstract: The review of synchrotron radiation (SR) research in the field of solid-state luminescence spectroscopy in the vacuum ultraviolet (VUV) spectral range is presented. The historical aspect of VUV spectroscopy with SR is briefly discussed. Some highlights of VUV luminescence spectroscopy are outlined, including such fields of activity as time-resolved spectroscopy of scintillator materials, crossluminescence, VUV spectroscopy of rare earth ions in solids. The possible future developments in VUV luminescence spectroscopy of solids, such as VUV spectroscopy with free electron laser, are discussed.

Key words: Vacuum ultraviolet; time-resolved spectroscopy; wide band-gap crystals; crossluminescence; rare earth ions

1.   INTRODUCTION

Synchrotron radiation (SR) is now routinely used for different kinds of spectroscopy, including spectroscopy in the vacuum ultraviolet (VUV) spectral range. During early years of SR research the VUV spectroscopy of solids was one of the main direction of activity in many SR Laboratories. A special interest of VUV spectroscopy was to investigation of wide band-gap crystals for which the edge of intrinsic absorption lies in the VUV spectral range. The reviews of early years of activity in the field of VUV spectroscopy of solids can be found e.g. in Refs. 1,2. Except for traditional optical spectroscopy such as absorption and reflection spectra measurements the luminescence spectroscopy of solids with SR became a

349

Vasili Tsakanov and Helmut Wiedemann (eds.), Brilliant Light in Life and Material Sciences, 349–359.
© 2007 *Springer.*

powerful tool for the studies of electronic structure of solids. Luminescence spectroscopy experiments started almost simultaneously at several SR centers near the end of 60'th. However, the first paper, where the potential of SR as an excitation source for luminescence spectroscopy was demonstrated, has been published in 1970.[3] That paper was based on pioneering luminescence spectroscopy experiments performed at the storage ring TANTALUS, Wisconsin. In early 70'th several publications appeared in scientific journals and conference Proceedings in which the experiments on luminescence spectroscopy with SR have been described. One of the pioneering works with SR on VUV luminescence spectroscopy of KCl and NaCl crystals has been done at the S-60 electron synchrotron at Lebedev Physical Institute in Moscow.[4]

Nowadays the main activity in the field of solid-state spectroscopy is concentrated in the X-ray spectral region. However, there are several directions of research where VUV spectroscopy still remains an attractive tool for the studies of optical and electronic properties of solids. In the present paper some of these directions will be described and perspectives of VUV spectroscopy with SR will be discussed. Most of experimental data presented here were obtained by the author in collaboration with many research groups at the S-60 electron synchrotron in Moscow, at the SRS storage ring in Daresbury and probably at the best set-up in the world for luminescence spectroscopy with SR, SUPERLUMI,[5] which is now in routine operation at the DORIS storage ring in Hamburg.

2.   TIME-RESOLVED SPECTROSCOPY OF SCINTILLATOR MATERIALS

An important aspect of luminescence spectroscopy with SR is time resolution, based on the pulsed nature of SR. A lot of work has been done during recent years on time-resolved spectroscopy with SR of different kinds of scintillator materials (see, e.g. Refs. 6-8). This kind of activity was driven mainly by the needs in the fast and efficient scintillators for applications in high-energy physics, nuclear medicine and security systems.

It is obvious that for the better time resolution of a scintillation detector the scintillation pulse should be short, i.e. emission decay time of the luminescence center in the scintillator should be small. For Ce^{3+} doped scintillator materials, which are at present widely studied and discussed as the mostly promising fast scintillators, the decay time of emission due to allowed $5d - 4f$ transitions lies in the nanosecond range. The decay time is shorter for wide band-gap materials because of shorter emission wavelengths and accordingly faster decay (in rough approximation, decay time is proportional to the third power of wavelength).

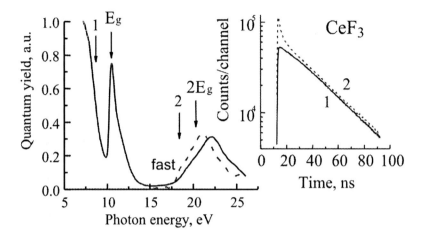

Figure 1. Excitation spectra of fast component and time-integrated luminescence (l = 300 nm) from CeF3. Decay kinetics of CeF3 luminescence under excitation by photons of different energies: (1) intracenter excitation; (2) impact excitation.

So, the wide band-gap (especially fluoride) compounds doped with Ce^{3+} may be considered as very promising fast scintillator materials. However Ce^{3+} containing fluoride crystals seem always to produce less light than is expected and in many cases the real light yield is much less than the theoretical fundamental limit. This is the case, e.g., for the well-known and widely studied scintillator CeF_3. The reason of this effect is that the conventional recombination mechanism of Ce^{3+} emission centers excitation is not efficient in fluoride crystals because the high-lying position of the Ce^{3+} ground state above the valence band prevents efficient hole trapping. The dominant mechanism of Ce^{3+} excitation in such systems is so-called impact mechanism when fast photoelectrons excite directly (by impact) the Ce^{3+} ions.[9,10] The presence of this mechanism was directly demonstrated from the measurements of excitation spectra of CeF_3 emission in VUV region (Fig. 1), where the clear threshold is observed for the Ce^{3+} emission excitation at photon energies slightly below the value of two band-gaps. Besides, the studies of emission decay kinetics clearly indicated that at excitation energies just above the energy corresponding to the threshold for the impact excitation the quenching of Ce^{3+} emission due to interaction of closely spaced excitations comes into play.[6,7,10]

During past years a lot of activity was directed towards the studies of scintillation mechanisms in different scintillator materials. In many cases the deeper understanding of these mechanisms resulted in both improving the properties of the well-known scintillators such as widely used $PbWO_4$ and developing new scintillators for different applications (see, e.g., Refs. 11-15).

3.   CROSSLUMINESCENCE

One more example of successful using VUV SR spectroscopy is a series of works on the studies of a special kind of intrinsic luminescence in ionic crystals so-called crossluminescence (CL). This type of luminescence is also called as core-valence luminescence or Auger-free luminescence. All these terms are used in the literature since each of them reflects some important property of this luminescence, as will be shown below.

The scheme of energy bands and electronic transitions in BaF_2 crystal as an example describing the mechanism of CL is shown in Fig. 2. CL is due to radiative recombination of electrons from the valence band with the holes in the uppermost core band created in the crystal by VUV radiation with the energy exceeding the ionization edge of the uppermost core band. CL is observed in ionic crystals, in which the Auger-decay of the holes in the uppermost core band is energetically forbidden, i.e. if the energy separation between the tops of the valence and uppermost core bands is less than the band-gap of the crystal.

The key experiment for the explanation of the nature of this luminescence has been made at Lebedev Physical Institute on BaF_2.[16] A clear threshold for CL emission excitation has been found at an energy corresponding to the energetic distance between the top of the uppermost core band and the bottom of the conduction band. Just the presence of this threshold is the criterion of the CL existence in the particular crystal. CL is observed only in 6 binary ionic crystals (BaF_2, CsF, CsCl, CsBr, RbF, KF), but is also observed in many ternary and multi-component crystals based on these 6 binary compounds.

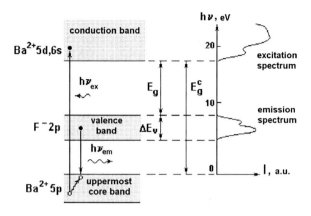

Figure 2. Simple model of crossluminescence.

The spectrum of CL consists usually of a few wide bands situated mainly in the UV and VUV regions and corresponds in the energy scale to the spectrum of transitions from the whole valence band to the top of the uppermost core band, since it is supposed that the core hole relaxes to the top of the uppermost core band before the radiative transition takes place. The characteristic decay times for CL are ~1 ns that corresponds to radiative transition probabilities for allowed electronic transitions with the energies typical for the CL spectrum. An important property of CL is its very high temperature stability, namely for the most of CL-active crystals the CL does not show thermal quenching up to temperatures near 600-700 K.

Since radiative lifetime of the core hole in CL crystals is relatively long ($\tau \sim 10^{-9}$ s), not only the complete relaxation of the electronic system takes place before radiative transition (thermalization of the core hole to the top of the uppermost core band) but also relaxation (deformation) of the lattice should occur near the core hole resulting in localization of the hole near some (cation) site in the crystal. Therefore CL should be treated as radiative recombination of the valence electrons with the localized core holes.

One manifestation of the local nature of CL is thermal broadening of CL emission bands. This broadening is well described in the framework of the model of phonon broadening for the local optical center in the crystalline environment in the limit of strong electron-lattice coupling. The FWHM for emission band of such local optical center can be represented as:[17]

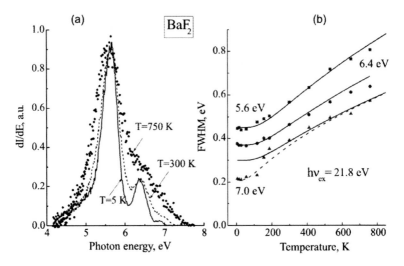

Figure 3. (a) Crossluminescence spectra of BaF2 measured at different temperatures. Spectra are normalised with respect to their maxima; (b) Temperature dependencies of the width of the bands in the crossluminescence spectrum of BaF2. Dots are experimental data, lines are calculated curves with hw = 38 meV (solid lines) and 18 meV (dashed line). Excitation energy is 21.8 eV.

$$W(T) = W_0 \times [coth(\hbar\omega/2k_BT)]^{1/2}, \tag{4}$$

where $W(T)$ and W_0 are the spectral widths of the emission band at temperatures T and $T = 0$, respectively, ω is the phonon frequency, k_B is Boltzmann constant.

This broadening is shown in Fig. 3 for the CL spectrum of BaF$_2$ crystal as an example.[18] Temperature broadening can differ for different emission bands in the CL spectrum of the same crystal and is almost independent of temperature in some cases. So, more complicated model of optical center should be applied for the description of temperature broadening for CL emission bands. However, from the comparison of the features of phonon broadening for CL and luminescence of so-called self-trapped excitons one can conclude that the lattice structure of the emitting center for CL has most probably the on-center character, i.e. lattice deformation is symmetric around the localized core hole.[19]

4. VUV SPECTROSCOPY OF RARE EARTH IONS IN SOLIDS

Another field of VUV spectroscopy actively developing in recent years is the studies of $4f^n$ - $4f^{n-1}5d$ transitions in rare earth (RE) ions doped into wide band-gap hosts. The interest to VUV spectroscopy of RE ions has reappeared recently due to commercial demand of high-efficiency phosphors for mercury-free fluorescence lamps (due to environment requirements) and plasma display panels for flat television. In both kinds of devices the phosphor is excited by VUV radiation from discharge in Xe gas. The VUV-excited phosphors should have quantum efficiency exceeding 100% to be competitive in energy efficiency with mercury lamps.

The main principle of such high-efficiency phosphor is two-photon emission due to cascade transitions. The first observation of cascade luminescence was reported many years ago for Pr^{3+} ion.[20,21] Such cascade transitions are also called as quantum cutting, or quantum splitting. The quantum efficiency for such process in Pr^{3+} was obtained to be about 140%. However, since the wavelength of the first photon in the cascade is about 400 nm, i.e. lies close to UV region, Pr^{3+} doped materials cannot be used in practice as phosphors. Although the cascade radiative transitions are also observed in many other RE ions, it has been shown that quantum efficiency of visible emission higher than 100% can be obtained only in systems doubly or triply doped with RE ions. The first proposed system was the Gd-based host doped with Eu^{3+}.[22] In such system the internal quantum efficiency of red emission from Eu^{3+} ion can be close to 200% if the phosphor is excited by a photon with the wavelength near 200 nm. However, this wavelength is far from the maximum of intensity in the spectrum of Xe discharge. Moreover, the absorption of 200 nm photons on Gd^{3+} $4f$ - $4f$ transitions is very weak, and accordingly the real external quantum efficiency of such Eu doped Gd-based phosphor is much less than 100%.

Nevertheless, the proper co-doping of Gd^{3+}-based fluoride crystals with two different RE ions can provide a phosphor with a quantum efficiency of visible emission (under VUV excitation) higher than 100%.[23] However, by now only the $LiGdF_4$ crystal co-doped with Er^{3+} and Tb^{3+} has been studied as a promising VUV-excited green phosphor for which quantum efficiency up to 130 % has been achieved.

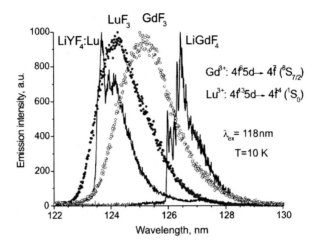

Figure 4. VUV emission spectra due to 5d – 4f transitions in Lu3+ or Gd3+ taken from LiYF4:Lu3+(5.0%), LuF3, GdF3 and LiGdF4.

Up to recently, luminescence in VUV spectral range has been detected only from three RE^{3+} ions, namely from Nd^{3+}, Er^{3+}, and Tm^{3+}.[24] However, our recent studies[25-27] have revealed that VUV luminescence at around 10 eV, which is due to $5d – 4f$ transitions, is observed also from Gd^{3+} and Lu^{3+} ions incorporated into some fluoride hosts with wide enough band-gaps (Fig. 4). The Lu^{3+} ion has no excited $4f$ energy levels available for non-radiative relaxation and subsequent quenching of Lu^{3+} $d{\rightarrow}f$ luminescence, i.e. the existence of $d{\rightarrow}f$ emission from Lu^{3+} was really expected. The reason why it has never been detected before may arise from the fact that its emission energy is the highest among all RE^{3+} ions and for spectroscopic research at such high photon energies it is necessary to use a special window-free optics and detectors in combination with VUV SR.

More surprisingly is that VUV luminescence has been detected also from several Gd-containing compounds. It was generally supposed that radiative $d{\rightarrow}f$ transitions in Gd^{3+} would be completely quenched because of the considerable number of closely spaced $4f^7$ levels in the same energy region as the $5d$ radiative states, that should enable an efficient non-radiative relaxation from the lowest Gd^{3+} $5d$ level to lower-lying $4f$ levels. However, at low temperature the VUV luminescence around 10 eV has been detected from several gadolinium-containing fluoride crystals. The possible reason for the existence of efficient radiative decay from the Gd^{3+} $5d$ level is that the competing non-radiative transitions are heavily spin-forbidden because the multiplicity of the lowest $5d$ level of Gd^{3+} is eight, whereas the excited $4f$ levels of Gd^{3+} closest to this $5d$ level are doublets or quartets.

5. VUV SPECTROSCOPY WITH FEL

Although the "conventional" VUV spectroscopy with SR is still rather attractive tool for many directions of research, the future of VUV solid-state spectroscopy will be definitely based on using undulators and free electron lasers (FEL). Undulator radiation being much brighter than SR from bending magnets provides a possibility for measurements with much smaller spots of excitation light on the samples enabling spectroscopy with high spatial resolution and spectroscopy in small high-pressure cells[28] or with magnetic field[29] as well as two-photon spectroscopy when VUV undulator radiation is used in combination with (UV/visible) laser.[30]

However, in order to investigate the dynamics of relaxation processes in solids in a full time domain, a femto-second pulsed excitation is necessary. Besides the excitation source is needed with tunable energy and with variable intensity in a large dynamic range. In the VUV spectral range, only FEL radiation is suitable for this purpose. First experiments on luminescence spectroscopy with VUV-FEL at HASYLAB have been already performed.[31] Those experiments have demonstrated the possibility of two-photon excitation of CL in BaF_2 as well as the effect of shortening for Ce^{3+} $5d$ – $4f$ emission from YAG:Ce crystal under high-power excitation due to the interaction of closely spaced excited Ce^{3+} ions. Very promising direction of research with FEL seems to be pump‑ probe experiments in the femto-second time domain for the studies of relaxation phenomena in solids.

ACKNOWLEDGEMENTS

The author would like to thank all co-workers from various Institutions for fruitful collaboration when performing joint experiments at different SR sources. The support by RFBR Grant 05-02-17306 of experiments on VUV spectroscopy of RE ions is gratefully acknowledged.

REFERENCES

1. *Vacuum Ultraviolet Radiation Physics*, edited by E.E. Koch, R. Haensel, and C. Kunz, (Pergamon-Vieweg, Braunschweig, 1974).
2. *Synchrotron Radiation – Techniques and Applications, Topics in Current Physics*, edited by C. Kunz (Springer, Berlin, Vol. 10, 1979).

3. W.M. Yen, L.R. Elias, and D.L. Huber, Utilization of near and vacuum ultraviolet synchrotron radiation for the excitation of visible fluorescence in ruby and MgO:Cr^{3+}, *Phys. Rev. Lett.* **24**, 1011 (1970).

4. S.N. Ivanov, E.R. Il'mas, Ch.B. Lushchik, and V.V. Mikhailin, Photon multiplication in KCl and NaCl crystals, *Sov. Phys.– Solid State* **15**, 1053 (1973).

5. G. Zimmerer, Status report on luminescence investigations with synchrotron radiation at HASYLAB, *Nucl. Instrum. & Meth. A* **308**, 178 (1991).

6. V.N. Makhov, Time-resolved luminescence studies of fast scintillators using synchrotron radiation, *Proc. Int. Workshop "Crystal 2000" on Heavy Scintillators for Scientific and Industrial Applications, Chamonix, September 22-26, 1992,* "Editions Frontieres", p.167.

7. I.A. Kamenskikh, M.A. MacDonald, V.N. Makhov, V.V. Mikhailin, I.H. Munro, and M.A. Terekhin, Fast crystalline scintillators for high counting rate X-ray detectors, *Nucl. Instrum. & Meth. A* **348**, 542 (1994).

8. V.V. Mikhailin, SR study of scintillators, *Nucl. Instrum.& Meth. A* **448**, 461 (2000).

9. Yu.M. Aleksandrov, V.N. Makhov, and M.N. Yakimenko, Impact excitation of impurity centers in LaF$_3$ crystals activated with rare-earth elements, *Sov. Phys. – Solid State* **29**, 1092 (1987).

10. C. Pedrini, B. Moine, D. Bouttet, A.N. Belsky, V.V. Mikhailin, A.N. Vasil'ev, and E.I. Zinin, Time-resolved luminescence of CeF$_3$ crystals excited by X-ray synchrotron radiation, *Chem. Phys. Lett.* **206**, 470 (1993).

11. I.N. Shpinkov, I.A. Kamenskikh, M. Kirm, V.N. Kolobanov, V.V. Mikhailin, A.N. Vasil'ev, and G. Zimmerer, Optical functions and luminescence quantum yield of lead tungstate, *phys. stat. sol. (a)* **170**, 167 (1998).

12. A.N. Vasil'ev, Relaxation of hot electronic excitations in scintillators: account for scattering, track effects, complicated electronic structure, *Proc. Fifth Int. Conf. on Inorganic Scintillators and Their Applications SCINT'99, Moscow, August 16-20, 1999 (Moscow State University Press, 2000),* p. 43.

13. V.N. Makhov, J.Y. Gesland, N.M. Khaidukov, N.Yu. Kirikova, M. Kirm, J.C. Krupa, M. Queffelec, T.V. Ouvarova, and G. Zimmerer, VUV scintillators based on d-f transitions in rare earth ions, *Proc. Fifth Int. Conf. on Inorganic Scintillators and Their Applications SCINT'99, Moscow, August 16-20, 1999 (Moscow State University Press, 2000),* p. 369.

14. I. Kamenskikh, E. Auffrey, N. Gerassimova, C. Dujardin, P. Lecoq, V. Mikhailin, C. Pedrini, A. Petrosyan, G. Stryganyuk, and A. Vasil'ev, LuAP:Ce and LuYAP:Ce crystals: relaxation channels competing with cerium emission, *Proc. 8th Int. Conf. on Inorganic Scintillators and their Use in Scientific and Industrial Applications SCINT 2005, Alushta, Crimea, Ukraine, September 19-23,* 2005, p. 11.

15. V. Nagirnyi, P. Dorenbos, E. Feldbach, L. Jönsson, M. Kerikmäe, M. Kirm, E. van der Kolk, A. Kotlov, H. Kraus, A. Lushchik, V. Mikhailik, R. Sarakvasha, and A. Watterich, Conduction band structure in oxyanionic crystals, *Proc. 8th Int. Conf. on Inorganic Scintillators and their Use in Scientific and Industrial Applications SCINT 2005, Alushta, Crimea, Ukraine, September 19-23,* 2005, p. 36.

16. Yu.M. Aleksandrov, V.N. Makhov, P.A. Rodnyi, T.I. Syrejshchikova, and M.N. Yakimenko, Intrinsic luminescence of BaF$_2$ excited by synchrotron radiation pulses, *Sov. Phys. – Solid State* **26**, 1734 (1984).

17. V.N. Makhov, M.A. Terekhin, I.H. Munro, C. Mythen, and D.A. Shaw, Temperature dependence of crossluminescence bandwidth, *J. Lumin.* **72-74**, 114 (1997).

18. V.N. Makhov, I. Kuusmann, J. Becker, M. Runne, and G. Zimmerer, Crossluminescence at high temperatures, *J. Electron Spectrosc. Relat. Phenom.* **101-103**, 817 (1999).

19. V.N. Makhov, V.N. Kolobanov, M. Kirm, S. Vielhauer, and G. Zimmerer, Phonon broadening of emission spectra for STE and Auger-free luminescence, *International Journal of Modern Physics B* **15**, 4032 (2001).

20. J.L. Sommerdijk, A. Bril, and A.W. de Jager, Two photon luminescence with ultraviolet excitation of trivalent praseodymium, *J. Lumin.* **8**, 341 (1974).

21. W.W. Piper, J.A. DeLuca, and F.S. Ham, Cascade fluorescent decay in Pr^{3+}-doped fluorides: achievement of a quantum yield greater than unity for emission of visible light, *J. Lumin.* **8**, 344 (1974).

22. R.T. Wegh, H. Donker, K.D. Oskam, and A. Meijerink, Visible quantum cutting in $LiGdF_4:Eu^{3+}$ through downconversion, *Science* **283**, 663 (1999).

23. R.T. Wegh, E.V.D. van Loef, and A. Meijerink, Visible quantum cutting via downconversion in $LiGdF_4:Er^{3+},Tb^{3+}$ upon Er^{3+} $4f^{11} \rightarrow 4f^{10}5d$ excitation, *J. Lumin.* **90**, 111 (2000).

24. K.H. Yang and J.A. DeLuca, VUV fluorescence of Nd^{3+}, Er^{3+} and Tm^{3+}-doped trifluorides and tunable coherent sources from 1650 to 2600 Å, *Appl. Phys. Lett.* **29**, 499 (1976).

25. M. Kirm, J.C. Krupa, V.N. Makhov, M. True, S. Vielhauer, and G. Zimmerer, High resolution vacuum ultraviolet spectroscopy of 5d-4f transitions in Gd and Lu fluorides, *Phys. Rev. B* **70**, 241101(R) (2004).

26. M. Kirm, V.N. Makhov, M. True, S. Vielhauer, and G. Zimmerer, VUV-luminescence and excitation spectra of the heavy trivalent rare earth ions in fluoride matrices, *Physics of the Solid State* **47**, 1416 (2005).

27. V.N. Makhov, J.C. Krupa, M. Kirm, G. Stryganyuk, S. Vielhauer, and G. Zimmerer, "VUV luminescence of Gd^{3+} and Lu^{3+} ions in fluoride matrices", *Russian Physics Journal*, **4** (supplement), 86 (2006).

28. J.W. Wang, R. Turos-Matysiak, M. Grinberg, W.M. Yen, and R.S. Meltzer, Mixing of the $f^2(^1S_0)$ and 4f5d states of Pr^{3+} in $BaSO_4$ under high pressure, *J. Lumin.* **119-120**, 473 (2006).

29. V.N. Makhov and V.N. Kolobanov, A conceptual design of the set-up for solid state spectroscopy with free electron laser and insertion device radiation, *Nucl. Instrum. & Meth. A* **467-468**, 1537 (2001).

30. T. Tsujibayashi, M. Itoh, J. Azuma, M. Watanabe, O. Arimoto, S. Nakanishi, H. Itoh, and M. Kamada, Two-photon spectroscopy of core excitons in barium fluoride using synchrotron radiation and laser light, *Phys. Rev. Lett.* **94**, 076401 (2005). M. Kirm, A. Andrejczuk, J. Krzywinski, and R. Sobierajski, Influence of excitation density on luminescence decay in $Y_3Al_5O_{12}:Ce$ and BaF_2 crystals excited by free electron laser radiation in VUV, *phys. stat. sol. (c)* **2**, 649 (2005).

X-RAY DIFFRACTION IMAGING OF NANOSCALE PARTICLES

Ruben A. Dilanian[1], Andrei Y. Nikulin[1], and Barrington C. Muddle[2]
[1]School of Physics, Monash University, Clayton, Victoria 3800, Australia; [2]Department of Material Engineering, Monash University, Clayton, Victoria 3800, Australia

Abstract: An experimental "momentum transfer" X-ray diffraction imaging technique is suggested and tested for nondestructive determination of nanoscale particles and clusters in three-dimensions with a spatial resolution of few nanometers. The advantage of the proposed approach is that it does not require coherent X-ray source and therefore is suitable for almost any beamline and many laboratory sources. A simple and robust quantitative technique to reconstruct an average nanoparticle randomly dispersed over a large volume within a sample should enable researchers to study nanoparticles using conventional laboratory X-ray diffraction equipment with resolution up to a few nanometers.

Key words: Nano-particles, 3-D imaging, X-ray diffraction

1. INTRODUCTION

In many fields of science the ability to visualize nanoscale objects (nanoparticles and nanoclusters) embedded in metal, ceramic or polymer materials is proving crucial to our understanding of the real structure of objects and mechanism involved in growth processes. The detailed, high-resolution, 3D study of the microstructure of the nanoparticles puts in the forefront new special requirements to experimental conditions, data acquisition, and data analysis. From this point of view, X-ray diffraction technique offer unique opportunity for non-destructive 3D characterization of nanoscale objects, such as nanoparticles, nanoclusters, and nanotubes.

Vasili Tsakanov and Helmut Wiedemann (eds.), Brilliant Light in Life and Material Sciences, 361–370.
© 2007 *Springer.*

X-ray diffraction and imaging techniques are powerful diagnostic tools for the non-destructive analysis of matter, because of the greater penetration length of X-rays. At the present time various nondestructive X-ray imaging techniques are used to determine the structure, shape and size of non-periodic objects and nanoscale crystalline materials.[1-6] Most of them adopt so-called direct or "real space" imaging technique where the data is collected in real space and the intensities are measured as a function of real space coordinates, Fig. 1 (a). X-ray diffraction microscopy, combining coherent X-ray diffraction with the oversampling approach to recovery of phase information has recently been demonstrated capable of 3D image reconstruction of non-crystalline nanostructure at nanoscale resolution.[5,6] This technique, however, requires a highly coherent X-ray source that makes a systematic study of nanoscale materials possible only by using high-end third generation synchrotrons.

This article presents an application of the Gerchberg-Saxton (G-S) technique[7-9] to reconstruct a 3D image of various specimens with dispersed nanoparticles and carbon nanotubes of the size of 10-500 nm. To perform this reconstruction we collected X-ray diffraction data from the sample as a 2D function of the diffraction angle, which is the main difference of the present approach from the previously used direct imaging methods.[10,11]

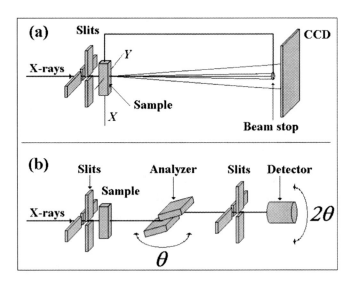

Figure 1. Sketch of the setup for an image reconstruction experiment: (a) direct imaging method; (b) momentum transfer method.

2.   MOMENTUM TRANSFER DATA ACQUISITION

An alternative solution to the aforementioned problem is to record the diffracted from the object intensity in the reciprocal (momentum transfer) space coordinates.[12,13] The methodology is based on the measurement of a high angular resolution X-ray Fraunhofer diffraction pattern. In this case the diffraction pattern of the object is measured as a 2D function of the crystal-analyzer and detector angular positions. Mapping of reciprocal space is a widely used experimental layout for obtaining 2D information about the crystal structure. For instance, such technique was used to map the 2D elastic strains in silicon crystals after ion implantation.[13] The experimental arrangement for the momentum transfer technique is shown schematically in Fig. 1 (b). The required X-ray radiation energy is selected using the primary double-crystal Si beamline monochromator (for synchrotron experiments only). Further angular collimation can be performed by a double-crystal channel-cut Si monochromator. Then the beam is spatially collimated by a pair of slits. The sample is placed downstream just after the slits. The Si crystal analyzer and a detector with a pair of collimated slits just before it are placed downstream from the sample. The resulting 2D intensity distribution is collected from the parallelogram area around the selected Bragg reflection of the crystal analyzer. In such experiments the spatial resolution is determined by conventional Fraunhofer diffraction relationship:

$$\Delta z = \frac{\lambda}{\Delta\theta},$$

where $\Delta\theta$ is the total angular aperture of the experimental data and λ is the wavelength[10,12].

The advantage of the method presented here is that it is not sensitive to the sample vibration and does not require unusual mechanical stability of the source – optics – sample – detector arrangement, unlike the real space (direct) imaging technique. Moreover, use of a single photon counting detector (e.g. APD) allows one to record data within a linear dynamic range of 8-9 orders of magnitude. The crucial central peak of the diffraction pattern is therefore readily measurable in this case. Accordingly, the intensity distribution measured as a function of angular direction in the reciprocal space does not require coherent radiation.

The major benefit of the momentum transfer X-ray diffraction imaging experimental technique is that it's suitable for almost any beamline and many laboratory sources. The method does not allow one to reconstruct shape and size of an individual particle since the diffraction pattern of objects is measured as a function of the angular direction. However, this method does allow to non-destructively and *in-situ* analyze large, over a several cubic millimeters, volume of material and determine an average shape and size of the dispersed particles and closely evaluate the distribution of the particle sizes.

3.   RECONSTRUCTION ALGORITHM

At the present time various nondestructive x-ray imaging techniques are used to determine the structure, shape and size of non-periodic objects and nanoscale crystalline materials. One of these techniques, the so-called Gerchberg-Saxton (G-S), method,[7,8] has been shown to reconstruct a two-dimension (2D) image of a scattering object from experimentally collected diffraction data. The problem of uniqueness of the reconstruction for a two-dimensional case has evolved from the work of Barakat and Newsam.[15] According to this work, in general the two-dimensional reconstruction problem has a unique solution if one exists. The idea of the G-S algorithm is that, if partial information about the magnitude of the object density, $|g(r)|$, as well as about the magnitude of the object's Fourier transform, $|F(q)|$, can be supplied, the phase information can be recovered. Later Fienup modified the G-S algorithm by using finite supports and positivity constrains in real space instead of the magnitude of the object density to reconstruct a shape of the real and positive object[8,9]. The objective of this algorithm is to produce an input whose Fourier transform's phase is the same as that of the true image. The problem is solved by an iterative approach which algorithm consists of the following four steps:

1. Fourier transform $g_k(r)$ to get a new Fourier pattern, $G_k(q)$:

$$G_k(q) = |G_k(q)| \times \exp[i\psi_k(q)] = \Im[g_k(r)].$$

2. Generate a new Fourier pattern by replacing the modulus of the resulting Fourier transform with the measured Fourier modulus (Reciprocal-space constraint):

$$G'(q) = |F(q)| \times \frac{G_k(q)}{|G_k(q)|}.$$

3. Inverse Fourier transform $G'(q)$ to get a new object function, $g'(r)$:

$$g'(r) = |g'(r)| \times \exp[i\varphi'(q)] = \Im^{-1}[G'(q_k)].$$

4. Generate a new object function, $g_{k+1}(r)$, using the real-space constraints:

$$g_{k+1}(r) = |f(r)| \times \frac{g'(r)}{|g'(r)|} \quad \text{(Gerchberg-Saxton algorithm)}$$

$$g_{k+1}(r) = \begin{cases} g'(r), r \in S \\ g_k(r) - \beta \times g'(r), r \notin S \end{cases} \quad \text{(Fienup algorithm)}.$$

Here the vector position $r = (r_x, r_y)$ represents a two-dimensional spatial coordinates and q is a spatial frequency, β is a constant feedback parameter between 0.5 and 1.0, and S is a support function. The convergence of the algorithm can be monitored by computing the error function,[8] E_k:

$$E_k = \left(\sum_{r \notin S} |g_k(r)|^2 \Big/ \sum_{r \in S} |g_k(r)|^2 \right)^{1/2}.$$

Fienup indicates the importance of positivity of the original real function, $g(r)$, and *a priori* information (constrains) for the successful reconstructing of the original image[8]. For example, for sky brightness objects is a real, nonnegative function. In most cases, however, the function $g(r)$ is a complex and the reconstruction algorithm can not be applied in form presented in this paragraph. Thus, in order to use the G-S algorithm some additional assumptions must be made. For example, Fienup demonstrated the possibility of reconstruction some special complex objects in a two-step approach by using additionally a low-resolution intensity image from a telescope with a small aperture.[9] In the case of X-ray diffraction, the complex-valued object density can be expressed by using the complex atomic scattering factor, $f_1 + if_2$,[1] where f_1 is the effective number of electrons that diffract the photons and is positive for X-ray diffraction (this assumption is note applicable in the case of neutron diffraction) and f_2 is the attenuation and should be positive for ordinary matter. It has been proposed that the positivity constraints on the imaginary part of complex-valued objects can be used as internal constraints for phase retrieval. In [3] authors suggested other positivity constraints for the complex-valued object function. The complex image (step 3) was made real by setting its phase to zero and then made positive. Strictly speaking, both algorithms utilize measurements of the attenuation of the incident beam to determine the form of the object by using direct imaging methods. When diffraction and scattering effects become appreciable, such approach generally will give unsatisfactory results.

In case of small objects, such as nanoparticles, the diffracted intensity profiles can be well approximated as a modulus squared of a Fourier transform of the complex transmission fuction[16]:

$$g(r) = \exp(i\varphi) = \exp\left\{i\frac{2\pi}{\lambda}n(r)T(r)\right\},$$

where $n(r) = \delta(r) + i\beta(r)$ is the complex refractive index, and $T(r)$ is a function of the object thickness. If the magnitude of φ is small compared to unity, $\varphi << 1$, then one can simplify:

$$g(r) \approx 1 + i\Delta\varphi \approx 1 + i\frac{2\pi}{\lambda}n(r)T(r) = g_1(r) + ig_2(r)$$

$$g_1(r) = \left[1 - \frac{2\pi}{\lambda}\beta(r)T(r)\right]$$

$$g_2(r) = \left[\frac{2\pi}{\lambda}\delta(r)T(r)\right].$$

This substitution is possible only if the following condition is satisfied:

$$T << \frac{\lambda}{2\pi|n|}.$$

For a homogenous object we can also assume that the complex refractive index, n, is a constant and its value depends only on chemical composition of the object and radiation energy[17]. Consequently, the "shape" of the real and imaginary parts of the expanded complex transmission function should be the same – the real and positive function, $T(r)$. This means that the positivity constraints[1] can be used for reconstruction of shapes of scattering objects. In this case, the real-space constraints in the G-S reconstruction process can be written as

$$T_{k+1}(r) = \begin{cases} T'(r), r \in S \\ T_k(r) - \beta \times T'(r), r \notin S \end{cases}$$

In addition, since in case of hard X-rays the imaginary part of a complex refractive index is generally much smaller than the real part, it is also very useful to deliberately use the X-rays energy just above the absorption edge of the dominated in the scattered object chemical element to be able to compare the results of reconstruction focused on either of the parts of the generally complex function.

4. RESULTS AND DISCUSSION

Fig. 2 (a) represents experimentally recorded 2D map of the diffraction intensities for Al_2O_3 nanoparticles dispersed in a polymer matrix. The spatial resolution in the reconstructed image was thus 2.5 nm. The central vertical line in the maps is the main peak due to Bragg reflection of the monochromator. The initial complex transmission function $g_o(r)$ of the nanoparticles was estimated using experimental data for the size of the average nanoparticle and a randomly distributed phase of the diffracted wave. The support function was selected as a circle with a diameter $D \approx 60$ nm. The shape of the reconstructed Al_2O_3 nanoparticle yielded the average size of the nanoparticles to be approximately 50 nm, which is in good agreement with the expected value.

Fig. 3 represents 3D rendering of a straight section of a carbon nanotube dispersed in a polymer. The angular range of the analyzer where chosen to allow the reconstruction of the average nanotube with spatial resolution of 1.0 nm. The shape of the resulting image yielded the average size (cross-section, in this case) of the carbon nanotube to be approximately 10 nm, which is in good agreement with the expected value.

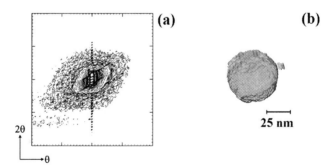

Figure 2. Image reconstruction of an Al_2O_3 nanoparticle from the experimental diffraction data: (a) experimentally recorded x-ray diffraction intensity (contour map, logarithmic scale); (b) 3D representation of an average nanoscale Al_2O_3 particle.

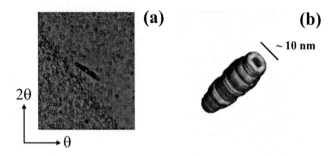

Figure 3. Image reconstruction of a carbon nanotube from the experimental diffraction data: (a) experimentally recorded x-ray diffraction intensity (logarithmic scale); (b) 3D rendering of a straight section of a carbon nanotube dispersed in a polymer matrix.

To understand how the reconstructed, using the momentum transfer method presented here, $g(r)$ function correlates with an average size and shape of nanoscale particles, let us consider the scattered intensity using the conventional concept of the so-called form-factor function.[18] The intensity diffracted from a sample containing nanoscale particles can then be written as:

$$I(q) \approx |F(q)|^2 \cdot |P(q)|^2,$$

where $F(q)$ is the structure amplitude and $P(q)$ is the form-factor function which depends on the size and shape of the particle. In the present case, the form-factor function $P(q)$ is the Fourier transform of the complex function $g(r)$, which is determined by the average shape of the scattering object. For the particular case when all particles are of the same shape and size, the reconstruction yields the actual shape of the particle. In case of differently sized particles contributing to the diffraction process, the G-S reconstruction approach presented here would yield a function of distribution of particles by their size. In some cases we can estimate the average size of the scattering objects before the G-S calculations. For spherical particles, randomly distributed inside amorphous matrix, a good approximation for the form-factor function is the spherical zeroth-order Bessel function,

$$I \approx |P(q)|^2 \approx |j_0(2qR)|,$$

where R is the radius of the particle.[18] This function has zero intensity values at points $2qR = n$ ($n = 1, 2,...$). In general, for randomly distributed particles, $P(q)$ can be replaced by the well-known Gunier approximation function,

$$I \approx |P(q)|^2 \approx \exp(-\overline{R}^2 q^2 / 3),$$

where $\overline{R}$ is the average radius of the particle.[18] It should be noted that this approximation is correct only if $q\overline{R} \leq 1$. The Gunier approximation function defines that a linear relationship exists between $\ln(I)$ and q^2, and thus a plot of $\ln(I)$ vs. q^2 obtained from experimental data can be interpolated by a linear function. The gradient of this line gives the value of $\overline{R}^3 / 3$ and, consequently, the average size of the cluster/particle can be estimated.

5. CONCLUSION

In summary, we have reconstructed 3D-images of the nanoscale particles embedded in an amorphous/polycrystalline matrix from experimental 2D data collected using momentum transfer method. The limitation of the presented method is that it is applicable only to study of structures with dispersed nanoparticles. The method does not allow one to reconstruct shape/size of individual particle. However, a simple and robust technique has a dramatic advantage for material science as it does allow non-destructive and *in-situ* analysis over large volumes of material (several mm^3) and determination of the average shape and size of the dispersed particles. It should also be emphasised here that presented method does not require coherent X-rays and can be implemented in a laboratory or at a "low-end" synchrotron.

ACKNOWLEDGMENT

The work was supported by the Australian Synchrotron Research Program (ASRP) and the Australian Research Council (ARC) Discovery grants.

REFERENCES

1. J. Miao, D. Sayre, and H. N. Chapman, Phase retrieval from the magnitude of the Fourier transforms of nonperiodic objects, J. Opt. Soc. Am. A **15**, 1662-1669 (1998).

2. J. Miao, P. Chalambous, and D. Sayre, Extending the methodology of x-ray crystallography to allow imaging of micrometer sized non-crystalline specimens, Nature **400**, 342-344 (1999).
3. K. Robinson, I. A. Vartanyans, G. J. Williams, M. A. Pfeifer, and J. A. Pitney, Reconstruction of the shape of gold nanocrystals using x-ray diffraction, Phys. Rev. Letters **87**, 195505-195508 (2001).
4. Y. Nishino, J. Miao, and T. Ishikava, Image reconstruction of nanostructured nonperiodic objects only from oversampled hard x-ray diffraction intensities, Phys. Rev. B **68**, 220101-220104 (2003).
5. S. Marchesini, H. He, H. N. Chapman, S. P. Hau-Riege, A. Noy, M. R. Howells, U. Weiertall, and J. C. H. Spence, X-ray image reconstruction from a diffraction pattern alone, Phys. Rev. B **68**, 140101-140104 (2003).
6. S. Marchesini, H. N. Chapman, A. Barty, A. Noy, S. P. Hau-Riege, J. H. Kinney, C. Cui, M. R. Howells, R. Rosen, J. C. H. Spence, U. Weiertall, D. Shapiro, T. Beetz, C. Jacobsen, A. M. Minor, and H. He, Progress in three-dimensional coherent x-ray diffraction imaging, arXiv:physics/0510032, v2 (2005).
7. J. R. Fienup, "Reconstruction of an object from the modulus of its Fourier transform". Optics Letters **3**, 27-29 (1978).
8. J. R. Fienup, Phase retrieval algorithm: a comparison, Applied Optics **21**, 2758-2769 (1982).
9. J. R. Fienup and A. M. Kowalczyk, Phase retrieval for a complex-valued object by using a low-resolution image, J. Opt. Soc. Am. A **7**, 450-458 (1990).
10. R. A. Dilanian and A. Y. Nikulin, X-ray diffraction imaging of Al_2O_3 nanoparticles embedded in an amorphous matrix, Appl. Phys. Letters **87**, 61904-161906 (2005).
11. A. Y. Nikulin, R. A. Dilanian, N. A. Zatsepin, B. A. Gable, B. C. Muddle, A. Y. Souvorov, Y. Nishino, and T. Ishikawa, Unpublished.
12. A. Y. Nikulin, A. V. Darahanau, R. B. Horney, and T. Ishikawa, High-resolution x-ray diffraction imaging of non-Bragg diffracting materials using phase retrieval x-ray diffractometry, Physica B **349**, 281-295 (2004).
13. A. Y. Nikulin, O. Sakata, H. Hashizume, and P. V. Petrashen, Mapping of two-dimensional lattice distortins in silicon crystals at submicrometer resolution from x-ray rocking-curve data, J. Appl. Cryst. **27**, 338-344 (1994).
14. C. Hammond, The basics of crystallography and diffraction. (IUCr, Oxford, 249 p., 1997)
15. R. Barakat and G. Newsam, Necessary conditions for a unique solution to two-dimensional phase recovery, J. Math. Phys **25**, 3190-3193 (1984).
16. M. Born and E. Wolf, Principles of Optics, Pergamon Press, Oxford, 1970.
17. X-ray Server, http://sergey.gmca.aps.anl.gov
18. A. Guinier, X-ray crystallographic technology, Dunod, Paris, 1956.

X-RAY SPECTROSCOPY

E. O. Filatova and A. S. Shulakov
V.A.Fock Institute of Physics, St. Petersburg State University, St. Petersburg, RUSSIA

Abstract: The hierarchy of X-ray spectroscopic methods is presented. Principles, basic and modified methods and the main information which one can extract employing one or another method, are discussed. Example of the application of X-ray reflection and emission spectroscopic analysis of Al_2O_3/Si structure synthesized by atomic layer deposition (ALD) method is presented, too.

Key words: X-ray, core levels, spectroscopic methods, valence and conduction bands, radiative and non-radiative decay.

1. INTRODUCTION

X-ray spectroscopy is one of the effective methods of the nondestructive analysis of the electronic structure, atomic concentration and chemical phase composition of materials.[1-5] Near edge spectral dependencies of the X-ray absorption coefficient and spectral distributions of the intensity in the characteristic X-ray emission bands reflect the energy distribution of the density of empty electronic states of the conduction band (CV) and occupied electronic states of the valence band (VB), respectively. X-ray absorption and emission processes have a local character (associated with hole localization in the core shell) and dipole selection rules for the transitions between the initial and the final state have been worked out. Thus the possibility to obtain the information about local and partial (allowed for certain angular momentum symmetry) density electronic states of the conduction and the valence band is appeared. Such unique information does not posses a single method.

Vasili Tsakanov and Helmut Wiedemann (eds.), Brilliant Light in Life and Material Sciences, 371–381.
© 2007 *Springer.*

The spectral resolution in the X-ray region is associated with the whole excited state lifetime τ, $\Gamma = \hbar / \tau$ that is determined by a sum of the radiative and non-radiative decay $\Gamma = \Gamma\text{rad} + \Gamma\text{nonrad}$ and spectrometer resolution.[6] Advantages offered by synchrotron radiation, such as high brightness, coherency, polarization control and photon-energy tunability[7] give a new push to the development of the X-ray Spectroscopy and now the X-ray Spectroscopy includes a whole complex of methods. The purpose of this paper is to classify all methods of X-ray Spectroscopy and to present principles, basic and modified methods as well as the main information which one can extract employing one or another method.

2.   BASIC PRINCIPLES

A high energy photon impinges on a sample and, via the photoeffect an electron from the core shell is liberated. The system pass into the excited state, which is characterized by a core-hole and photoelectron escaped into an empty final state just above the Fermi level or into the vacuum. The core-hole has an effect on the system as a huge scattering centre leading to the excitation of electron-hole pairs, phonons and plasmons. As a result a reconstruction of the electronic states arranged near the scattering centre is occurred. Subsequently the spontaneous X-ray transition through the excited intermediate state when a photon is emitted and the hole in the valence band is appeared takes place. The other channel of such core excited state decay is non-radiative Auger-process. Thus the energy of the absorbed high energy photon is transformed into the kinetic energy E_{kin} of photoelectrons and auger – electrons and into the secondary photon energy.

At this point, it is convenient to classify X-ray Spectroscopic methods taking into account the product of the interaction of the photon with the sample by following way:
1. Methods where the products of excitation are analyzed;
2. Methods where the products of relaxation are analyzed;
3. Methods where both the products of excitation and relaxation are analyzed.

Figs. 1-5 illustrate schematically the principles of the basic methods of X-ray Spectroscopy.

2.1 X-ray absorption spectroscopy (XAS)

When X-rays pass through mater, their intensity is attenuated. The relationship between the intensities before $I_o(E)$ and after $I(E)$ passing a layer of thickness d is

$$I = I_o \exp[-\mu(E)d] = I_o \exp[(-\mu/\rho)\rho d], \tag{1}$$

where ρ is the density of material, μ is the linear absorption coefficient, μ/ρ is the mass absorption coefficient, which is determined as $\mu/\rho = \Sigma \, (\mu_i/\rho_i)C_i$, where C_i are the weight concentration of the atoms making up the matter. $\mu(E)$ shows edges at certain energies which are characteristic for the absorber. The shape and position of the edge depend on the chemical bonding of the absorbing substance.

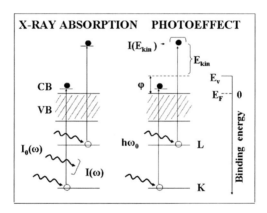

Figure 1. Energy diagram for X-ray Absorption Spectroscopy (XAS) (left) and X-ray Photoelectron Spectroscopy (PES) (right).

Fig. 1 illustrates the principle of XAS. Photon with energy $\hbar\omega$ impinges on a sample and photoabsorption takes place in a core shell with binding energy E_B (with respect to E_F). Depending on the photon energy the electron can be excited into empty states with small kinetic energy or can escape into the vacuum with positive kinetic energy ($\hbar\omega - E_B$). $\mu(E)$ shows oscillations above the edge. When $\hbar\omega \geq E_B$ the oscillating contribution in $\mu(E)$ called NEXAFS (near X-ray absorption fine structure) is appeared in the vicinity of the absorption edge. The shape and the position of NEXAFS depend strongly on the chemical bonding of the absorbing atom. When $\hbar\omega \gg E_B$ the oscillating contribution in $\mu(E)$ called EXAFS (extended X-ray absorption fine structure) is caused by interference of scattering photoelectron wave on the atoms surrounding. This structure contains information on the geometric structure (interatomic distances, coordination numbers) around the absorbing atom.

The XAS in transmission mode is a bulk probe. The alternative but surface sensitive method is X-ray quantum yield spectroscopy (XQYS). A fact of photoionization is investigated in this method. The atomic photoionisation inside the sample is followed by appearance of a budge of secondary scattered low energy electrons together with primary photo- and auge- electrons penetrated over the surface in vacuum. The main characteristic of this process is a quantum yield of external photoefffect in the current regime $æ_c(\omega) = n_e/N_o(\omega)$, where n_e is number of emitted electrons and $N_o(\omega)$ is a number of photons in primary beam.

2.2   X-ray photoelectron spectroscopy (PES)

In this method a photon interacts with a sample and via the photoeffect an electron escapes into the vacuum (Fig. 1). During this process the photon transfers all its energy to the bound electron and the identification of the element is carried out via measuring of the kinetic energy of the electrons which escape the sample without energy loss (photoelectron). Thus, if photoabsorption takes place in a core level with binding energy E_B ($E_B = 0$ at E_F) the photoelectrons can be detected with kinetic energy $E_{kin} = \hbar\omega - \varphi - E_B$ (φ is work functions) in the vacuum. The kinetic energy of the electron in vacuum is measured with respect to E_v. The energy of incoming photon can be in the soft X-ray region from 100 eV to 1000 eV or in the X-ray region more than 1 keV. The electron escape depth in this energy region is only of 10-20 Å. This means that PES registrates predominantly electrons from near the surface range. In the X-ray region the photoabsorption takes place in the core levels. Their energies depend on the atomic number and chemical state of the atom that allows carrying out the atomic chemical analysis and estimation of effective atomic charge. Thus, the important parameters to be measured in PES are the kinetic energy of the photoemitted electron. The polarization of the light is a useful property, which gives an additional opportunity for method in an angle-resolved PES experiment. The kinetic energy of the photoemitted electron and its angle with respect to the impinging light and the surface is analyzed in this case.

Photoelectron spectra allow obtaining information on the nature of excited states and on the channels of their non-radiative (Auger-effect) decay. A special interest presents Resonance PES (RPES). "Resonance" PES means that the photoemission is excited with photon energies very near to the absorption edge of a core level (index c). In this case the direct photoemission of valence band electrons (v) can interfere with Auger CVV-electrons that are emitted in a Super Koster-Kronig process. The coherent superposition photoemitted electrons and the (CVV)-Auger electrons is only possible exactly at resonance.

2.3 Inverse photoemission spectroscopy (IPES)

The radiative capture of electrons is a process studied in Inverse Photoemission Spectroscopy (IPES). In this spectroscopy, which is inverse to PES, an electron with energy E_o impinges on the sample (Fig. 2) and being captured by atom emits Bremsstrahlung which is detected.

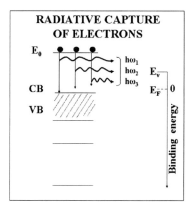

Figure 2. Energy diagram for Inverse Photoemission Spectroscopy (IPES).

In this case the electron makes a transition from initial state with energy E_o defined by the kinetic energy of the incoming electron into an empty final state E_f, that depends on the properties of the studied system. A photon with energy $(E_o - E_f)$ is emitted and detected. From the experimental point of view there are two possibilities to obtain spectra of radiative capture of electrons. If the electron energy is fixed and the energy distribution of photons is detected the technique is called IPES. If detector for photons of a fixed energy is used and the energy of the incident electrons is varied we talk about Bremsstrahlung Isochromat Spectroscopy (BIS). The information about the density of unoccupied states (CB) one can obtain with help of these methods.

In contrast to XAS the capture of electrons in a common case does not connect with creating of a strong spatial localized core-hole state. Although the dipole selection rules for the transitions between the initial and the final state worked out in this case, delocalized character both of the initial and final states of transition makes impossible to get information about energy structure of local density of unoccupied electronic states. Nevertheless, when the kinetic energy of impinging electrons is near to ionization potential of the core level the conditions for the competition between channel of direct radiative capture and other radiative processes connected with creating of localized core level excited state are appeared. Under this condition one can wait a manifestation of the effect of dynamical screening of the direct radiative capture by virtual excitations of atom, that are inducted by impinging primary electrons. The obtained spectra include the information about local electronic structure. Resonant inverse photoemission spectroscopy (RIPES) is used in this case.

The techniques of PES and IPES are complimentary to one another. PES connects the energy levels below the Fermi energy and above the vacuum level. IPES connects the energy levels above the E_F.

2.4 X-ray emission spectroscopy (XES)

X-ray normal (characteristical) emission can be described as a spontaneous emission of X-ray photons in a dipole transition between two electron states when the core-hole is filled (Fig. 3). The core hole can be created by either photon or electron beams. Thus, this method investigates the products of relaxation of interaction of photon or electron beams with substance. X-ray emission does not depend on the primary photon energy or primary energy of electron beams and reflects the distribution of the local partial density of occupied states. Dependence of the shape of the emission bands on the chemical state of the emitting atom gives an opportunity of reliable determination of the phase chemical composition of substance.

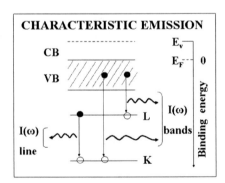

Figure 3. Energy diagram for X-Ray Emission Spectroscopy (XES).

Different modified methods of X-ray emission spectroscopy (XES) are used now for the investigation of electronic structure of solids. Depth Resolved soft X-ray Emission Spectroscopy (DRXES – depth resolved XES) is based on the energy dependence of a penetration depth of primary electron beam, applied for excitation of analyzed X-ray characteristic emission. The characteristic X-ray emission bands are created by valence band electrons spontaneously filling the vacancies produced by electron impact in the inner atomic shells.

In resonance emission, the primary radiation energy is near the ionization threshold. In this case the core electron is promoted into empty state of the conduction band and the intermediate state of X-ray emission corresponds to the final state of X-ray absorption. Under resonant excitation, a core hole may be filled by both a valence electron and the electron excited to the conduction band, which results in an elastic peak. The resonance X-ray emission often called resonance inelastic X-ray scattering (RIXS) is very useful tool for the investigation of magnetic dichroism, selective excitation of inequivalent atoms in layered compounds, band dispersion in crystals, and the electronic structure of half-metallic systems.

2.5 Auger electron spectroscopy (AES)

The excited atom can liberate its energy by radiative or non-radiative decay. Both channels of decay contain the information on the excited states of the structure. Auger-decay is non-radiative decay and is accompanied by escape of the Auger-electron into the vacuum. Fig. 4 shows different non-radiative auger-processes in which a two-hole state is produced. First index means the initial hole-state and second and third indexes show the final hole-states. In Auger Electron Spectroscopy (AES) the energy of the emitted electron is defined by the difference between the whole energy before and after transition. The energy of auger-lines does not depend on the energy of impinged photons and is defined by the sort of atoms emitted the auger-electrons. Because each element is characterized by its own auger-lines this spectroscopy can be used for the quantitative atomic chemical analysis of chemisorbed or physisorbed species.

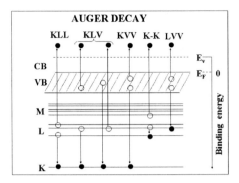

Figure 4. Energy diagram for X-Ray Auger Spectroscopy (AES). K-K means Koster-Kronig decay.

2.6 X-ray reflection spectroscopy (R)

X-ray total external reflection (specular reflection) is an integrated process and is defined by both the photoabsorption process and the polarizability of the material. The primary photon $\hbar\omega$ impinges on a sample and photoionisation takes place; locally the photon is absorbed and an electron is excited into empty states. Besides, the interaction of the photon $\hbar\omega$ with sample leads to the polarization of the substance. As a result the system comes back into the undisturbed state when the core-hole is filled and the photon is emitted (Fig. 5). The intensity of the "reflected" photons is detected. The reflection spectra R(E) exhibit "edges" due to the elements

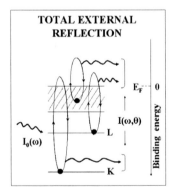

Figure 5. Energy diagram for X-Ray Reflection Spectroscopy (R).

present in the specimen and are very sensitive to the nature of the absorbing atoms, their chemical state and local coordination environment. Because the angular dependence of a penetration depth of the reflected beams, X-ray reflection spectroscopy is in-depth characterization tool of the local atomic structure of materials. It is evident that real surface is not absolutely smooth even after quite a perfect finishing process, but it is a 3D relief. Consequently, the scattering occurs along with reflected and refracted waves. Investigation of the radiation scattered by the surface allows obtaining a surface roughness characteristics.

3. EXAMPLE OF APPLICATION

X-ray reflection and emission spectroscopic methods were used to characterize the 30 nm-thick Al_2O_3 film grown on a silicon substrate by atomic layer deposition (ALD) method (Figs. 6-8).[5] According to our investigations ALD process results in the interfacial layer between film and substrate containing besides Al_2O_3 oxide a considerable amount of the silicon oxide SiO_2 and elemental Si. In the case under consideration the extent of this interface layer no less than 10 nm. The concentration of silicon oxide increases as one approaches the c-Si substrate interface because the interdiffusion of the oxygen and silicon, followed by silicon oxidation. The oxygen diffuses most probably over grain boundaries of the polycrystalline layer, although this requires supporting evidence.

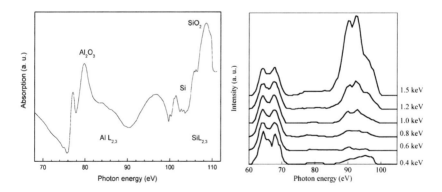

Figure 6. Left-Absorption spectrum of Al_2O_3 film calculated from the reflection spectrum, measured for glancing angle of 35° in the vicinity of Al L2,3- and Si L2,3-absorption edges, Right-Al L2,3 and Si L2,3 X-ray emission bands of the Al_2O_3 film for different primary electron energy between 0.4 and 1.5 keV. The spectra are normalized to Al L2,3 band maximum.

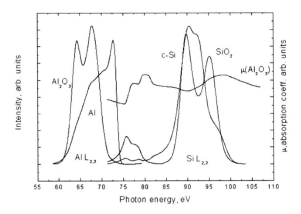

Figure 7. X-ray L2,3 emission bands used for analysis of spectra shape: metal Al and Al$_2$O$_3$, c-Si and SiO$_2$. The absorption coefficient μ(Al$_2$O$_3$) is shown too (dotted line).

4. SUMMARY INFORMATION

All afore-cited information is summarized in Fig. 9 and Table 1. Fig. 9 shows the main information that one can obtain employing one or another X-ray spectroscopic method. The table presents the hierarchy of X-ray spectroscopic methods.

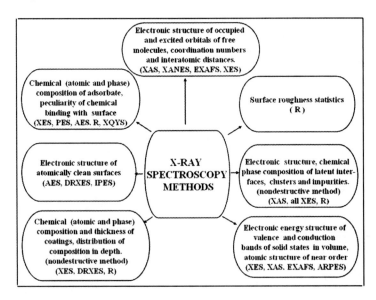

Figure 8. Information obtained by modern X-ray spectroscopic methods.

REFERENCIES

1. S. Hüfner, Photoelectron Spectroscopy: principles and applications, Springer Series in Solid State Science 82 (1996)
2. L. C. Feldman, J. W. Mayer, Fundamentals of surface and thin film analysis (1986)
3. E. Z. Kurmaev, X-Ray Fluorescence Spectroscopy of Novel Materials, Inorganic Materials 41 (1) S1-S23 (2005)
4. A. S. Shulakov, V. A. Fomichev, Threshold effects in Ultrasoft X-Ray Emission, Phys. Scr. 41 99-104 (1999)
5. E. O. Filatova, E. Yu. Taracheva, A. A. Sokolov, S. V. Bukin, A. S. Shulakov, P. Jonnard, J.-M. André, V. E. Drozd, Ultrasoft X-ray reflection and emission spectroscopic analysis of Al2O3/Si structure synthesized by ALD method, X-Ray Spectrometry (2006), in press
6. L. G. Parrat, Electronic band structure of solids by X-ray spectroscopy, Rev. Mod. Phys. 31, 616-645 (1959)
7. Workshop on X-Ray Science in the next Millennium: The Future of Photon in/Photon out Experiments, Pikeville, 2000

HARD X-RAY HOLOGRAPHIC METHODS
Instrumentation and Experimental Technique

G. Faigel, G. Bortel and M.Tegze
Research Institute for Solid State Physics and Optics, H1525 Budapest, POB 49 Hungary

Abstract: In structural studies local methods play an increasing role. The reason is the appearance of nanoscale systems, in physics, chemistry and also in biology. There is also a need to understand the local environment of impurities or dopant sites etc. Hard x-ray holography based on the inside reference point concept is a local probe of the atomic order in solids. It gives the 3D real space image of atoms without the phase ambiguity inherent to diffraction methods. Therefore it is a potentially powerful method to attack problems difficult for traditional diffraction. In this paper the basics of atomic resolution x-ray holography are given and recent theoretical and experimental developments are discussed. The capabilities of hard x-ray holography are illustrated by examples. Future possibilities and new directions are also briefly outlined.

Key words: structure; hard x-ray; holography; imaging

1.   INTRODUCTION

The atomic and molecular structure is the starting point in understanding and explaining many properties of solids. Therefore structure determination is fundamental in physics, chemistry and biology. Beside the interest in basic research, there is also an increasing need from the high tech industry to control the structure of materials at the atomic level. Therefore it is not surprising that large efforts were concentrated to the development of methods, capable of resolving the atomic structure of materials. Most of these techniques rely on the intensity measurement of waves elastically scattered by the sample. However, waves scattered in different directions have not only different magnitude (intensity) but different relative phases.

Vasili Tsakanov and Helmut Wiedemann (eds.), Brilliant Light in Life and Material Sciences, 383–393.
© 2007 *Springer.*

The phase relation is also determined by the spatial arrangement of the scattering objects. So the full information, which would be enough for the unambiguous reconstruction of 3D real space order of atoms, is the intensity and the phase of the scattered waves together. Without any of these, the inversion of measured data to 3D real space is not straightforward. In practice we use additional knowledge on the sample (such as composition, part of the atomic structure, closely related structures etc.) to replace the missing phase information and allow successful structure solution. However, this is not feasible in all cases. Therefore methods giving direct phase information or data allowing the retrieval of phases are valuable. Holography using local reference point is one of the methods. In this paper we overview the most significant results in the area. First, we briefly describe the basics of atomic resolution holography, then we discuss the experimental and evaluation related problems and the solutions to these. In the next part we illustrate the power of the method by selected applications. At the end of the paper future directions are described and different variants of the method are mentioned.

2.   PRINCIPLES OF HOLOGRAPHY

The holographic principle was suggested by Denes Gabor[1]. According to his idea, a known reference wave is mixed coherently to the unknown scattered wave, (called "object" wave in optics) and the resulting interference pattern (called hologram) is recorded on a surface in the far field. The phase information of the object wave is coded in the intensity distribution of the interference pattern. We can decode this by illuminating the recorded interference pattern with the reference wave, or equivalently we can apply the Helmholtz-Kirchhoff (HK) transformation on the digitally stored pattern[2]. In practice the decoding is direct, in the sense that we directly restore the object wave, the 3D real space image of the object.

The two most often used arrangements for making holograms are Gabor or in-line holography and Fourier holography. In first case, the reference wave is a plane wave. Part of the reference wave scatters on the sample, which becomes the object wave. At the photographic plate, we detect the interference pattern of the unscattered part of the reference wave and the object wave, which is the hologram.

In the second arrangement the reference beam is a spherical wave emitted by a point source. The formation of the hologram is analog to the Gabor case. The difference between the two cases is practical. In the first case the real space resolution is determined by the spatial resolution of the detector, while in the Fourier arrangement the limiting factor is the size of the point source. Of course the resolution is also limited by the wavelength of the probe beam, similarly to any method based on the interference of waves. It is clear that for atomic resolution imaging, we need a probe beam with angstrom wave length and either angstrom size sources or detectors. Let us consider hard x-rays as probe beam. In this case it is easy to satisfy the first requirement, since x-rays in the 10 keV range are readily available. However, producing angstrom size sources or detectors seems a difficult task. This problem was circumvented by the idea of Szoke, who suggested using the atoms of the sample as point sources[3]. He also gave those criteria, which makes this suggestion practically useable for solids. These are the following:

1. The environment of every source atom has to be the same, and oriented in the same way
2. The emitted radiation has to be monochromatic
3. The waves emitted by different atoms have to be incoherent
4. The size of the sample has to be much smaller than the sample-detector distance.

These criteria ensure that all the holograms produced by individual source atoms are identical and can be added in intensity at the detector. As a result we measure a single hologram characteristic to the atomic environment of a single source atom. The simplest type of materials, which satisfy these criteria are small single crystals, containing some element with emission lines in the 10 keV range. In this case the individually emitted fluorescent photons serve as the spherical reference beam. The first successful X-ray holography experiment was done on $SrTiO_3$[4]. The Sr atoms were used as point sources. The 3D picture of the Sr atoms was reconstructed[4]. This method is called X-ray fluorescent holography (XFH) in the normal mode.

Beside the Fourier type method, holograms can be taken in analogy to the Gabor method. This can be understood by the optical reciprocity theorem. In this case the atoms serve as local point detectors and the hologram is formed by the interference between the incident plane waves and the waves scattered by the atoms surrounding the detector atom[5]. The detection is through the absorption of the photons of the interference field by the detector atom. Therefore the energy of the incident X-ray beam has to be larger then the proper absorption edge of the detector atom. The response of the detector atom is given by the angular integrated fluorescent radiation. The hologram is formed by measuring the intensity of the fluorescent radiation as a function of the incident beam direction relative to the sample. This method is often called XFH in the inverse mode.

3.   EXPERIMENTAL CONSIDERATIONS

In an XFH experiment one has to measure a single energy (the fluorescent energy) on a spherical surface with angular resolution in the degree range (normal XFH) or detect the fluorescent radiation in a given fixed (and as large as possible) solid angle while changing the direction of the collimated incident beam (inverse XFH). Since the holographic effect is in the 10^{-3} range for a medium heavy element, one has to collect a few times 10^6 photons in every pixel. The size of a pixel is about 10^{-3} sr (corresponding to degree step size) which leads to a few thousand pixels in a hologram. Therefore the integrated number of fluorescent photons in a hologram is about 10^{10} or more. At a laboratory source 10^4 to 10^5 s^{-1} fluorescent yield can be reached at the required angular resolution. This leads to measuring time in the week range. So laboratory experiments are not very practical. For this reason X-ray holography measurements are mostly done at synchrotrons. Therefore we discuss here the synchrotron setups only.

A straightforward realization of normal XFH measurements would be a small spherical stationary sample illuminated from a fixed direction and moving the detector on a sphere. However, this arrangement is not easy to realize, since it is quite complicated to maintain the reproducible motion of a heavy detector on a spherical surface. The same straightforward thinking leads to an even more complicated solution in the case of inverse measurements, namely, to the motion of the source on a sphere. Since this cannot be done easily, one moves the rigidly fixed sample–detector assembly together, in order to reach all possible directions of the incident beam relative to the sample. This also asks for a complicated set-up similar to that used in normal measurements. One can simplify the mechanical motions by using a flat sample (Fig. 1.).

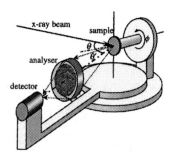

Figure 1. Sketch of experimental setup for holography measurements working both in normal and inverse mode.

In normal measurements the incident beam hits the sample at a fixed θ_0 angle. The detector moves on a circle which defines a plane perpendicular to the sample flat surface. The sample is rotated about an axis (ϕ) perpendicular to its surface and it crosses the rotation axis of the detector. This arrangement imitates the motion of the detector on a sphere[4]. A similar solution can be found for the inverse measurements. In this case the detector is fixed to the sample, so that the angle θ does not change during measurements[5]. However, the sample detector assembly moves together about the θ_0 axis relative to the incident beam. In addition to this the sample is rotated about the axis (ϕ). There is one drawback of the above simplified motions: while we measure the normal (inverse) hologram we also change the orientation of the crystal lattice relative to the incident beam (detector direction) by rotating about ϕ. This way we measure a part of the inverse (normal) hologram while taking the normal (inverse) hologram. This can cause serious distortion of the image, but it can be corrected by measuring the contribution of the unwanted hologram, and subtract it in the evaluation.

Now we turn to the optics and detector. In the case of normal holography the exciting beam could come from any direction and with any energy above the absorption edge of the source atom. However, in practice a slightly collimated and monochromatized incident beam is used. The cause of this is that the use of white radiation results in many Bragg peaks like those in a Laue photograph. We could filter out the Bragg peaks at the detector side; but it does not work since the high intensity leads to high detector dead time.

In inverse measurements the incident beam has to be collimated in the degree range and monochromatized with $\Delta E/E \sim 10^{-3}$. The fluorescent radiation is detected in as large a solid angle as possible. This leads to higher intensity than in the normal case, therefore the limitation on the dead time becomes even more serious. Considering all the above, and the spectral and angular characteristics of synchrotron beam one arrives at the same optics for both cases. The difference is in the detector, which should accept much larger solid angle in inverse holography.

Since we do not need high angular and energy resolution, a simple Si(111) channel cut or similar monochromator suffices. A precise incident beam monitor is essential. The fluorescent intensity could reach 10^7-10^9 1/s range, so fast detectors like avalanche photo diodes (APD) or plastic scintillators are needed. However, these do not have good enough energy resolution to suppress photons with unwanted energy. Therefore one has to use crystal analyzers. Since flat crystal analyzers have small angular acceptance, it is advisable to use some kind of sophisticated focusing system. One of the simplest solutions is a cylindrical curved graphite analyzer[6]. More sophisticated designs, like logarithmic shape[7] or toroidally bent[8] graphite analyzers were also developed. These yield higher acceptance angle, but they work only for a single energy.

4.   EVALUATION PROCEDURE

The evaluation consists of three basic steps:

a. removal the contribution of the reference beam. This is complicated by the non uniform angular intensity distribution caused by the combined effect of the absorption and sample shape. In the case of a flat sample, which is used in all of the experiments done so far, the angular dependence of the intensity distribution can be given in analytical form[9]. This allows the removal of this background by fitting a surface to the measured data.

b. removal the contribution of the inverse (normal) hologram caused by mechanical construction of the experimental setup (see fig.1.). This correction can be done by measuring very precisely the intensity of one (ϕ) circle and normalize with it[9]. At this point we arrive at the hologram without background.

c. The last step is the reconstruction. Traditionally this is done similarly to the high wavelength case by the application of the HK transformation[2]. However, after this procedure many problems may arise. First of all, there is a truncation error because the intensity cannot be measured on the full hemisphere about the sample. Secondly, the twin images of the atoms appear. Most of the efforts were concentrated to avoid these problems. Truncation error can be minimized if the hologram could be extended to the full sphere. In special cases this can be done by using the measured symmetries of the Kossel lines, and apply these symmetries to the hologram[10]. Another possibility to reduce the spurious oscillations is to use no direct transformation of the measured intensities to real space image, but to apply some iterative algorithm and fit the measured data. A Japanese group worked out a method called Scattering Pattern Matrix method (SPM), which works this way[11]. The distortions caused by the twin image can be eliminated by taking holograms at many energies and combine these in a multiple energy reconstruction[9,12]. Various combinations of the above procedures can be applied to obtain the best results.

5.   EXAMPLES FOR APPLICATIONS

In this section we selected three applications of very different types to illustrate the possibilities given by XFH.

The first one is imaging of the atomic decoration in a PdAlMn quasicrystal[13]. As it is well known, quasicrystals are nonperiodic in 3-D, but in 6-D they can be described as a periodic lattice. However, projecting this lattice to 3-D does not give the atomic decoration. This has to be modeled starting from chemical considerations. Local methods, such as electron microscopy or AFM, could give a picture of the atomic order. However, they probe only a small area on the surface of the sample. To get a bulk picture of the atomic order, x-ray methods have to be used. Although traditional crystallographic measurements show strong peaks in well-defined directions, the atomic positions cannot be derived. Using holography we could image selected shells about the Mn atoms. The hologram of the environments of Mn atoms is shown in the left panel of Fig. 2. and the reconstructed image is given in the right panel of Fig. 2. Comparing the result to model structures, one could validate theoretical predictions[14].

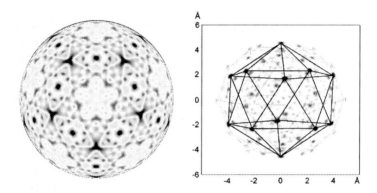

Figure 2. Hologram of an $Al_{70.4}Pd_{21}Mn_{8.6}$ quasicrystal taken at 16 keV energy showing the non-crystallographic five fold symmetry (left panel), reconstructed real space atomic arrangement around the Mn atoms (right panel).

The second example is a study of the local order in FePt thin layers. A Japanese group examined the effect of the growth conditions on the local order. The measurement itself is difficult because of the small amount of sample material on a bulk substrate. Further, the effect what the researchers wanted to see is masked by the artifacts inherent to the traditional evaluation process. For this reason a new iterative evaluation method the SPM[12] was used. The authors examined the order of Pt atoms about the Fe site[15]. A clear difference in the relative weights of the first and second neighbors between the high and low temperature grown samples were found.

The third application is on the origin of Kondo like behavior in the ThAsSe system. This compound becomes ferromagnetic below 110 K[16], but the Kondo like behavior of the resistivity persists below this temperature[17]. This indicates a nonmagnetic origin of the Kondo type anomaly. One of the possibilities, which could explain this feature, is the existence of two level systems (TLS)[18]. The most likely origin of TLS is structural: site occupation or positional disorder. To underpin the reason behind this unexpected Kondo anomaly, diffraction studies were carried out on ThAsSe and related compounds[19]. These studies revealed high values of the anisotropic displacement factors and also slight anion disorder (As/Se~12% site disorder). However, these traditional methods relying on the long range periodicity could not give an unambiguous connection between these features and the possible TLS. This prompted us to investigate the ThAsSe by a local method: hard x-ray holography. We chose the As as the central atom and took holograms at 4 energies (Fig. 3. left panel). This compound has a tetragonal crystalline structure (P4/nmm, Fig. 3. middle panel).

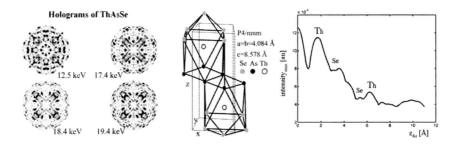

Figure 3. Holograms of ThAsSe about the As sites (left panel), crystal structure of ThAsSe (middle panel), intensity of atomic layers along "z" (right panel)

The flat surface of the sample was perpendicular to the (001) axis, which was parallel to the ϕ rotation. The four fold symmetry of the holograms clearly reflects this. Since we expect some As atoms on Se sites, the simple HK reconstruction gives a very complicated figure showing all the As environments and also their twin images at the same time. Therefore direct observation of the 3D image is not too informative. Since, the symmetry of the sample allows a variation in the z coordinate of Th and Se only, we concentrated to this. In the right panel of Fig. 3. we plotted the intensity maxima of the x-y planes as a function of "z" (z=0 is the As plane). We can identify the consecutive crystalline planes, which we marked on the figure. Note, that the Th plan is very wide. This is in accord with the large anisotropic displacement factors obtained from diffraction. However, the Th distribution is asymmetric and its maximum is shifted towards the As planes. The second interesting feature is the appearance of a second up shifted Se peak, at z=3.6 Å. This suggests two possible Se positions, which might be the origin of the expected TLS. We also examined the As/Se site disorder. This cannot be derived directly from the 3D real space data. We did model calculations of the holograms starting from the known structure with the modified z coordinates of Se and Th, and compared this to the measured holograms. We allowed to change the r = As/Se ratio until the best fit. We got r = 10%, which is in good agreement with the diffraction measurements. Our conclusion is that the origin of the two level systems in the ThAsSe compound is probably connected to the doubling of the Se site along the z direction. This might be in connection with the shift of Th atoms along "z".

6. CONCLUSION AND OUTLOOK

In the last ten years atomic resolution holography became a practically useable method. The basic theoretical and experimental tools for holography using inside reference points have been worked out.

Among the many variants of holography the most advanced is the x-ray fluorescent holography. This was applied to problems for which holography gave unique structural information. Wider applications are expected when further experimental and evaluation related problems will be solved. We outline here the most probable advances we expect in the near future. Though measurements can be done relatively fast (in the hours range per hologram) it would be useful to increase data collection speed. This is essential in doing multiple energy holography. For shorter collection times we either use higher intensity incident beam or larger solid angle of collection or both. Higher intensity of the incident beam can be realized by using directly an undulator harmonic without monochromator (called pink beam). If we can get the pink beam cleanly (without high harmonic content) and with high temporal stability, experiments can be done in less then a minute. We did experiments at ESRF in this direction but no success so far. This development would also be essential to solve another problem: the sample shape. One could use an arbitrary shape small sample instead of a large flat sample if the total absorption of the sample were negligible. Of course we would loose many photons but this could be compensated by the higher incident intensity. A few words about evaluation related problems. It is very likely that beside the SPM other iterative algorithms will appear. This basically depends on the developments of computer technology. Having PC-s with few tens of Gigabyte memory and a bit faster processors will open this direction. This will lead to cleaner real space images and higher sensitivity of the method. Another direction is the use of anomalous scattering of the atoms around the central atom. This involves measurements close to absorption edges and also development of the evaluation to combine these data. There were some works in this direction[20,21] but more should be done to prove the usefulness of this approach.

At last we would like to mention that not only fluorescent photons but also photons emitted by externally injected accelerating electrons and by nuclei can serve as sources. The first is called bremsstrahlung X-ray holography[22,23] and the second gamma ray holography[24,25]. There are other related methods like angular integrated elastic scattering[26], which does not use inside reference point but apply similar evaluation procedure (the HK transformation), and gives similar local structural information as holography.

Of course the above developments are the most trivial ones, probably as the use of holography gets wider more advances will appear.

ACKNOWLEDGMENT

This work was supported by OTKA T043237 and T048298, GB was supported by the Bolyai scholarship, and we are indebted to Dr. Cichorek for the sample.

REFERENCES

1. D. Gabor *Nature* (London). **161**, 777 (1948).
2. JJ. Barton *Phys. Rev. Lett.*, **61**: 1356 (1988).
3. A. Szöke In *Short Wavelength Coherent Radiation: Generation and Applications; AIP Conference Proc. No. 147*; Attwood DT, Boker J (eds). AIP: New York, 1986; 361.
4. M. Tegze and G. Faigel, *Nature*, **380**, 49 (1996).
5. T. Gog et al. *Phys. Rev. Lett.*, **76**, 3132, (1996).
6. S. Marchesini, et al., *Nuclear Instr. and Methods* **A457**, 601 (2001).
7. M. Tilman Donath, Diplomarbeit am Fachbereich Physik der Universit"at Hamburg 2002, Construction and Test of a Fluorescence Analyzer for Use in Reciprocal X-Ray Holography
8. Sekioka T et al. *J. of Synch. Radiaton*, **12**, 530 (2005).
9. G. Faigel and M. Tegze. *Rep. Prog. Phys*, **62**, 355. (1999).
10. M. Tegze, et al.. *Phys. Rev. Lett.*, **82**, 4847, (1999).
11. T. Matsuhita, A. Agui and A. Yoshigoe, *Europhys. Lett.*, **65**, 207 (2004).
12. JJ. Barton, *Phys. Rev. Lett.* **67**, 3106. (1991).
13. S. Marchesini, et al. *Phys. Rev. Lett.*, **85**, 4723 (2000).

14. M. Boudard et al. *J. Phys. Condens. Matter*, **4**, 10149 (1992).
15. Y. Takahashi et al. *Appl. Phys. Lett.*, **87**, 234104 (2005).
16. A. Zygmunt, M. Duczmal, *Phys. Stat. Sol.*, **A9**, 659 (1972).
17. Z. Henke, R. Fabrowski and A. Wojakowski, *J. Alloys Comp.*, **219**, 248, (1995).
18. D.L. Cox, A. Zawadowski, *Adv. Phys.*, **47**, 599 (1998).
19. Z. Henke et all. *J. Phys. Chem. Solids*, **59**, 385. (1998).
20. S. Marchesini et al. *Phys. Rev.* **B66**, 094111 (2002).
21. Y Takahashi, K. Hayashi and E Matsubara. *Phys. Rev.* **B71**, 134107. (2005).
22. GA. Miller, LB. Sorensen, *Phys. Rev.* **B56**, 2399, (1997).
23. SG. Bompadre, TW. Petersen and LB. Sorensen. *Phys. Rev. Lett.*, **83**, 2741 (1999).
24. P. Korecki, J. Korecki and T. Slezak, *Phys. Rev. Lett*, **79**, 3518 (1997).
25. G. Faigel, *Hyperfine Int,*, **125**, 133 (2000).
26. G. Faigel, M. Tegze, G. Bortel and L. Koszegi, Europhys. Lett. **6**, 1201 (2003).

NATURAL ZEOLITES AND APPLICATION IN LIQUID WASTE TREATMENT

H. N. Yeritsyan, A.A. Sahakyan, V. V. Harutunyan, S.K. Nikoghosyan,
E. A. Hakhverdyan, N. E. Grigoryan
Yerevan Physics Institute, 2, Alikhanyan Bros. str., Yerevan, 375 036, Armenia

Abstract: Zeolite-clinoptilolite minerals crystal structures and their main properties are
              presented depending on property structure relation particular attention is
              given on irradiated clinoptilolite properties in connection w3ith application
              for radioactive waste water treatment. It was shown the high effiency for
              reducing the radioactivity of waters by electron irradiated Armenian
              clinoptilolite more than 1700 times.

Key words: zeolite, clinoptilolite, radiations modification, liquid waste

1.   INTRODUCTION

"Rarely in our technological society does the discovery of a new class
of inorganic materials result in such a wide scientific interest and
kaleidoscopic development of applications as has happened with the zeolite
molecular sieves", declared Donald W.Breck in 1974 [1].

Minerals have been investigated for many year sand it is evident that the
study of their formation, crystal structure and reactivity is an endless source
of inspiration for designing novel materials. Zeolites are heterogeneous
microporous high-internal-surface-area crystalline minerals with an open,
three-dimensional framework consisting of tetrahedral AlO_4^{-5} and SiO_4^{-4}
units linked through shared oxygen atoms and acting as catalysts and their
active sites are located within the internal cavities of the structure [2]. The
knowledge of the structure of a given material is an important step for
understanding its properties.

Vasili Tsakanov and Helmut Wiedemann (eds.), Brilliant Light in Life and Material Sciences, 395–401.
© 2007 *Springer.*

Physical and chemical properties of zeolites depend on type and concentrations of cations existing in their elementary cell. It is at this point that much chemistry and other disciplines begin: new experiments and theoretical approaches, which may yield related materials with improved properties, may be proposed. It was shown the importance of combining synchrotron powder X-ray diffraction and spectroscopic methods in the study of guest molecules in zeolite hosts. The reactivity of the active sites is analyzed via the adsorption of probe molecules monitored by infrared and Raman spectroscopies.

There are 40 known naturally occurring zeolites and more than 150 synthetic ones. Zeolite structures can be visualized by taking a neutral SiO_2 framework and isomorphously substituting AlO_2^- for SiO_2. The resulting structure (Fig.1) exhibits a net negative charge on the framework aluminum. This negative charge is balanced by cations (for instance, Na+, K+, or NH4+) that reside in the interstices of the framework. Many of these cations are mobile and free for exchange. This ion-exchange property accounts for the greatest volume use of zeolites today.

Some zeolite properties that are important include (Table 1):
- structure;
- silica-to-alumina ratio;
- pore size; framework density (that is, atoms per unit cell).

The idealized formula of zeolites is given by

$$M_{x/n}[Al_xSi_yO_{2(x+y)}]\, pH_2O$$

where M is (Na,K,Li) or (Ca, Mg, Ba, Sr), n-charge of a cation, $y/x = 1\text{-}6$; $p/x = 1\text{-}4$. The oxidic formula of zeolites is $M_{2/n}O \cdot Al_2O_3 \cdot xSiO_2 \cdot yH_2O$.

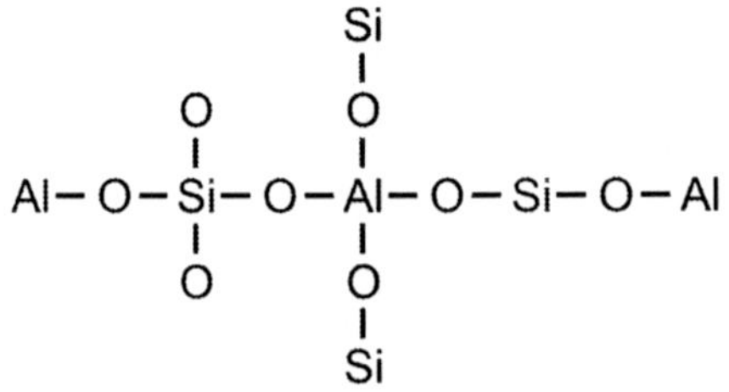

Figure 1. Zeolite structure.

The pore size is two-dimensional opening of the zeolites and is determined by the number of tetrahedral atoms joined together. The structure is built up further by connecting the tetrahedral atoms in a three-dimensional array. This array can lead to larger inner cavities connected by pore openings. In some zeolites, there are no cavities, but a series of one-, two-, or three- dimensional channels through the structure.

Because of this ability to customize properties zeolites are commercially valuable as adsorbents and molecular sieves – selectively admitting some molecules while excluding other whose size, shape, or polarity preclude adsorption.

Table 1. Zeolite properties

Property	Range
Channels	2.2-8 Å
Cavities	6.6-11.8 Å
Thermal stability	500-1.000°C
Ion-exchange capability	Up to 700 milliequivalents/100g
Surface area	Up to 900 m²/g

2.   CLINOPTILOLITE

Among mentioned 40 types of natural zeolites the three of the most significant economic important are clinoptilolite. mordenite and chabazite. Clinoptilolite has unique properties of high cation exchange capacity and stability to set attrition, which make it highly effective for the removal of toxic pollutants from water and soil. There are about 150 million metric tons of natural clinoptilolite in different regions of Armenia, the most known of them is Noyemberyan region, the samples of which have the most useful properties. (Tables 2,3)

The oxidic formula of clinoptilolite:
$\{K,Na,1/2Ca\}_2O \cdot Al_2O_3 \cdot 10SiO_2 \cdot 8H_2O$

Composition:
Crystal data: Space group:C12/ml (#12)
 a=17.662 Å b=17.911 Å c=7.407 Å
 $\alpha=90°$ $\beta=116.40°$ $\gamma= 90°$
 X-ray single crystal refinement, Rw=0,088
Comment: unique axis b, cell choice 1
Denseness: 2,16g/cm³
Hardness on the Mohs: 3,5-4

There are many articles concerning the zeolite- clinoptilolite properties investigations and application[3-5]. The present paper focuses on the Armenian Clinoptilolite study and application for radioactive waste water treatment from Armenian Nuclear Power Plant (ANPP). Moreover, radiation modified clinoptilolite samples were used for this purpose at the first time, which presented very effective method for the reduction of summary radioactivity of the water.

Table 2. Chemical Composition of Zeolites from Noyemberyan region of Armenia (in weight %).

SiO_2	TiO	Al_2O_3	Fe_2O_3	FeO	MgO
67.11	0.20	11.69	1.43	0.36	1.28
CaO	Na_2O	K_2O	H_2O	SO_3	Other loss
4.9	0.79	2.22	3.01	0.10	6.91

Table 3. Specification and Results of Analysis of Armenian Zeolite.

	Specification		Results
1.	Description Off-wite (normally pale green)		Off-white Fine powder
2.	Identification		
	Test for silicate	Positive	Positive
	Test for Aluminum	Positive	Positive
	Test for Barium Absorption	Positive	Positive
3.	Loss of Drying (105 deg C)	2-6%	4.04%
4.	Loss on ignition (800 deg C)	13% max	10.78%
5.	Arsenic (As)	1 ppm max.	< 2ppm
6.	Heavy metals	10 ppm max.	< 2ppm
7.	Oxidisable substances	0.3 ml 0.02M KMn04/g	0.16ml
8.	Calcium Ion Exchange	4mg/g minimum	5.9 mg/g
9.	Lead Ion Exchange	60 mg/g minimum	75 mg/g
10.	Apparent weight/ml	0.6 – 0.9 g/ml	0.72 g/ml
11.	Sieve Test		
	Passing 75 micron sieve	35% minimum	47.5%
	Retained on 125 micron sieve	35% maximum	26.7%
12.	Microbial		
	Total aerobic count (Bacteria)	<1,000 cfu/g	<10
	Total aerobic count (Fungi)	<100 cfu/g	<10
	Enterobacteriacae Count	<10 cfu/g	<10
	Presence of E. coli, Ps Aeruginosa and Staph.Aureus (all in 1 g)	All absent	ND*
	Presence of Salmonellae	All absent	ND*

ND* - Not detected

Since the liquid radioactive wastes from nuclear plants contain much more Strontium and Cesium ions, it seems interesting to study zeolites with these elements which are absorbed by zeolite structural cages during the cleaning process.

Radiation modified natural zeolites (mainly clinoptilolite) were tested for the treatment of waste water from Armenian Nuclear Power plant (ANPP). An electron irradiation with 8Mev energy was used for the zeolite radiation modification.

The main part of waste water radioactivity is caused by the Cs^{+1} element hence special attention was given to the content of its isotopes (^{134}Cs and ^{137}Cs) although the radioactivity for secondary elements (I^{-1}, Ag^{+1}, Co^{+3}) was measured too which present in waste water.

An automatic installation consisting of 3-column stainless steel tube system was used which was filled in with natural granular clinoptilolites (column1) and radiation modified (columns 2-7) ones. Radioactive water enters into column (containing clinoptilolite as a sorbent) from the bottom of column and after cleaning enters into an intermediate tank where a little pump is located. When the water reaches a certain level the pump automatically switches on and transfers the cleaned water to a storage tank for further processing through the same or other column.

This equipment acts in two regimes: step-by-step and apparently as autonomous cleaning system. The parameters are:

1. Three column system which can be operated separately and together. Each column has following sizes: length - 50cm. diameter - 4 cm.
2. The multiplicity and velocity of radioactive water flow was defined by the time intervals per 1 hour, after which radioactivity and chemical tests came carried out.
3. Different radiation modified clinoptilolite samples are applied.
4. The granule sizes of clinoptilolite are 1,8 – 2 mm in diameters.

Radioactive water with low concentration of Na and K was chosen for determination of processing efficiency. The chemical impurities in water were the following: Cl-0,15 mg/kgr, ammonium-0,5 mgr/kgr, Na-0,15mgr/kgr, Ka-0,2 mgr/kgr; B-12,6 gr/L.

The pH-was 5,9; it was increased to 12 by adding of NaOH.

Transparency was 90.

Radioactivity of initial water was:

^{137}Cs –2,9x10^4 Bk/L, ^{134}Cs - 2,4x10^4 Bk/ L, ^{60}Co - 4,7x10^3Bk/L.

The measurement results are given below in tables and graphically. The sampling of water was performed after each cycle.

In order to determine the efficiency of the whole 3-column installation all 3 columns were connected consecutively: columns N° 1,2,3.

Table 4. Water cleaning scheme – 8 cycles thorough columns №1,2,3 consequently.

Nuclid	Energy keV	Intensity	Accuracy %	Radioactivity		Relation 0/n
				Cu/l	[Bk/l]	
Initial activity of water						
^{137}Cs	662	3,19E+1	4%	$14,46 \times 10^{-7}$	54390	
^{134}Cs	796	2,58E+1	6%	$13,26 \times 10^{-7}$	49050	
^{60}Co	1173	1,74E+0	8%	$1,09 \times 10^{-7}$	4051	
Cycle 8						
^{137}Cs	662	1,62 E-1	13	$0,002 \times 10^{-7}$	72	755
^{134}Cs	796	1,87E-2	15	$0,0096 \times 10^{-7}$	35	1427
^{110m}Ag	885	6,32E-2	16	$0,04 \times 10^{-7}$	152	
^{60}Co	1173	9,62E-2	18	$0,058 \times 10^{-7}$	217	

It is seen from given results that the maximum result is observed at the defined degree of radiation modification, i.e. the reduction of radioactivity of ^{137}Cs and ^{134}Cs was about 1750 times. Note, that even natural row untreated clinoptilolite reduces radioactivity about 75 times which is higher than literature data.

The maximum value of initial radioactivity of the water was $2,9 \times 10^4$ Bk/L, after processing this value decreased to 30 Bk/L which is only 10 times higher than acceptable level (3Bk/L). So, it is quite possible to get „ pure" results thorough further investigations.

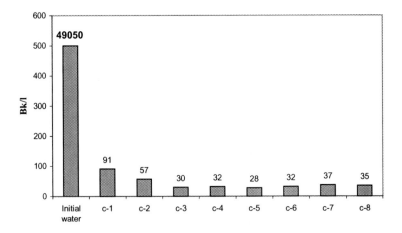

Figure 2. The cleaning dynamics of 134Cs thorough columns №1, 2, 3 connected consequently (water specific activity).

3. CONCLUSIONS

One can conclude from the presented results that the radiation modification of zeolites leads to considerable enhancement of their sorption properties. This can be explained by the increase of the sorption activity of granular surface of zeolites due to radiation modification. At the same time this has stronger effect for elements with higher ion radius ($r_{Cs} \sim 1,67\text{Å}$) than for elements with low radius ($r_{Co} \sim 0,63\text{Å}$; $r_{Ag} = 1,13\text{Å}$; $r_I = 2,2\text{Å}$). The latter has negative valence.

The radiation enhancement of the sorption surface may be explained by the following phenomena: radiation shaking leads to the removal of different inclusions from the zeolite cage pores, and variation of charge state of cations or sublattices which can capture for radioactive Cs. Future experiments are necessary to elucidate this mechanism and continue electro physical investigations.

Current results correspond to our previous findings for electro-physical and optical properties of zeolites, in particular, the reduction of Cs^{+1} radioactivity with the radiation modification degree has antibatic character with respect to the same dependence of zeolite conductivity.

The scientific and practical significance of presented results allows continue current studies to develop and design pilot industrial installation taking into account some factors: granule sizes of zeolite, diameter and length of column, water flow rate, optimum pH, i.e. final cleaning technology for radioactive liquid waste.

REFERENCES

1. Breck D. W. Zeolite Molecular Sieves: Structure, Chemistry and Use, Wiley. New York , (1974).
2. Th. Armbruster, M. E. Gunter, Crystal Structure of Natural Zeolites, Reviews in Mineralogy and Geochemistry, V. 45, Natural Zeolites: Occurrence, Properties, Applications. p. 1-57, 2001.
3. Denes Kallo, Applications of Natural Zeolites in Water and Waste Water Treatment Reviews in Mineralogy and Geochemistry, V. 45, Natural Zeolites: Occurrence, Properties, Applications. p. 519 – 550, 2001.
4. L. M.Wang,, S. X. Wang and R. C. Ewing. Radiation Effects in Zeolite. Proceedings of the 9-th Annual International Radioactive Waste Management., American Nuclear Society, Las Vegas, May 11-14, 1998), p. 772-775.
5. Hrant Yeritsyan, Aram Sahakyan, Sergey Nikoghosyan, Vachagan Harutiunian, Volodia Gevorkyan, Norair Grigoryan,Eleonora Hakhverdyan, Yeghis Keheyan, Rudolf Gevorgyan,Hakob Sargisyan. Dielectric properties and specific conductivity of armenian natural clinoptilolite irradiated by electrons. CEJP 3(4) 2005 610–622.

STUDY OF MEMORY IN SOLIDS STATES UNDER SYNCHROTRON RADIATION EXCITATION

V.V. Harutunyan[1], E.A. Hakhverdyan[1], G.A. Hakobyan[2]

1Yerevan Physics Institute after A.I. Alikhanian,Yerevan,375036,Armenia, 2Yerevan State University, Yerevan, 375001,Armenia

Abstract: A mechanism of "radiation memory" in corundum single crystals is discussed consisting in restoration of some optical absorption bands within 200-650 nm range after irradiation and thermal treatment at high temperatures with subsequent X-ray irradiation.

Key words:

1. INTRODUCTION

The problem of increase of corundum crystals (α-Al_2O_3) radiation resistance is of permanent interest in respect of the formation mechanisms of radiation defects (both point and complex ones), a part of which are the color centers (CC), i.e. the defects capable to absorb or radiate quanta in ultraviolet (UV), vacuum ultraviolet (VUV), visual and infrared spectral ranges[1-6].

Radiation induced defects in α– Al_2O_3 have been intensively studied by many investigators[1-3] lately. The defects were obtained by electron, neutron or ion bombardment with electron irradiation the defects are predominantly isolated, while with neutron bombardment the defects may aggregate, and with ion implantation dense cascades of defects are obtained[7-9].

Vasili Tsakanov and Helmut Wiedemann (eds.), Brilliant Light in Life and Material Sciences, 403–409.
© 2007 *Springer.*

The study of radiation influence on corundum resulted in revealing of the "radiation memory" phenomenon in the irradiated corundum single crystals.

The purpose of this work is to reveal, at what stage of structural transformations the color centers are formed responsible for the "radiation memory" phenomenon, as well as to broaden the existing ideas on explanation of new and earlier obtained experimental results.

2. EXPERIMENTAL TECHNIQUES

The subjects of study were samples of nominally pure (undoped) corundum (α-Al_2O_3) single crystals grown by various ways, namely, by horizontal-oriented crystallization (HOC), Verneil methods. Concentrations of uncontrollable impurities in the reaction mixture were (in mass percent): $3 \cdot 10^{-3}$ for Cr_2O_3, 10^{-4} for 10^{-4} Ti_2O_3, 10^{-3} for Ca, Fe and Ni.

The samples of corundum (plane-parallel plates and cubes with C_3 optical axes parallel to the large side to within $\pm$ 3°) used for the measurement of optical absorption, photoluminescence and excitation of luminescence, were manufactured from specially selected perfect boles. The side surfaces of all samples were carefully mirror-finished using the AM-1 diamond paste.

These corundum single crystals were irradiated using 50 MeV "ARUS" linear electron accelerator, 2 MeV reactor neutrons, X-rays and "white" beam of synchrotron radiation (hν~12 keV). The samples were heat-treated at various temperatures in air. The optical absorption spectra were investigated within 190-640 nm by means of a double lattice monochromator. The luminescence excitation measurements in UV and VUV ranges (4.5 – 30 eV) at various temperatures were carried out using time-resolved SUPERLUMI (HASYLAB, DESY) experimental device[8].

3. EXPERIMENTAL RESULTS AND THEIR DISCUSSION

The study of radiation-optical properties of corundum single crystals in a wide spectral range has allowed to reveal and interpret the "radiation memory" phenomenon.

Within the investigated range of optical absorption (Fig. 1) the corundum single crystals are characterized by rather high transparence which decreases at higher energies. It is clear from Fig. 1 that the optical absorption spectra (OAS) of unirradiated and irradiated corundum samples differ. The sample irradiation by high-energy electrons (50 MeV, dose of 10^{17} electrons/cm^2) results in the increase of absorption constant within the whole spectral range. The spectra are composite curves, the resultants of all components of the induced absorption (IA) bands (i.e. the difference between the absorption constants before and after irradiation).

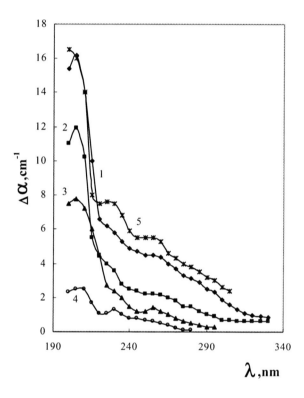

Figure 1. Corundum optical absorption spectra.(Verneil method). Verneil crystals irradiated with the dose of $3 \cdot 10^{17}$ electrons / cm^2. 1 – untreated sample; 2 – sample treated at 225°C; 3 - sample treated at 325°C; 4 - sample treated at 600°C; 5 - untreated, measured at 77 K.

Evidently, the most intensive is the 6.05 eV (205 nm) band in the spectrum. It is necessary to note that this absorption band is analogous to similar bands in the IA spectra of corundum grown by other methods. For example, after heat-treatment of Verneil crystals their IA spectra display distinct 5.4 eV (230 nm) and 4.86 eV (255 nm) bands. These bands are well manifested when measuring OAS's at 77 K (Fig. 1, curve 5). It is probably related to unequal thermal stability of various bands forming the IA spectrum.

It is known from other publications[1-5, 7-13] that the detected bands are caused by anion centers: F-centre (6.05 eV - anion vacancy with two localized electrons) and F^+-centre (5.4 eV, 4.86 eV - anion vacancy localized by one electron).

It follows from the obtained results and other publications that these bands well coincide in their location, half-width and thermal stability.

To reveal the nature of the detected CC induced by electrons, some samples irradiated by reactor neutrons with energy 2 MeV, dose of 10^{17} neutrons/cm^2 and treated at 700°C, were investigated.

The analysis of experimental data presented in Fig. 1, 2 clearly proves the existence of absorption bands belonging to F- and F^+-centers, as well as to other CC centers. But the most important is the fact that the detected absorption bands are directly connected with the state of other CC. It is known that in real crystals there also exist uncontrollable impurities of some metals which stimulate formation of growth defects to maintain the charge composition which can not be annihilated even as a result of high-temperature treatment.

Such defects are potential traps for point radiation defects such as interstitial ions and their vacancies. A part of these defects can become color centers (F, F^+, F^{2+})[14]. The depth of above-mentioned potential wells (traps) sometimes can be so substantial that the interstitial ions are not capable to be released from these traps even at high temperatures (1000°C). Besides, there are inelastic interactions which modify the charge state of pre-irradiation defects according to the following reaction: $F^2 + e \rightarrow F^+$; $F^2 + 2e \rightarrow F$, resulting in the formation of F – и F $^+$-centers as well as other CC centers.

The study of thermal stimulated processes for both electron and neutron irradiated crystals have shown that when temperature increases to 1000°C, the bands intensity decreases within the whole investigated spectral range (Fig. 1 and 2). The mentioned 6.2 eV absorption band (Fig. 2) was also detected by authors of [14] and ascribed to F^{3+} ions, but actually it is caused by oxygen vacancies as F^{2+}–center. Appearance of peaks in UV spectral range in 205, 230 and 255 nm bands specifies an incomplete annihilation of these centers at these high temperatures and, hence, the existence of a "threshold" for coloring.

The process of thermal ionization for color centers should be experimentally verified. With that end in view, a series of experiments was carried out to reveal the state of some CC (Fig. 2, curves 2 - 6). It is seen from Fig. 2 that after X-ray irradiation, synchrotron radiation (SR) with hν ~12 keV during various periods of time results in changes in absorption bands intensities. This experiment has shown that decrease in $F^\pm$-center (230, 255 nm) band intensity in comparison with that of F –center takes place due to the formation of free electrons and holes in corundum at irradiation by high-energy SR photons. A part of electrons and holes are captured by single and complex CC (bands in near UV and visual spectral ranges), the others – by F – and F^+-centers. It is found that in the case of neutron irradiation with energy 2 MeV in result of a cascade of elastic collisions, the concentration of knocked-on atoms is more than in the case of electron irradiation and should be even more taking into account the losses of the basic part of electron energy at inelastic scattering due to the Coulomb interaction.

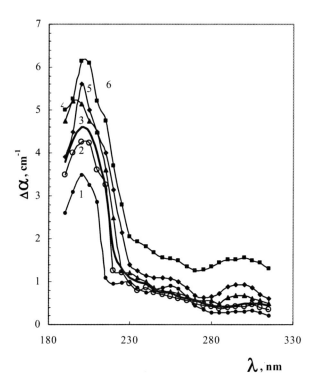

Figure 2. Optical absorption spectra of the irradiated (dose- $3 \cdot 10^{17} n/cm^2$)and thermal treated corundum crystals (HOC method). 1 - treated at 700°C; 2 –irradiated by photons during – 90 sec; 3 -150 sec; 4 -300 sec; 5 - 450 sec; 6 – 600 sec.

Before the collision with the atoms of substances the accelerated electrons lose a considerable part of their energy due to ionization, that is why the number of generated vacancies, oxygen (O_i) and aluminum (Al_i) interstitial ions in the case of electron irradiation is much less than in the case of neutron irradiation.

The CC responsible for the 302 nm absorption band (4.1 eV) and the ions of uncontrollable impurities (Fig. 2) can be one of the reasons of $\Delta\lambda$ increase within the 290-320 nm spectral range. Revealing of the nature of this band is of much interest because the CC responsible for this band play considerable role in the corundum "radiation memory" phenomenon. According to[15], the 302 nm band is caused by F_n - type complex centers. In[7] we have shown that the process of F –centers aggregation cannot be carried out even at temperatures above 1000°C. Migration of anion vacancies at the temperatures below 1800°C and their subsequent aggregation are impossible[15].

It follows from the obtained experimental results that the existence of F - centre and CC (302 nm band, see Fig. 2,curves 4-6) enables to present it as complexes. The CC energy states are related, basically, to the aluminum charge state and can only enter into the complexes with anion vacancies as a $[Al_i^+F]$ center.

4.   CONCLUSIONS

The experimental results are presented confirming clearly the effect of thermal treatment on the behavior of the absorption bands, i.e. decrease in the absorption spectra intensities within the investigated area. Based on these experimental regularities (irradiation, heat treatment), a mechanism of the "radiation memory" phenomenon as a result of repeated irradiation by X-ray quanta is offered.

REFERENCES

1.   G.W. Arnold, W.D. Compton. Phys. Rev. Lett, v. 31, 130 (1961).
2.   J.H. Graword Ir., Nucl.Instrum.Methods phys. Res. Sect. B1, 159 (1984).
3.   F.W. Clinard Jr. J.Nucl. Mat. v. 85-86, 393 (1979).
4.   E.R. Hodgson. Cryst. Latt. Def. Amorph. Mater. v. 18, 169 (1989).
5.   G.P. Pells and G.J. Hill. J. Nucl. Mater. v. 141-143, 375 (1986).
6.   K.J. Caulfield, R. Cooper. Phys. Rev. B. v. 47, 55 (1993).
7.   V.V. Harutunyan, G.N. Eritsyan, R.K. Ezoyan, V.A. Gevorkyan. Phys. Stat. Sol., (b), v. 149, # 77 (1988).
8.   V.V. Harutunyan, V.A. Gevorkyan, V.N. Makhov. The European Phys. Journal B, v. 12, 31 (1999).
9.   V.V. Harutunyan, M. Kirm, V.N. Makhov, G. Zimmerer. Preprint DESY, Hasylab, Part I, 601 (2002).
10. J.P. Batra. J.Phys. C. Sol. Stat. Phys., v. 15, 5399 (1982).
11. S.I. Choi, T. Takeyehi. Phys. Rev. Lett. v. 50, 19, 1474 (1983).
12. A.I. Surdo, V.S. Kortov, I.I. Milman. Optics and Spectroscopy, v. 62, 4, 801 (1987).

13. A. Lushchik, E. Feldbach, M. Kirm, G. Zimmerer. J. of Electron spectroscopy and related phenomena. v. 101-103, 587 (1999).
14. R.R. Atabekyan, R.K. Ezoyan, V.A. Gevorkyan, V.L. Vinetskii. Cryst. Latt. Def. Amorph. Mat., v. 14, 155 (1987).
15. A.I. Surdo, V.S. Kortov, I.I. Milman. Optics and Spectroscopy, v. 64, 6, 1363 (1987).

PART 5

Instrumentation and Experimental Technique

HOM DAMPED CAVITIES FOR HIGH BRILLIANCE SYNCHROTRON LIGHT SOURCES

Ernst Weihreter, Frank Marhauser
BESSY GmbH, Albert-Einstein-Str. 15, 12489 Berlin, Germany

Abstract: Cavities with damped higher order modes (HOMs) are an essential ingredient
             for state of the art high brilliance synchrotron radiation sources to avoid
             degradation of the beam quality due to coupled bunch instabilities. In the last
             15 years the development of such cavities has been driven by the
             requirements of high luminosity meson factories and only since a few years a
             significant design effort was made for HOM damped cavities optimised for
             the specific needs of storage ring based synchrotron light sources. This paper
             gives an overview on the design concepts of existing superconducting and
             normal conducting HOM damped cavities with special emphasis on a normal
             conducting 500 MHz cavity developed for low and medium energy high
             brilliance synchrotron light sources.

Key words: RF-cavities, synchrotron radiation sources, high brilliance, HOM damping,
             cavity technology, higher order mode impedance, electron storage rings.

1.   WHY DO WE NEED HOM DAMPED CAVITIES?

The generation of high brilliance photon beams with state of the art storage ring based synchrotron radiation (SR) sources requires two conceptual ingredients: low emittance optics and undulators. Both ingredients are rather expensive for the following reason: The equilibrium emittance scales as $\varepsilon \sim \gamma^2 \theta^3 <H>$, where γ is the relativistic energy, θ the bending angle per dipole and $<H>$ the average of the Snyder invariant H over the dipole[1]. For low emittance we therefore need i) a large number of dipoles (many lattice cells) to reduce θ, and ii) a low $<H>$ which can be

413

Vasili Tsakanov and Helmut Wiedemann (eds.), Brilliant Light in Life and Material Sciences, 413–427.
© 2007 *Springer.*

achieved only by rather strong horizontal focusing. Strong focusing generates high chromaticities with the need for strong sextupols for chromatic correction, leading to reduced dynamic apertures which often must be compensated by additional (achromatic) sextupoles in the dispersion free straight sections. Also, the undulators and wigglers must be matched to the lattice which often asks for additional quadrupoles. For these reasons storage ring based high brilliance SR sources have complex lattice structures with a large number of lattice cells and many magnets per cell.

To make sure that the "expensive" low emittance is not spoiled by beam instabilities, it is wise to take adequate countermeasures to keep the emittance increase within tolerable limits. With the use of HOM damped cavities it is possible to avoid (or at least to minimize) the degrading effects of a full class of beam instabilities, the coupled bunch oscillations driven by the small band HOM impedances of the accelerating cavities.

The narrow band beam coupling impedance of modern synchrotron radiation sources is dominated by the higher order modes of the accelerating cavities thanks to the smooth structure of state of the art vacuum chambers. The beam induced HOMs of the RF cavities are therefore the essential driving terms for the excitation of coupled-bunch oscillations if they coincide with the synchrotron and betatron sidebands of the Fourier spectrum of the beam current. These oscillations can severely limit the photon beam brilliance of undulator dominated 3rd generation SR sources by

- increased effective emittance in both transverse directions caused by transverse coupled-bunch oscillations,
- increased energy spread due to longitudinal coupled-bunch oscillations,
- potential beam current limitations caused by very large oscillation amplitudes.

As an example Fig. 1 shows a simulation of the brilliance degradation for the U49 undulator beam in BESSY II assuming a HOM induced increase in effective emittance and energy spread by a factor of 2 and 5 respectively. The reduction in brilliance is particularly strong for higher undulator harmonics, which are increasingly used in modern facilities. Thus to reach full performance it is evident that such instabilities must be avoided.

Various cures have been applied in the past to increase the thresholds for the onset of coupled-bunch instabilities, e.g. by

i. detuning of the most dominant HOM away from its driving beam frequency using a second tuner and/or by changing the cavity temperature[2],

ii. using small band damping antennas to damp one or a few dominating HOMs[3,4],

iii. using higher and lower harmonic RF systems for Landau damping and decoupling of the synchrotron tune of neighbouring bunches respectively[5,6], and

iv. using broadband bunch to bunch feedback systems.

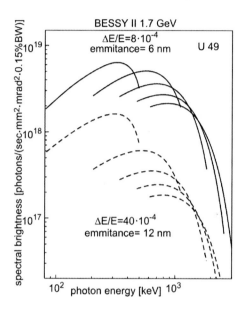

Figure 1. Brilliance spectra calculated up to the 9th odd harmonics for the U49 undulator in BESSY II. Full line: nominal emittance and energy spread. Dotted line: HOM driven increase in emitance and enery spread by a factor of 2 and 5 respectively.

All these methods, however, have there own specific limitation and the most straight forward way to avoid cavity driven coupled-bunch instabilities is to reduce the HOM impedances below threshold.

The beam current thresholds for the excitation of such instabilities are inversely proportional to the total HOM impedance Z_{tot} in a ring given by

$$Z_{tot} = N_c \left(\frac{R}{Q_0} \right)_{HOM} Q_{ext} \qquad (1)$$

for $Q_0 \gg Q_{ext}$, assuming NC identical single cavity cells, where $(R/Q_0)_{HOM}$ is the shuntimpedance / quality factor ratio of the strongest HOM, and Q_{ext} is the external quality factor if the cavity is coupled to an external load.

Thus Z_{tot} can be minimised by using a small number of cavities, by a reduction of $(R/Q)_{HOM}$, by a reduction of Q_{ext} or a combination thereof.

2.   CONCEPTS FOR HOM DAMPED CAVITIES

Cavity resonators have higher order modes up to very high frequencies. Above the cut-off frequency of the vacuum chamber (typically in the range of 3 GHz for 3rd generation SR sources) the HOMs are not trapped anymore. They propagate down the vacuum chamber and are damped by the surface resistance of the chamber material. Therefore cavity HOM damping is relevant in the frequency range from the first HOM above the fundamental mode to the vacuum chamber cut-off. The principles of HOM damping have been reviewed by E. Haebel[7], and in several PAC and EPAC conferences overviews were given on the cavity design and RF issues for meson factories[8-11]. Figure 2 illustrates the two most popular concepts, waveguide couplers and beam tube damping loads, which are also applied for broad band damping of HOMs in cavities used for high brilliance SR sources.

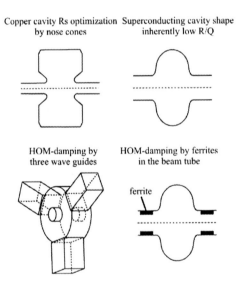

Figure 2. Concepts for HOM damped cavities.

For normal conducting NC cavities, the nose cones necessary to maximise the fundamental mode shuntimpedance lead to rather large (R/Q) also for the HOMs. Therefore external waveguides are attached to the cavity in radial direction to couple out the HOMs and dump their energy in an rf-load to provide a low Q_{ext}. The transverse size of the waveguide becomes rather large if the cut-off frequency is chosen to guarantee propagation of all lower frequency HOMs, and the length of the guide must be long enough to allow the fundamental mode evanescent field to decay sufficiently before reaching the absorber. A minimum of three waveguides is necessary to couple to the different polarisations of dipole modes[12]. In the case of superconducting SC cavities the cavity shape is preferably elliptical to avoid multipacting, and openings in the equator region have the risk to enhance quenching at high field gradients. Therefore the (R/Q) of all modes is minimised using a large beam tube diameter. This is welcome for the HOMs, but only affordable for the fundamental mode in the case of SC cavities. Thanks to the large beam tube diameter the HOMs propagate down the beam tube and can be absorbed at a proper location outside the cavity cell.

2.1 Superconducting cavities

There are two SC cavities as a spin-off from particle colliders which are now also used in SR sources. The CESR cavity[13] developed at Cornell University has been adopted for the Canadian Light Source, the NSRRC facility in Taiwan, DIAMOND in England, and the Shanghai Light Source in China, and the LEP cavity, developed in collaboration between CERN and CEN-Saclay is used in a modified version for the SOLEIL facility[14] in France. In a few rings[15-17] 3rd harmonic SC cavities are used for bunch lengthening to improve the Touscheck lifetime and provide Landau damping for collective oscillations. These cavities are scaled versions of the CESR and CERN/CEN cavity and will not be considered in the following.

The CESR cavity[13] (see Fig. 3) is a single cell cavity made of niobium sheet metal with only the 500 MHz TM_{010} mode being trapped and all HOMs propagating out through the beam tubes. To ensure propagation of the lowest dipole modes the pipe cut-off at one cavity side is lowered by the addition of 4 grooves, and the HOM power is absorbed in a cylindrical array of ferrite tiles outside the cold area, one on each side, reducing the HOM Q-values down to the order of 100. These HOM dampers have been tested up to 7 kW. In acceptance tests acceleration gradients up to 12 MV/m have been demonstrated. The input coupler is a waveguide coupler with a planar ceramic window and a power capability of 500 kW in cw operation. Thermal losses of 100 W at 4.3 K must be compensated by the cryogenic system. A conceptually similar 509 MHz cavity has been developed for KEKB[18].

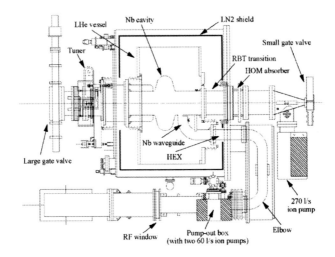

Figure 3. Superconducting RF cavity for CESR B at Cornell University.

The technique of sputter spraying niobium onto copper developed at CERN provides increased stability against thermal breakdown due to the higher thermal conductivity of copper, and the SOLEIL cavity[14] has adopted the CERN experience with Nb/Cu cavities. A single cryostat houses two 352 MHz cavity cells with a beam tube of 400 mm diameter between the two cells to allow HOM propagation (see Fig. 4).

Four coaxial loop type HOM couplers are connected to the central tube to extract up to 5 kW of HOM power. The couplers are connected to external loads through coaxial lines and ceramic vacuum windows. Each of the two coaxial input couplers can transmit 200 kW. An acceleration voltage of 2.5 MV can be generated per cell. The Nb/Cu structure is immersed in a LHe bath, and the cryogenic system must provide 100 W of refrigeration power at 4.5 K and 20 l/h liquefaction of He.

2.2   Normal conducting cavities

Several room temperature single cell copper cavities have been developed for meson factories, e.g. the PEP II cavity[19], the DAPHNE cavity[10], the ARES cavity[20] for KEK-B, and the cavity for the VEPP2000 ring[21]. So far, only a few cavities have been built specifically for SR sources, e.g. the cavity for the DUKE-FELL ring[22] and the BESSY cavity.

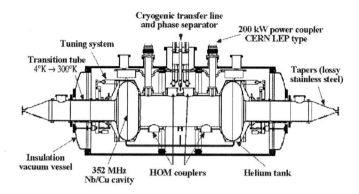

Figure 4. Schematic of the SOLEIL SC cavity based on Nb/Cu technology.

The PEP II cavity has a spherical shape with nose cones optimized for a high impedance of the accelerating mode. Three rectangular waveguides are strategically placed on the cavity body to maximize coupling to the worst HOMs. Under typical operation conditions 103 kW are dissipated in the walls generating 850 kV of RF voltage. This gives a peak thermal surface power density of about 80 W/cm². The thermal layout is based on a power level of 150 kW leading to a complex hydraulical and mechanical design as shown in Fig. 5.

The HOM loads are designed for a maximum power of 10 kW each using in vacuum AlN-40%SiC ceramics as the lossy material. RF-power is fed into the cavity through a circular alumina vacuum window designed for 500 kW input power. Beam based measurements of the HOM damping characteristics are in agreement with numerical simulations of the HOM impedances as well as with bench measurements[23]. Of all room temperature cavities mentioned above only the PEP II design has been adopted also for a high brightness SR source, the upgrade of the SPEAR III ring. Suggestions how to reduce the complexity of the mechanical PEP II design have been made for damping ring cavities[24] which may also be beneficial for future SR source cavities.

The DAPHNE cavity[10], designed for a thermal power level of 16 kW, has long tapered beam tubes, three rectangular damping waveguides placed on the cavity body and two waveguides on the beam tubes. These waveguides are ridged with broad-band coaxial transitions, coaxial vacuum feedthroughs and external rf-loads. This design avoids ferrites in vacuum and permits to sample the HOM power. Power levels of 1 kW can comfortably be handled by the waveguide/load assemblies.

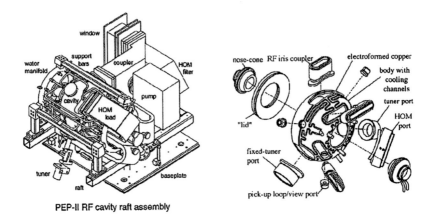

PEP-II RF cavity raft assembly

Figure 5. The PEP II cavity, left: schematic view of the cavity assembly, right: exploded view of the cavity cell.

With the ARES cavity[20] (Accelerator Resonantly coupled with Energy Storage) a very special structure has been developed for KEK-B to reduce the R/Q of the fundamental mode in order to significantly increase the threshold also for the Robinson instability. The structure is a three cavity system operating in the $\pi/2$ mode, where the HOM damped accelerating cavity is coupled with an energy storage cavity via a coupling cavity. The accelerating cavity is loaded by a coaxial waveguide equipped with a notch filter to reject the fundamental mode, and 16 bullet shaped SiC absorbers are inserted in the waveguide for HOM damping.

At the Budker Institute for Nuclear Physics (Russia) several HOM damped cavities have been designed and built for frequencies in the range of 170 MHz. The VEPP-2000 single mode cavity[21] is a coaxial structure as shown in Fig. 6 with two cylindrical loads, a coaxial and a waveguide load, both made of SiC. A choke filter for fundamental mode rejection is placed in the coaxial line before the load. This load is matched to the coaxial line with a VSWR < 1.5 for frequencies up to 3.5 GHz. The waveguide load is placed in the vacuum chamber of 187 mm diameter with a cut-off frequency of 1.23 GHz far above the fundamental mode. A modified version of this concept has been adopted for the DUKE-2 cavity[22] of the DUKE-FELL ring (see Fig. 6). The cavity is made of copper clad stainless steel and HOMs are damped by a cylindrical load connected to the cavity through a circular waveguide of 702 mm diameter. Since 2004 the cavity is in routine operation in the DUKE-FELL ring. For both cavities, VEPP-2000 and DUKE-2, the loads are made of RF absorbing ceramic cups bolted to the cooled copper wall. The HOM power to be absorbed in the VEPP2000 cavity is about 300 W.

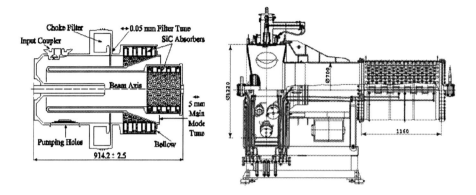

Figure 6. Schematic drawing of the VEPP-2000 cavity (left) and the cavity for the DUKE FELL ring (right).

2.3 Superconducting vs normal conducting cavities

The essential performance parameters of the cavities reviewed above are summarized in Table 1. Since the surface resistivity of SC cavities is extremely low, power dissipation in the structure is low and higher RF voltages can be produced per cell, which allows to minimize the number of cells, i.e. reduced overall impedance. The higher voltages also allow to increase the energy acceptance of optimized higher energy SR source lattices to a level which can not be reached economically with NC cavities. Also RF power consumption is inherently lower compared with NC cavities, however the power necessary for the cryogenic system must be included in the total wall plug power balance. This makes SC cavities a preferred choice for storage rings operating at higher energies. On the other side there is a prize to be paid for the higher complexity related with the technology of SC RF systems: reliability. The operation mode of SR sources is different from meson factories and a high mean time between failure (MTBF) is a crucial precondition. Although SC RF systems came in use in SR sources only recently, a MTBF of one week has already been reached at NSRRC[26]. This is acceptable for a ring with one cavity. For several cavities this figure should still be improved. The LEP collider at CERN with 288 cavities has reached a MTBF of 23 days for a single cavity which indicates that high reliability can be realised.

Table 1. Performance parameters of several cavities (Rs = Vcy2/2Pcy , L insertion length).

	f_0	V_{cy}	R_s/Q	Q_0	P_{cy}	L	f_{HOM} II	RII	f_{HOM} ⊥	R⊥
	MHz	MV	Ω		kW	m	MHz	kΩ	MHz	kΩ/m
CESR	500.	2.5	44.5	-	-	2.9	2253.	0.18	715.	32.
SOLEIL	352.	2.5	45.	-	-	3.65	699.	2.1	504.	49.
		V_{cy} kV	R_s MΩ							
PEP II	476.	850.	3.8	32400	103.	~1.5	1295.	1.83	1420.	144.
DAPHNE	368.2	250.	2.	33000	16.	1.9	863.	259.	-	-
ARES	509.	500.	1.75	118000	72.	~1.1	696.	1.35	989.	10.
VEPP2000	172.1	120.	0.23	8200	29.	0.95	2460	0.4		<10.
DUKE-2	178.5	730	3.46	39000	77	3.16	-	-	-	-
BESSY	500.	780.	3.1	26700	100.	0.5	670.	1.6	1072.	54.

Room temperature cavities are based on well proven and cost efficient fabrication techniques. Several decades of experience with NC cavities led to a high level of operational reliability. The maximum HOM impedances of advanced NC cavities are of the same order of magnitude as for SC cavities (see Tab.1). So for SR sources with only a smaller number of cavities NC cavity technology is a good choice. For higher energy storage rings in the range above 3 GeV SC cavities show increasing advantages.

In 3rd generation SR sources the straight sections are primarily foreseen for undulators, thus the insertion length of the RF cavities should be small to make economic use of this costly resource. Most of the normal conducting cavities built for meson factories are longer than 1 meter and therefore are not always perfectly suited for SR sources. Furthermore there are more than a dozen of low and medium energy SR sources worldwide operating with 500 MHz RF systems which could take advantage from NC HOM damped cavity technology. To provide a suitable solution for these cases a NC HOM damped 500 MHz cavity has been developed at BESSY[26] in collaboration with Daresbury Lab (England, DELTA) Germany and Tsing Hua University (Taiwan) in the frame work of an EC funded project.

3.   THE BESSY HOM DAMPED CAVITY

Conceptually the cavity is of re-entrant type with nose cones to improve the fundamental mode impedance and with three circular waveguides, which are equally spaced in azimuth by 120° (see Fig. 7). Circular waveguides offer considerable engineering advantages compared to rectangular waveguides as the mechanical joining of a circular tube to a rotationally sym-metrical cavity body is simpler. Simulations have shown that the fundamental mode thermal power density in the critical transition region between the waveguide and the cavity body is lower as compared to rectangular waveguides. To reduce the diameter of the circular waveguide (cut-off frequency 615 MHz) two symmetrically arranged ridges are introduced, and the waveguides are tapered to provide a circular waveguide to coaxial transition[27] (CWCT) at the end. HOM energy is removed from the cavity via a broadband coaxial RF vacuum window and absorbed in a matched external coaxial load. This design avoids ferrites in vacuum and permits to sample the HOM power. The cavity is designed for a thermal power of 100 kW.

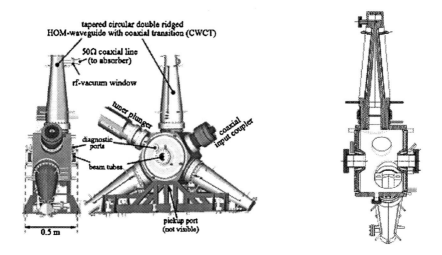

Figure 7. The BESSY 500 MHz HOM damped cavity.

A numerical method[23] has been used to evaluate the frequency dependence of the cavity impedance by the Fourier transform of the wake-field computed with the 3D time domain solver of the MAFIA code[28]. The cavity geometry has been optimized for minimum HOM impedances by iteration of all relevant parameters: length and radius of the cavity, nose cone angle, accelerating gap length, waveguide cutoff-frequency and position of the waveguides.

The cavity impedances are shown in Fig. 8 for the relevant frequency range below the beam pipe cutoff, indicating that impedances can be reduced down to the level of 2 kΩ and 60 kΩ/m for the longitudinal and transverse case respectively, assuming a perfectly matched homogenous waveguide with constant cross-section. For tapered waveguides the peak HOM impedances are in the range of 5 kΩ and 220 kΩ/m respectively in good agreement with bead pull measurements. A detailed analysis has shown that this difference can be attributed to a stronger coupling of the homogenous waveguides to the cavity modes.

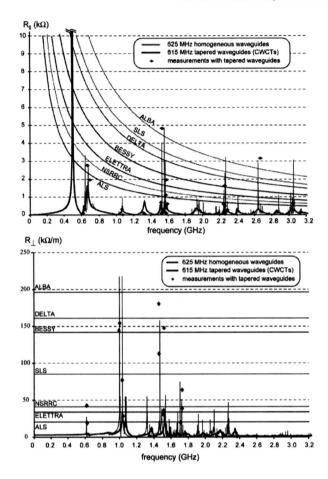

Figure 8. Longitudinal (top) and transverse (bottom) impedance spectra of the BESSY cavity and critical impedances for several machines.

The relevance of a given cavity impedance spectrum for the excitation of coupled bunch instabilities in a storage ring is best described by the threshold impedance, Z^{thresh}, which can be obtained by equating the radiation damping time with the respective multi-bunch instability rise time, giving

$$Z_{\|}^{thresh.} = \frac{1}{N_c} \cdot \frac{1}{f_{\|,HOM}} \cdot \frac{2 \cdot E_0 \cdot Q_s}{I_b \alpha \, \tau_s} \tag{3}$$

$$Z_{x,y}^{thresh.} = \frac{1}{N_c} \cdot \frac{2 \cdot E_0}{f_{rev} I_b \beta_{x,y} \tau_{x,y}} \tag{4}$$

for the longitudinal and the transverse case respectively (E_0 beam energy, I_b average beam current, Q_s synchrotron tune, α momentum compaction, NC number of cavities, $f_{\parallel,HOM}$ long. HOM frequency, f_{rev} revolution frequency, $\tau_{x,y,s}$ damping times, $\beta_{x,y}$ beta function at the cavity). Here the conservative assumption has been made, that every HOM coincides with a Fourier component of the beam current and that the NC cavities have identical impedances. With these expressions and the machine parameters given in Tab. 2 the threshold impedances have been calculated for several SR sources with 500 MHz RF systems. As shown in Fig. 8, for the case of homogenous waveguides all considered machines can be operated without being affected by longitudinal multi-bunch oscillations, whereas a few rings may still be subject to transverse multi-bunch instabilities under these rather conservative assumptions.

Table 2. Parameters for several SR sources using 500 MHz RF systems.

	ALBA	ALS	BESSY II	DELTA	ELETTRA	NSRRC	SLS
E_0 / GeV	3.	1.5	1.7	1.5	2.	1.5	2.4
I_b / mA	400	400.	250.	220	300.	200.	500.
$Q_s \times 10^3$	8.34	9.12	6.16	6.9	9.9	10.6	7.21
N_c	6	2	4	1	4	2	4
f_{rev} / MHz	1.12	1.52	1.25	2.61	1.156	2.498	1.04
$\alpha \times 10^3$	0.859	1.59	0.73	5.3	1.6	6.78	0.7
τ-s / ms	3.106	13.58	8.	3.7	7.9	5.67	4.5
τ-x / ms	4.077	15.3	16.	9.	10.	6.96	9.
τ-z / ms	5.289	21.5	16.	9.	13.	9.37	9.
β-x / m	2.	11.	1.	3.6	6.5	10.5	3.
β-z / m	1.4	5.	1.2	2.5	6.5	2.9	3.

To take advantage from the reduced HOM impedance levels achievable with homogenous waveguides, a suitable waveguide[29] compatible with the BESSY cavity has been developed which terminates in a wedge shaped ferrite load (see Fig. 10). The load is designed for less than 20% reflection and a power level of 2.5 kW per waveguide.

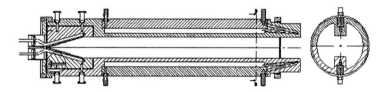

Figure 9. Homogenous ridged circular waveguide with 2 ferrite loaded absorber elements on the left hand side.

A prototype cavity with tapered waveguides has been installed in the DELTA ring at the University of Dortmund one year ago and beam instability studies indicate that there are no coupled bunch oscillations driven by the cavity HOMs[30]. A second cavity with homogenous damping waveguides is under test for the Metrology Light Source in Berlin and six cavities will be installed next year in the ALBA ring in Spain.

4.   CONCLUSIONS

The first HOM damped cavities were a spin-off from high intensity collider rings. In the meantime a few 3rd generation SR sources have been equipped with cavities of this origin and demonstrate that their advantages can also be exploited by state of the art SR sources to avoid brilliance degradation caused by coupled bunch instabilities. Superconducting cavities show their virtues for higher energy rings above 3 GeV whereas room temperature cavities are a good choice for rings operating at medium and lower energies. Recently normal conducting HOM damped cavities have been designed and built for the specific boundary conditions of SR sources (e.g. short insertion length) which can also be used for upgrading existing facilities.

ACKNOWLEDGEMENTS

The authors want to thank the members of the EC funded project team at BESSY, Daresbury Lab, DELTA/ Dortmund University and Tsing Hua University/Taiwan for their dedicated collaboration and S. Belomestnykh, A. Gallo, P. Marchand, R. Rimmer, and V. Volkov for providing figures and technical information on their work.

REFERENCES

1.   H. Wiedemann, Particle Accelerator Physics, Springer, Berlin, New York, p.352, (1993)
2.   M. Svandrlik, G. D'Auria, A. Fabris, E. Karantzoulis, A. Massatotti, C. Pasotti, C. Rossi, Investigation of the higher order modes in the ELETTRA cavities, Proc. EPAC 1994, 2136 (1994)
3.   B. Dwersteg, E. Seesselberg, A. Zolfhagari, Higher order mode couplers for normal conducting DORIS 5-cell cavities, Proc. PAC 1985, 2797 (1985)
4.   D. Boussard, P. Baudrenghien, T. Linnecar, G. Rogner, W. Sinclair, The 100 MHz RF system for the CERN collider, Proc. EPAC 1988, 991 (1988)

5. E. Weihreter, H. Hoberg, W. Klotz, P. Kuske, R. Maier, G. Mülhaupt, Instability studies and double rf-system operation at BESSY, IEEE Trans. NS No. 5, 2317 (1985)
6. P. Marchand, Possible upgrading of the SLS RF system for improving the beam lifetime, Proc. PAC 1999, 989 (1999)
7. E. Haebel, Higher order mode suppression in accelerators, Proc. EPAC 1992, 307 (1992)
8. J. Kirchgessner, Review of the development of RF cavities for high currents, Proc. PAC 1995, 1469 (1995)
9. K. Akai, RF issues for high intensity factories, Proc. EPAC 1996, 205 (1996)
10. R. Boni, HOM – free cavities, Proc. EPAC 1996, 1223 (1996)
11. H. Padamsee, Review of HOM damped cavities, Proc. PAC 1997, 184 (1997)
12. P. Arcioni, G. Conciauro, Feasibility of HOM free accelerating resonators, Particle Accelerators 36, 177 (1991)
13. H. Padamsee et al., Accelerating cavity development for the Cornell B-factory, CESR-B, Proc. PAC 1991, 786 (1991)
14. P. Marchand et al., Successful RF and cryogenic tests of the SOLEIL cryomodule, Proc. 2005 PAC, 3438 (2005)
15. A. Fabris, C.Pasotti, P. Pittana, M. Svandrlik, D. Castronovo, Design of a 3rd harmonic superconducting cavity for ELETTRA, Proc. EPAC 1998, 1879 (1998)
16. M. Svandrlik et al., The super 3HC project, Proc. EPAC 2000, 2052 (2000)
17. P. vom Stein, M. Pekeler, H. Vogel, W. Anders, S. Belomesnykh, J. Knobloch, H. Padamsee, A SC Landau cavity for BESSY II, Proc. PAC 2001, 1175 (2001)
18. Y. Funakoshi, K. Akai et al., Beam test of a SC damped cavity for KEKB, Proc. PAC 1997, 3087 (1997)
19. R. Rimmer, M. Allen, K. Fant, A. Hill, M. Hoyt, J. Judkins, J. Saba, H. Schwarz, M. Franks, High power testing of the first PEP II cavity, Proc. EPAC 1996, 2031 (1996)
20. T. Kageyama, K. Akai et al., The ARES cavity for the KEK B-factory, Proc. EPAC 1996, 1243 (1996)
21. V. Volkov, A. Bushuev, E. Kendjebulatov, N. Mityanina, D. Myakishev, V. Petrov, I. Sedlyarov, A. Tribendis, VEPP-2000 single mode cavity, Proc. EPAC 2000, 2008 (2000)
22. V. Volkov, N. Gavrilov, E. Gorniker, O. Deichuli, E. Kenjebulatov, I. Kuptsov, G. Kurkin, L. Mironenko, N. Mityanina, V. Petrov, E. Rotov, I. Sedlyarov, A. Tribendis, 178 MHz cavity with HOM damping for the DFELL ring, Problems of Atomic Science and Technology, 2 (43), 64 (2004)
23. R. Rimmer, J. Byrd, D. Li, Comparison of calculated, measured, and beam sampled cavity impedances, Phys. Rev. S. T. Acc. And Beams, Vol.3 , 102001 (2000)
24. R. Rimmer, D.Li, Design considerations for a 2nd generation HOM-damped rf-cavity, PAC 1999,907 (1999)
25. C. Wang, L. Chang, S. Chang, F. Chung, F. Hsiao, G. Hsiung, K. Hsu et al., Successful operation of the 500 MHz SRF module at TLS, Proc. PAC 2005, 3706 (2005)
26. F. Marhauser, E. Weihreter, C. Weber, Impedance measurements of a HOM damped low power model cavity, Proc. PAC 2003, 1189 (2003)
27. F. Schönfeld, E. Weihreter, Y.S. Tsai, K.R. Chu, Layout of a broadband circular waveguide to coaxial transition, Proc. EPAC 1996, 1937 (1996)
28. MAFIA code, CST, Darmstadt, Germany
29. E. Weihreter, V. Dürr, F. Marhauser, A ridged circular waveguide ferrite load for cavity HOM damping, to be presented at EPAC 2006
30. R. Heine, T. Weis, to be presented at EPAC 2006

REVIEW OF ELECTRON AND PHOTON BEAM DIAGNOSTICS IN ADVANCED LIGHT SOURCES

S.G. Arutunian
Yerevan Physics Institute, Alikhanian Brothers St 2., 375036 Yerevan, Armenia

Abstract: Improvement of advanced light source photon beams quality requires corresponding improvement in electron and photon beam diagnostics. Typical methods of electron and photon beam diagnostics are reviewed. Attention is devoted to the features that provide high quality of photon beams in outlets of storage rings and insertion devices. The new tendencies in area of photon and electron beam diagnostics are also presented.

Key words: Beam diagnostics, storage ring, emittance,

1. THIRD-GENERATION LIGHT SOURCES

Synchrotron storage rings optimized for insertion devices are frequently used in present light sources (so called "third-generation" light sources). In Table 1 we present typical parameters of electron beams in synchrotron light sources[1,2].

Table 1. Beam parameters in some synchrotron light sources.

	Energy, GeV	Hor. Emit., nm*rad	Vert. Emit., pm*rad	σ_x, µm	σ_y µm
APS	7	2.5	25	271	9.7
ESRF	6	4.0	30	380	14
Spring-8	8	6	14	390	7.5
Bessy-II	1.72	6	180-240	290/76	27/17
CANDLE	3	8.4	84	128	41

Vasili Tsakanov and Helmut Wiedemann (eds.), Brilliant Light in Life and Material Sciences, 429–439.
© 2007 *Springer.*

Their figures of merit are strongly facility dependent. Providing filling pattern options and short pulses with excellent bunch purity for those having a strong time-structure user community (5 to 33 %), has a strong priority[3]. To achieve this aim very important is to decrease, if possible, the electron beam emittance. Some ways of achieving 1 nm*rad emittance (horizontal) are proposed in[4] (longitudinal gradient in dipoles; special damping wigglers proposed for NSLS II and PETRA 3; high periodicity of the lattice combined with the high gradients in the small gap dipole magnets, MAX-IV approach). Subpicosecond duration electron bunches have many potential applications including generation of femtosecond X-ray pulses for ultrafast phenomena studying, and as a means to pump and/or probe media with the electrons directly[1,5].

High stability of the photon beam has the highest priority for users. As a consequence the typical tolerances in synchrotron light sources are the following: orbit tolerances about 5% of bunch sizes and bunch sizes relative about 0.1%, energy spread 10^{-3}-10^{-4}.

The base of diagnostics in third generation light sources is the electron beam diagnostics. In this review we also present photon beam "accelerator" means that normally operate to correct the electron beam parameters via feedback system.

2.   ELECTRON BEAM DIAGNOSTICS

2.1 Beam current, particle losses

To measure the beam current one can capture the beam and let the current flow through some kind of meter[6]. In its simplest form it consists of a conducting metallic chamber or cup (Faraday Cup). The design will be significantly more complicated when it is necessary to make measurements of very short pulses or very high energy beams which may not be fully stopped in the thickness of the detector. Simulations and bench tests demonstrate that Faraday Cup with enough bandwidth should reliably measure features of the beam to less than 10 ps[7]. Modifications of Faraday Cups are commercial available[8].

Bunch structure of the beams is measured successfully by "beam transformers"[8]. In order to see the magnetic field of the beam the transformer is mounted over ceramic insert in the metallic vacuum chamber. The ferromagnetic core is wound of high permeability metal tape or made of ferrite, to avoid eddy currents. Bandwidths exceeding 100 MHz can be achieved. In case of length of a beam bunch is longer than the transformer's rise time and shorter than its droop time, the signal from its will be a good reproduction of the bunch shape.

This method can be used also for train of bunches when circulating time is shorter than the droop time. This condition is not satisfied in the storage rings where the beam circulates many hours. Here a true DC beam current measurements are needed. Such a device was developed by K.Unser for ISR (CERN)[9, 10]. The Direct Current Transformers (DCTs)[11] consists of a high permeability and very fast rise time toroidal cores made of cobalt-based nanocrystalline and amorphous alloys and a few-turn sense winding wound by the proprietary multithread technique[11,12]. Cores are switched between flux saturation levels by counter-phased windings powered by an external source (see Fig. 1). In the absence of any DC beam current and to the extent that the two cores exhibit matched and symmetric B-H characteristics, sense windings produce equal and opposite signals.

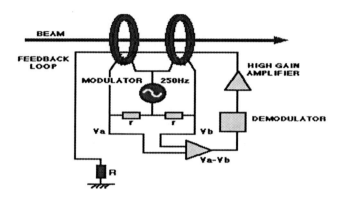

Figure 1. Direct Current Transformers schematic.

A DC beam current through the two cores biases each with flux of the same polarity. The net result is a flux imbalance between the two cores, producing an output signal at even harmonics of the excitation frequency. DCTs are normally designed to operate in a feedback configuration. The Parametric Current transformer is used on most particles accelerators in the world to measure the average beam current. The large dynamic range up to 10^5, the wide bandwidth more than GHz and high resolution about 1 µA[13] make it the ideal instrument to measure beam lifetime in storage rings. It is often the only truly calibrated beam instrument in an accelerator and serves as a reference to calibrate other beam diagnostics.

Fast protection systems with a response time in the µs range are required to shut off the beam in case of high losses. High losses at single points cannot only damage the undulator, but could also produce holes in the vacuum system[6]. Here we present one type of BLM so called Wittenburg's

Beam Loss Monitor[13]. Device is based on the difference between free path length in the media for electrons and synchrotron radiation photons. The charged particle crosses both PIN-diodes, which causes a coincidence. Synchrotron radiation photons, if stopped by the first PIN-diode, do not cause a coincidence (see Fig. 2).

Operating principle

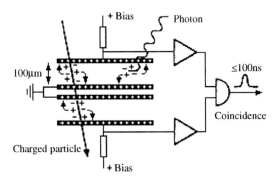

Figure 2. Beam loss monitor operating principle.

BLM have up to 10 MHz count rate with dynamic range[13] $> 10^8$.

The electrons lost during the injection process can be detected via the shower of secondary particles created due to their passage through the vacuum chamber, the magnet blocks and the detector shielding. With these diagnostic losses from the end of the transfer line, losses due to transverse phase space acceptance and then due to the longitudinal capture process can be distinguished. A Perspex rod 25 mm in diameter and 600 mm long is used to convert the high-energy radiation shower into visible scintillations[14]. The Perspex rod then guides this light to the sensitive face of a photo multiplier tube.

2.2   Beam position monitor for electrons

Diagnostics with most usage is the electron beam transversal monitoring. For this purpose beam position monitors (BPM) of different types are used. The principle of operation of widely used BPMs is[6]: as beam passes through the special chamber with few electrodes it will induce electric charges on the metallic electrodes. The induced charges can be carried away for measurement into low-impedance circuit or be sensed on high impedance as a voltage on a capacity between the electrode and surrounding vacuum chamber. The effective signal on the electrodes is almost linear. In electron and positron machines so-called "button" electrodes are used[6].

To measure the closed orbit in a circular accelerator many BPMs are arranged around the circumference. For example, in CANDLE electron beam position in the storage ring will be measured by 80 4-buttons BPMs.

The monitors for separate bunches position are to be separated out, especially for short bunches. Thus, the length of bunches stored in ESRF lies in the 30 ps to 120 ps range (FWHM). The observation of single bunch phenomena like transverse or longitudinal oscillations or bunch length variation requires the acquisition and analysis of signals at frequencies higher than 10 GHz. A set of microwave cavity pickups operating at 10 GHz and 16 GHz together with the appropriate electronics has been implemented on the ESRF storage ring[15]; it detects the wall currents on the vacuum chamber due to the electron beams circulation.

2.3 Longitudinal diagnostics

Diagnostics of longitudinal motion is especially important for short bunches accelerators. The bunch length measurement techniques generally are divided into measurements in time domain (streak camera, RF kicker cavity), measurements in frequency domain (coherent, incoherent, RF pickups), optical and laser techniques (laser-heterodyne, laser-micro-probe, optical-electro effect[16]).

The mentioned above[15] set of microwave cavity pickups operating at 10 GHz and 16 GHz together can obtain also the longitudinal oscillations or bunch length. This type of pick up is a good complement to the streak camera, particularly for the accurate study of the spectrum of the single bunch signals.

An electrooptical method was offered for measurement of the longitudinal profile of electron bunches in storage rings by the signal laser beam modulation in nonlinear crystal by the coherent synchrotron radiation electric field component which correlates with the bunch longitudinal profile[17,18]. Idea was realized in[19]. Further developments of proposed method are described in[20-22].

For example in[22] electro-optic sampling (EOS) method with optical anisotropy of a ZnTe crystal, induced by the electric field of the relativistic electron bunch and sampled with a 15 fs titanium-sapphire laser pulse.

2.4   Usage of synchrotron radiation

Interference and diffraction of synchrotron radiation (SR) in hard X-ray can be used for electron beam diagnostics. Simple optical schemes providing X-ray interference patterns highly sensitive to transverse size of the emitting electron beam are suggested[23]. For each scheme, the visibility of fringes in the pattern depends on transverse size of the electron beam. The schemes can be applied nearly at any X-ray beamline. The visual part of the SR radiation in SLS is transported to an optical lab, where the temporal profile of the storage ring bunches can be measured with a minimal time resolution of 2 ps using a dual sweep, synchroscan streak camera. A fast avalanche photo diode has been set up in parallel to monitor the filling pattern of the storage ring. Beam size and coupling is intended to be measured with a zone plate monitor at 1.8 keV photon energy overcoming diffraction limitations[24].

SR can be used also for measurement of beam transverse profile (see e.g.[25]). A beam profile monitor based on two Fresnel zone plates has been developed at the KEK-ATF damping ring. SR from electron beam is monochromatized and the transverse beam image is twenty-times magnified by two zone plates and detected on the X-ray CCD camera. This monitor can take real-time images and measure the beam profiles with a resolution less than 1 μm.

3.   PHOTON BEAMS DIAGNOSTICS

Chemical Vapor Deposition (CVD) techniques allow to develop diagnostics for photon beams. X-ray beam position monitors and profilers[26] based on the position-sensitive photoconductive detector using insulating-type CVD diamond as its substrate material. A thin CVD-type diamond disk is patterned on both surfaces with a thin layer of electrically conductive material. These coated patterns are connected to a biased current-amplifier circuitry (see Fig. 3). When the electrically biased CVD disk is subjected to X-ray beam, the photons activate the impurities in the CVD diamond causing a local conductivity change. Compared with other photoconductors, diamond is a robust and radiation-hardened material with high dark resistivity and a large breakdown electric field.

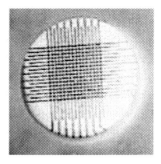

Figure 3. Sixteen aluminum strips are coated on both sides of the CVD-diamond disk creating a 16 x 16 pixel two dimensional array with 175 μm x 175 μm pixel size.

Another type of photon beam monitors is based on the scrapers, which react at touch with the beam [27].

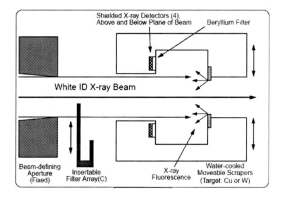

Figure 4. X-ray Beam Position Monitor.

The true performance of X-ray beam position monitors when installed on insertion device beamlines is severely limited due to the stray radiation that contaminates the insertion device photons (Fig. 4). Reduction in stray radiation background in[28,29] is proposed by decreasing the bend angle of the main dipole magnets.

4. WIRE SCANNERS

Conventional wire scanners with thin solid wires are widely used for beam size measurements in particle accelerators[30] (see Fig. 5).

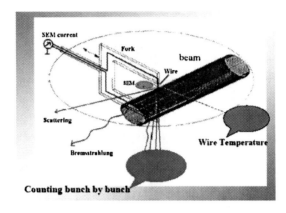

Figure 5. Conventional wire scanner schematically.

A wire is stretched between the two prongs of a fork, perpendicular to the beam and moved through the circulating beam, Secondary particles, produced by the interaction of the circulating particles with the wire hit a detector, which consists of a scintillator and photo multiplier. The output signal is sampled together with the wire transversal position[31]. The secondary emission (SEM) of electrons from the wire is also can be used to measure the number of the particles hitting the wire. The multiple scattering produces emittance blow-up. In order to minimize the blow-up the wire diameter is chosen as small as possible (to few microns) and the large feed velocity of the scanner is used (few m/s). The large velocity is also need by the reason that due to interaction with the circulating beam the temperature of the wire increases during transversal motion of the beam.

Thus transversal motion must be so fast that the wire during scan does not destroy. For scanning of low intensity beams or beam halo measurements low speed scanning is used. An example of slow scan (speed is 0.5 mm/s) is described in[32].

A new approach to measure the wire heating quantity is developed by us[33-37]. The wire heating measurement with great accuracy is provided by measurement of wire natural oscillations frequency. By the rigid fixing of the wire ends on the base an unprecedented sensitivity of the frequency to the temperature and to the corresponding flux of colliding particles is obtained. The first scanning experiments on a charged beam were done on an electron beam at the Injector of Yerevan Synchrotron with an average current of about 10 nA and electron energy of 50 MeV[37]. The special two wire thermocompensated monitor was developed for photon beamline at APS ANL and tested at low energy semiconductor laser. The resolution with respect to laser radiation power density is of order 1.3×10^{-4} W/cm^2 in air. In vacuum this value achieves 5×10^{-6} W/cm^2.

Also note the possibility of using the laser beam as a wire-like object[38] and atomic beams[6]. The idea of laser beam holds much promise and will be realized in PETRA3 project.

ACKNOWLEDGEMENTS

I am very grateful to J. Bergoz and K.Unser for support in development of Vibrating Wire Monitor for photon beam diagnostics and to G.Decker for kindly consent to test Vibrating Wire Monitor at APS ANL.

REFERENCES

1. G. Decker (2005) Beam stability in synchrotron light sources, Proceedings of DIPAC 2005, Lyon, France, pp. 233-237.
2. CANDLE design report (2002), Armenian Synchrotron Light Source CANDLE, July, 2002.
3. 37th ICFA Advanced Beam Dynamics (2006), Workshop on Future Light Sources, May 15-19, 2006, Hamburg, Germany.
4. H. Tarawneh, M. Eriksson, L.-J. Lindgren, S. Werin, L. B. Anderberg, U.E.J. Wallén (2004) Lattice Studies For The MAX-IV Storage Rings, Proceedings of EPAC 2004, Lucerne, Switzerland, 2155-2157.
5. W. D. Kimura (2001) Generation of Femtosecond Electron Pulses, - www.sunysb.edu/icfa2001/Papers/th1-2.pdf.
6. Koziol H. (1994) Beam diagnostics for accelerators, CAS CERN Accel. School, Fifth general accelerator physics course, Geneva 1994, 2, 565-599.
7. C. Deibele (2006) Fast Faraday Cup (SNS). - http://www.sns.gov/APGroup/Minutes/ColabMinutes/030211_FastCup.pdf
8. http://www.kimballphysics.com/detectors/support_PDF/FaradayCup_66_info.PDF.
9. Unser K.(1981) Part. Accel. Conf., Washington, 1981, IEEE Trans. Nucl. Sci., NS-28, N3.
10. K. Unser, (1969) Beam current transformer with D.C. to 200 MHz range, Presented at National Particle Accelerator Conf., Washington, D.C., Mar. 5-7, 1969. Published in IEEE Trans. Nucl. Sci. 16:934-938, 1969 (No. 3).
11. BERGOZ Precision Beam Instrumentation: Fast Current Transformer User's Manual (2006).
12. M. Wada, Y. Ishida, T. Nakamura, A. Takamine, A. Yoshida and Y. Yamazaki (2005) Nondestructive intensity monitor for cyclotron beams RIKEN Accel. Prog. Rep. 38.
13. Wittenburg loss monitor (2006), http://www.bergoz.com/products/BLM/BLM.html.
14. G.A. Naylor, B. Joly, U. Weinrich (2002) Detection of electron losses during injection into the ESRF storage ring, Proceedings of EPAC 2002, Paris, France, 1951-1953.
15. E. Plouviez (2001) Microwave pickups for the observation of multi GHz signals induced by the ESRF storage ring electron bunches, Proceedings DIPAC 2001, ESRF, Grenoble, France, 136-138.

16. X.J. Wang (2001) Review on the production of high-brightness short-pulse electron beams, Presented at 21st ICFA Beam Dynamics Workshop on Laser-Beam Interactions June 11-15, 2001 at Stony Brook USA.

17. Arutunian S.G. (1992) Bunch millimeter and submillimeter length measuring by coherent synchrotron radiation, DESY Internal Report, M-92-6, August 12, 1992.

18. Arutunian S., Dobrovolsky N., Karabekian S., Laziev R., Mailian M., Avakian R., Mailian A., Oganessian D., Tovmasian T., Vardanian A. (1995) Measuring of the short electron bunches longitudianal profile by electrooptical method, 2nd European Workshop on Beam Diagnostics and Instrumentation for Particle Accelerators (Travemuende, 28-31 May, 1995), 96-98.

19. X. Yan A. M. MacLeod W. A. Gillespie G. M. H. Knippels D. Oepts A. F. G. van der Meer W. Seidel (2000) Subpicosecond Electro-optic Measurement of Relativistic Electron Pulses, Phys. Rev. Let., 85, No. 16, 2000, 3404-3407.

20. Winter, M. Tonutti, S. Casalbuoni, P. Schmueser, S. Simrock, B. Steffen, T. Korhonen, T. Schilcher, V. Schlott, H. Sigg, D. Suetterlin (2004) Bunch length measurements at the SLS Linac using Electro Optical Techniques, Proceedings of EPAC 2004, Lucerne, Switzerland, 253-255.

21. Steffen, S. Casalbuoni, E.-A. Knabbe, H. Schlarb, B. Schmidt, P. Schmueser, A. Winter (2005) Electro-optic bunch length measurements at the VUV-FEL at DESY, Proceedings of 2005 Particle Accelerator Conference, Knoxville, Tennessee, 3111-3113.

22. T. Tsang, V. Castillo, R. Larsen, D. M. Lazarus, D. Nikas, C. Ozben, Y. K. Semertzidis, T. Srinivasan-Rao, L. Kowalski, (2000) Electro-optical Measurements of Ultrashort 45 MeV Electron Beam Bunches, arXiv:hep-ex/0012032 , 1, 11 Dec 2000.

23. O. Chubar, A. Snigirev, S. Kuznetsov, T. Weitkamp, V.Kohn (2001) X-ray interference methods of electron beam diagnostics, Proceedings DIPAC 2001 – ESRF, Grenoble, France, 88-90.

24. V. Schlott, M. Dach, C. David, B. Kalantari, M. Pedrozzi, A. Stre (2004) Using visible synchrotron radiation at the SLS, Proceedings of EPAC 2004, Lucerne, Switzerland, 2526-2528.

25. Nakamura N., Salai H., Takaki H., etc.(2004) Developments of the FZP beam profile monitor, Proceedings of EPAC 2004, Lucerne, Switzerland, 2353-2355.

26. D. Shu, P. K. Job, J. Barraza, T. Cundiff, T. M. Kuzay (1999) CVD-diamond based position sensitive photoconductive detectors for high-flux X-rays and gamma rays, www.aps.anl.gov/News/Reports/1999/alpe2.pdf

27. Decker G. (2005) Orbit Stabilization at the Advanced Photon Source, DOE-BES Review May 24, 2005.

28. G. Decker and O. Singh (1999) Method for reducing x-ray background signals from insertion device x-ray beam position monitors, Phys. Rev. Special Topics – Accelerators and beams, 2, 112801.

29. G. Decker, O. Singh H. Friedsam, J. Jones, M. Ramanathan, and D. Shu, (1999) Reduction of X-beam systematic errors by modification of lattice in the APS storage ring, Proceedings of the 1999 Particle Accelerator Conference, New York, 2051-2053.

30. Wittenburg Kay (2004) Beam tail measurements by wire scanners, http://www.sns.gov/workshops/20021023_icfa/presentations/KayWireScanner.pdf.

31. Agoritsas V., Falk E., Hoekemeijer F., Olsfors J., Steinbach Ch. (1995) The fast wire scanner of the CERN PS, CERN/PS 95-06.

32. R. Fulton, J. Haggerty, R. Jared, R. Jones, J. Kadyk, C. Field, W. Ozanecki, W. Koska (1988) A high resolution wire scanner for micron-size profile measurement at the SLC, SLAC-PUB-4605 LBL-25136 UM-HE-88-10 April 1988.

33. Arutunian S.G., Dobrovolski N.M., Mailian M.R., Sinenko I.G., Vasiniuk I.E. (1999) Vibrating wire for beam profile scanning, Phys. Rev. Special Topics - Accelerators and Beams, 1999, 2, 122801.

34. Arutunian S.G., Werner M., Wittenburg K. (2003) Beam tail measurements by wire scanners at DESY, ICFA Advanced Beam Dynamic Workshop: Beam HALO Dynamics, Diagnostics, and Collimation (HALO'03) (in conjunction with 3rd workshop on Beam-beam Interaction) (May 19-23, 2003 Gurney's Inn, Montauk, N.Y. USA).

35. Aginian M.A., Arutunian S.G., Hovhannisyan V.A., Mailian M.R., Wittenburg K. (2004) Vibrating wire scanner/monitor for photon beams with wide range of spectrum and intensity, NATO Advanced Research Workshop "Advanced Radiation Sources and Their Application" (Nor Amberd, Armenia, August 29 - September 02, 2004), NATO Science Series, II. Mathematics, Physics and Chemistry – Vol. 199, pp.335-342.

36. Arutunian S.G., Avetisyan A.E., Dobrovolski N.M., Mailian M.R., Vasiniuk I.E, Wittenburg K., Reetz R. (2002) Problems of Installation of Vibrating Wire Scanners into Accelerator Vacuum Chamber, Proc. 8-th Europ. Part. Accel. Conf. (3-7 June 2002, Paris, France), 1837-1839.

37. Arutunian S.G., Dobrovolski N.M., Mailian M.R., Vasiniuk I.E. (2003) Vibrating wire scanner: first experimental results on the injector beam of Yerevan synchrotron, Phys. Rev. Special Topics, Accelerators and Beams, 6, 042801.

38. G. Blair, T. Kamps, F. Poirier, K. Balewski, K. Honkavaara, Holger Schlarb, Peter Schmueser, S. Schreiber, N. Walker, M. Wendt, K. Wittenburg, M. Ross, I. Ross, D. Sertore (2001) Compton Scattering Techniques for the Measurement of the Transverse Beam Size at Future Linear Collider, www.pp.rhul.ac.uk/~lbbd/archiv/tk_iop2001.pdf.

PRODUCTION OF FEMTOSECOND-PICOSECOND ELECTRON AND X-RAY PULSES IN SYNCHROTRON RADIATION SOURCES

K.A. Ispirian
Yerevan Physics Institute, Brothers Alikhanian 2, 0036, Yerevan, Armenia

Abstract: A review of the femtosecond electron beam slicing methods used for production of soft and hard X-ray and low intensity high-energy electron femtosecond pulses on synchrotron radiation sources ALS and BESSY2 is given. Some new methods proposed for the same purposes are considered. Options for future higher energy SR sources ALBA, CANDLE etc are discussed.

Key words: X-ray pulses, laser, bunch slicing

1.   INTRODUCTION

The 2002, 2003 and 2005 workshops[1] on ultrafast time-resolved soft X-ray science witness that the femtosecond X-ray science with expected extension to shorter pulses produced on synchrotron radiation sources will revolutionize physics, chemistry and biology[2]. The state of art is shown in Fig. 1. To analyze a short time processes one can use either 1) fast detectors with time resolution better than ns and ps as photomultipliers and streak cameras, respectively, or 2) illuminate the objects by ps or shorter light pulses, the second method being more practical. In the synchrotron radiation light source domain one has simply down to a few of tens ps pulses of 1 eV – 10 keV X-ray photons. In the domain of pure lasers one has short flashes down to a few fs, allowing with the help of low energy, IR-UV photon to study atomic and molecular processes (rotation, vibration and dissociation). To be able to study shorter processes it is necessary either 1) to convert the laser photons into X-ray photons as in laser-induced

441

Vasili Tsakanov and Helmut Wiedemann (eds.), Brilliant Light in Life and Material Sciences, 441–453.
© 2007 *Springer*.

plasma and high-harmonic generation (HHG) domains or 2) to use high energy electron beams with or without fs laser pulses as in bunch slicing and free-electron laser (SASE) domains (see Fig. 1). In the case of 1) the brilliance of the X-beams is very low because the spectral and angular distribution of the radiation produced in laser-induced plasma is wide, while for HHG the efficiency is low and the spectrum has many harmonics. We shall not consider the proposed methods for production of ps and attosecond X-ray pulses and will discuss only the techniques for fs X-ray pulses realized with the help of fs lasers applicable at CANDLE.

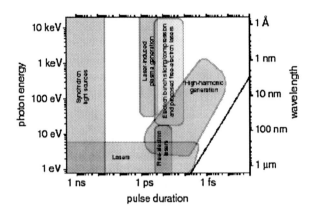

Figure 1. Pulse length and photon energy domains covered by various methods of production of nanosecond-attosecond optical - X-ray bunches. The dashed line shows the limits of single cycle radiation[2].

2. GENERATION OF FEMTOSECOND AND PICOSECOND X-RAY PULSES

2.1 Slicing and fs X-ray pulse separation methods via laser-electron beam interaction in insertion device

For the first time this method has been proposed[3] and realized at ALS, LBL[4,5], and now there are such beams at APS and BESSY2 too and is explained schematically in Fig. 2.

The energy exchange between the electron and laser beams in the wiggler is enhanced if the following resonance condition is satisfied

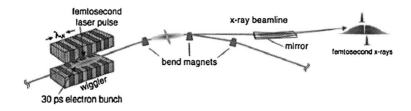

Figure 2. The arrangement for bunch slicing and fs X-ray pulse production at ALS. Due to the interaction of laser and electron beams after the wiggler one has energy modulated electron bunchs 1a). In the first dispersive bending magnet of a 3 magnet ALS sector one has space separation of the fs slices. In the second magnet fs and ps SR pulses are produced by the slices and bunches 1b). The X-ray mirror and slit separate the fs X-ray pulses 1c) [4].

$$\lambda_s = \frac{\lambda_W}{2\gamma^2}\left(1 + K^2/2 + \gamma^2\theta^2\right) = \lambda_L \,, \tag{1}$$

where λ_W is the wiggler period, λ_s and λ_L are the spontaneous radiation and laser wavelength,
$\gamma = E/mc^2$, θ is the radiation angle and

$$K = eB_0\lambda_W/2\pi mc \text{ is the undulator parameter.} \tag{2}$$

The total radiation energy after the wiggler is equal to:

$$A_{tot} = A_L + A_S + 2\sqrt{A_L A_S \Delta\omega_L/\Delta\omega_S}\, \cos(\varphi) \,. \tag{3}$$

where $\omega_i = 2\pi c/\lambda_i$ with spread $\Delta\omega_i$ (i = L,S) and $\varphi = \omega_L t_0$ is the electron-photon interaction phase with t_0 being the time of the electron interaction with the laser wave.
    Since[3,4]

$$A_S = \pi\alpha\hbar\omega_L \frac{K^2/2}{1+K^2/2}; \quad \frac{\Delta\omega_S}{\omega} \approx \frac{1}{M_W}; \quad \frac{\Delta\omega_L}{\omega} \approx \frac{1}{M_L} \,, \tag{4}$$

where M_W is the number of the wiggler periods and M_L is the number of the waves in the laser pulse.
    Assuming K>1 and Gaussian form for laser parameters the energy exchange or the electron energy modulation ΔE in (3) can be written in the form

$$\Delta E = 2(A_L A_W \frac{M_W}{\xi M_L} \eta_{emit})^{1/2} \cos \varphi \,, \tag{5}$$

where $\xi \approx 1.4$; $\eta \approx (1 + \sigma_{xe}^2 / 2\sigma_{xL}^2)^{-1} \approx 0.7$, σ_{xe} and σ_{xL} are the rms transversal sizes of the electron and laser beams.

Therefore after the wiggler one has E-modulation during fs laser pulse with

$$E_1 = E_0 + \exp(-t_0 / 4\sigma_{tL}^2)\Delta E(\varphi) / \sigma_E \,, \tag{6}$$

where σ_{tL} and σ_E are the widths of the laser time and electron energy.

This can be used for slicing and production of fs X-ray pulse in the following 2 ways:

a) i)-First using a bending magnet to separate horizontally the fs electron pulse (see Fig. 1b) then ii)- using an another bending magnet for production the X-ray beams and iii) mirror and slit to take the spacely separated fs X-ray beam produced by the fs electron pulse in the second magnet (see Fig.1c). This method has been realized at ALS still in 1999, and we shall call it ALS scheme.

b) i)-First using a bending magnet to separate horizontally the fs electron pulse then ii)- using an additional undulator downstream after the wiggler for production X-ray beams and iii) using angular cutoff to separate the fs X-ray beam. This method is still in consideration and preparation in ALS[6] and has been realized at BESSYII, still in 2004[7-9] and we shall call it BESSY scheme. Compared with the ALS scheme the undulator or BESSY scheme provides fs photon fluxes with a few order higher intensity and low background.

2.2 The ALS scheme[4]

Let us first consider the theoretical and simulation results of ALS scheme.

Taking a) the wiggler parameters: M_W =19 periods, λ_W =16 cm, K=13 and keeping (1) by adjusting the wiggler gap or B_0;

b) parameters of the chirping $TiAl_2O_3$ + Nd YAG laser with λ_L =800 nm, τ_L =50 fs, U=0.7 mJ, $\sigma_{xL} > \sigma_{xe}$ and repetition rate f_L =1000 pps; and using some of ALS electron beam parameters given in [4], one obtains from (5) $\Delta E \geq 6$ MeV which is greater than the ALS beam energy spread $\sigma_E \approx 1.2 MeV$.

Therefore after the wiggler one has E-modulation during fs laser pulse

$$E_1 = E_0 + \exp(-t_0 / 4\sigma_{tL}^2)\Delta E(\varphi) / \sigma_E, \tag{6}$$

where σ_{tL} and σ_E are the widths of the laser time and electron energy.

Prior to the construction of any fs slicing facility an extensive simulations are necessary. Fig. 2 shows the calculated x- and t-distributions of the electrons in the middle of the second magnet[4].

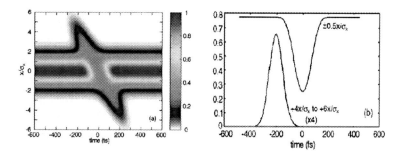

Figure 3. The calculated spatial (left) and integrated in the given domains temporal (right) distributions of electrons in the middle of the second magnet for approximately above given parameters[4].

The expected properties of fs SR pulse are calculated by standard formulas with the time characteristics of fs X-ray pulse given by

$$\sigma_{X-ray}^2 \cong 2\sigma_{tL}^2 + \tau_E^2 + \tau_x^2 + \tau_{x'}^2, \tag{7}$$

where the factor 2 comes from the slippage between the electron and laser photons and the last 3 factors due to electron energy, coordinate and angle are small. Meanwhile the flux and brightness are governed by 3 factors $\eta_1 = \sigma_{tL} / \sigma_{te}$, $\eta_2 = f_L / f_e$ and $\eta_3 \approx 0.2$, the last one is due to matching of e and L pulses.

Since the energy kick can be considered as random energy kick with certain relaxation time

$$\frac{1}{\tau_{rel}} \approx \frac{p^2 M_L \lambda_L}{2c\tau_b} \frac{f_L}{n}, \tag{8}$$

where

$$p = \frac{\Delta E}{\sigma_{eE}} \qquad (9)$$

one cannot increase the fs X-ray intensity increasing f_L, the laser repetition frequency or infinitely increasing the laser intensity, because the perturbation of the electron bunches must be recovered after one turn.

Finally, let us note that as the estimates (and further experiments) show the background in the fs X-ray pulse from the unperturbed pedestal of electrons far from central orbit is negligible.

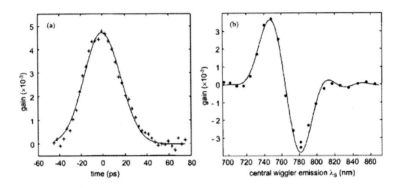

Figure 4. The dependence of the gain upon the delay time between the laser pulse and electron bunch (at optimal wiggler gap or wavelength) 4a) and upon wavelength (adjusted with the help of wiggler gap)[4].

Now consider the experimental methods and results at ALS. Fig.4 shows the delay time and frequency dependence of the measured and theoretically calculated gain[4]. The measured FEL gain in 2000 was 50% of the expected value (mainly due to laser phase-front distortions and other bad tunings) which means that a 30% loss in ΔE.

The visible SR ($\hbar\omega \approx 2eV$) after the slit allows the selection of correct horizontal region for X-ray SR production and helps to carry out cross-correlation measurements. As it follows from[4] about ~100 fs FWHM pulses has been obtained.

2.3   The Bessy scheme[7-9]

Let us first consider the theoretical and simulation results of BESSY scheme. They use

a) A planar "modulator" wiggler, U139, with parameters: M_w=10 periods, λ_w=13.9 cm, keeping (1) by adjusting the wiggler gap or B_0 or K.

b) The parameters of the chirping $TiAl_2O_3$ + Nd YAG laser with λ_L=780 nm, τ_L=30-50 fs, $\sigma_{xL} > \sigma_{xe}$ and U=2.8 mJ at repetition rate f_L=1 kHz or U=1.4 mJ at f_L=2 kHz.

One obtains from (5) $\Delta E / E \approx 1\%$ or $\Delta E \geq 17$ MeV which is greater than the ALS beam energy spread $\sigma_E = 1.2 MeV$. Then the electrons pass an elliptical "radiator" undulator, UE56, placed together with U139 and deflecting magnet B1, B2 and B3 in the same straight section. UE56 has the parameters d) M_W=30 periods, λ_W=5.6 cm.

The authors of[7-9] give the following results of numerical simulations sometimes accompanied with experimental data.

Laser induced energy deviation computed numerically in the following way:

$$\Delta E = -e \int_{-L/2}^{L/2} x'(s)\varepsilon(x, y, z)ds , \tag{10}$$

where $x'= dx / ds$ per infinitesimal small step ds over $L = M_U \lambda_U$ length of radiator-undulator for "macroelectrons" with various parameters.

$$\varepsilon(x, y, z)= \sqrt{2\rho(x, y, z)/\varepsilon_0}\ \sin[2\pi z / \lambda_L - \varphi_G(s_L)], \tag{11}$$

(x, y, z) is the t-dependent electron position relative to laser pulse center, $\rho(x, y, z)$ is the energy density, ε_0 is the free space permitivity, s_L is the laser pulse position relative to modulator center at $s = 0$,

$$\varphi_G(s_L) = \tan^{-1}\left(\frac{s_L \lambda_L}{4\pi\sigma_{x,y}(0)}\right), \tag{12}$$

is the Guoy phase shift with the beam waist at $s = 0$. Longitudinally and transversally the beam is Gaussian and

$$\sigma_{x,y}(s_L)= \sqrt{\sigma_{x,y}^2(0)+\left(\frac{M_{x,y}^2 \lambda_L}{4\pi\sigma_{x,y}(0)}\right)^2 s_L^2}\ , \tag{13}$$

The result of simulations for laser with 1.5 mJ and quality parameter $M_{x,y}^2 \leq 1.5$ is $\Delta E / E - 0.01$.

The laser induced $\Delta E / E$ results in longitudinal change of electrons position[10] z due to which the microbunching, slicing and production of terahertz coherent synchrotron radiation (THzCSR) can take place.

Microbunching takes place if the energy-dependent differences of the electron paths are less than $\lambda_L / 2$ which in its turn results in CSR at $\lambda > \lambda_L / 5$. At BESSY such microbunching takes place at certain places, however , the CSR can not be observed. One can show it can be observed with the help of terahertz coherent diffraction radiation (THzCDR).

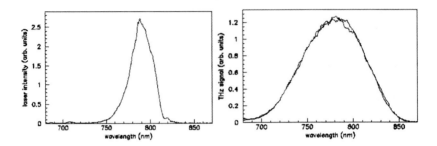

Figure 5. Laser spectrum (left) and intensity of THzCSR (right).

Slicing takes place if the energy-dependent differences of the electron paths are larger than $\lambda_L / 2$. Results of simulations for BESSY[11] show that a short dip of the order of 0.1 mm gives rise to terahertz coherent synchrotron radiation (THzCSR) with high intensity (see [10,11])

$$P_{coh}(\omega) = \left(N_e \sigma_L / \sigma_e\right)^2 P_1(\omega)\left|F(\omega)\right|^2 , \tag{14}$$

where $\sigma_{L,e}$ are the laser and electron pulse length, $P_1(\omega)$ is the intensity of SR from single electron, $\left|F(\omega)\right|$ is the bunch form factor connected with electron density longitudinal distribution. Laser spectrum (left) and intensity of THzCSR (right) are given in Fig. 5. Easily detecting this THzCSR with the help of InSb bolometer one can monitor the optimal resonant λ or gap (1), the optimal delay between the laser and electron bunches, the dependence of THzCSR on the delay between L and e-pulses.

3.   NEW METHODS OF BUNCH SLICING AND PRODUCTION OF FS X-RAY AND ELECTRON PULSES

As it has been proposed in[12] the fs transversal deflection of electron beams with the help of laser beams can be used for many purposes, in particular, for construction of femtosecond oscilloscopes and production of fs electron and photon beams. Indeed, as the results of theoretical calculations[13] show due to the finite length interaction of electrons with co propagating linearly polarized laser beams and then a motion in a long drift length the electrons oscillate as it is shown in Fig. 6.

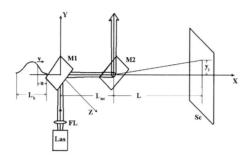

Figure 6. The experimental arrangement for obtaining transversal deflection. Lb is the length of the bunch of electrons each with distance from the bunch beginning u and velocity u. Las is the laser providing a photon beam passing the lens FL a being deflected by the mirrors M1 and M2. After the short interaction region Lint the electron pass the long drift region L and make transverse oscillations with a period of the laser photons. Using position sensitive detectors or having holes in the plane of the screen Sc one can measure the femtosecond length of the bunch of separate periodic fs electron pulses[12,13].

After the drift length L the electrons make oscillatory motion with laser photon frequency and the transversal deflection is equal to

$$y_2 \approx KL, \tag{15}$$

where p_{x1}, p_{y1} and y_1 are the projection of the momentum and coordinate of electrons at the end of the interaction region due to drift and oscillations is small, and the so called bunch scanning factor is equal to

$$K = \frac{p_{y1}}{p_{x1}} = \frac{A}{\sqrt{\gamma^2 - 1} + \frac{A^2}{2}\left[\gamma + \sqrt{\gamma^2 - 1}\right]}, \tag{16}$$

where

$$A = \frac{eE_0}{m_0 c\omega}\left[\sin(\eta_1 + \varphi_0) - \sin\varphi_0\right], \tag{17}$$

$\eta_1 = \omega t_1 - kL_{int}$ (φ_0 and t_1 are the phase and the time of electron fly through L_{int}).

Taking electron energy equal to 1.5 GeV, $L_{int} \sim$ 1cm, L =300 cm and CO_2 laser pulses with wave period $T_L = \lambda/c = 33.33$ fs, $E_0 = 2x10^8 \, V/cm$, with the help of above formulae one obtains $K = 2x10^{-5}$, $y_2 = 60\mu m$ which is greater than the ALS e-bunch vertical size in the dipole magnets, $\sigma_V = 12\mu m$. If the laser length τ_L is greater than the electron bunch length τ_e = 30 ps then after the slit allowing to separate slices with length 0.1 y_2 =0.012 cm one obtains trains of fs pulses consisting of $N_{fs} = \tau_e/T_L$ =30 ps / 33.3 fs = 900 pulses with length equal to $T_{fs} \approx 0.2T_L$ = 6.66 fs (see Fig.7). Using Nd:Glass lasers with wavelength =1.06 micron one obtains X-ray pulses with length 0.66 fs. The expected intensities in the case of this method which can be used with bending magnet or undulator, of course, are higher than in the case of the above-considered methods. There is no doubt that such truly fs pulses can find wide application.

Using the provided relatively large transversal deflection instead of SR in the bending magnet or in an undulator it is reasonable to use bent (for fs electron beam extraction) or not bent (for fs ChR and TR, respectively) single crystals, which are more effective, but destructive.

4.   COMPARISON WITH OTHER METHODS

A method for slicing and generation of subpicosecond X-ray pulses using RF orbit deflection has been proposed in[14]. The method is similar to the bunch crab deflection proposed for the increase of luminosity of colliders[15].

In the E_{110} field of RF cavity the bunch is "rotated or crabbed" and a second similar RF placed at half betatron wavelength downstream compensates all the introduced distortion. Just as above with the help of slit one can take SR form various point in the position A or use the electron angle in the position B. Compared with the above discussed fs slicing method the method has the advantages of providing higher intensities without any laser with certain repetition rate since all the bunches work. The main shortcoming of the in the first sight simple RF method is that the method cannot provide fs slicing and the length of the pulses is ps. To our knowledge the method has not still realized, and the R&D works are in progress[16,17].

As it is shown in the review[18] devoted to the tabletop production of intense quasimonochromatic X-ray beams using small 2-20 MeV accelerators the best mechanism among the bremsstrahlung, coherent bremsstrahlung, transition radiation, X-ray Cherenkov radiation, parametric X-ray radiation, channeling radiation and laser Compton scattering (LCS) the latter is the more effective (see the Table 1 in[18]). However, it is well known that the most effective "head on" collision of the laser and electron beams can not give fs LCS photons since in this case the length of the produced LCS pulse is equal to the long electron beam. To obtain fs X-ray pulses from LCS it is necessary to have large $\psi \approx 90^o$ crossing of laser and electron beams when the length of the produced LCS pulse is determined by the transit time of the laser pulse across the waist of the electron beam. For $\sigma_e \approx 40 \mu m$ *or* $90 \mu m$ (FWHM) the X-ray LCS length is ~130 fs. The characteristics of the fs X-ray can be described quantitavily by considering the laser field as a magnetic undulator (for the first time such an approach was done in the lectures[19] delivered after the first measurements of X-ray undulator radiation produced by 4.5 GeV electrons[20]) with periods number, M_L, equal to the number of laser waves in the laser pulse, and the undulator parameter is given by (2) where $B_0 = E_L$ and $\lambda_w = \lambda_L$. For typical lasers $K_L \approx 0.85 x 10^{-9} \lambda_L (\mu m) I^{1/2} (W/cm^2) \leq 1$. The length of X-ray LCS photons is given by the formula (1) divided by (1-cos ψ). One can calculate the number of fs LCS X-ray photons per laser electron collision assuming an undulator instead of laser pulse with the help of expression

$$ N_{LCS} \approx \frac{\pi\alpha}{2} \left(\frac{K_L^2}{1 + K_L^2/2} \right) M_L \frac{\Delta\lambda}{\lambda} N_e \frac{\sigma_{t-L}}{\sigma_{t-e}}, \qquad (19) $$

where N_e is the number of electrons in a single bunch, while σ_{t-L} and σ_{t-e} are the laser and electron bunch lengths.

The comparison of LCS and the discussed Zholents-Zolotarev fs methods is given in[4] (see Table 2 For calculated characteristics of 5 keV 100fs X-ray sources assuming 800nm laser).

Here we do not consider the Duke University method[21] for production fs X-ray pulses by creating and sustaining fs electron bunches in the storage ring with intensities more ~1000 higher than the method[2-4] provides. Unfortunately, there are no results on the method[21].

Table 1. calculated characteristics of 5 keV 100fs X-ray sources assuming 800nm laser

	LCS	Bend radiator	Undulator Radiator
Flux(ph/s/0.1%BW)	$5x10^3$	$3x10^5$	$3x10^6$
Brightness(ph/s/nm²/mrad²/0.1%BW)	10^7	10^8	$6x10^{10}$
Normal. beam emittance hor/ver	2	22/0.2	22/0.2
Laser pulse energy(mJ)	100	0.1	0.1
Laser repetition rate(Hz)	10	10000	10000
Electron beam Lorentz factor	40	3720	3720
Bunch length(ps)	1	30	30
Number of electrons/bunch	10^{10}	10^{10}	10^{10}

5.   CONCLUSION AND OPTIONS FOR ALBA AND CANDLE

The conclusion is: the realized methods for X-ray fs pulse production are good and their intensity though is low, but enough for various applications. There are some proposals providing higher intensities, however, the proposed methods needs in experimental investigation.

There is a lot of time for the optimal choice and testing new methods for CANDLE, ALBA and other future third generation SR sources before the fourth generation and energy recovery SR sources will be launched. CANDLE, ALBA. etc have higher 3 GeV energy instead of 1.5-1.7 GeV of ALS and BESSY resulting in some difficulties. Taking into account this circumstance and all the above discussed it is necessary to consider the following options:

Option 1. The simplest ALS scheme with modulator-undulator and after a large distance bending magnet radiator and space separation of fs X-ray pulses provides low intensities. However, one can compensate this drawback using instead of bending magnet, radiators of transition, channeling, and even very monochromatic parametric X-ray and X-ray Cherenkov radiation, since it is known that for single passage of electrons these mechanisms provide higher X-ray spectral distribution than SR. For the new ~3 GeV machines with their parameters (length of straight sections, transversal coordinate and angular beam spread, etc) it is necessary with the help of analytical calculations and simulations to study: Is it possible to use amorphous and/or crystalline radiators (with goniometer) placed in places where the fs pulse electrons are separated from the electrons main ps pulses? Can the fs electron pulses further be deflected by bent crystals? What is the physics interest for having fs pulses of MeV and even GeV gamma quanta produced by the above-mentioned mechanisms and coherent bremsstrahlung?

Option 2. The BESSY scheme with 2 modulator- and radiator-undulators, additional bending magnets and angular separation provides higher intensities. For production of fs pulses it is necessary to study: Can one use radiators for above-mentioned radiation instead of scrappers? Is the fact that fs electrons are flying under larger angles than the ones of ps pulses enough for separating the new types radiation or channeling only the fs electrons and their bending. The last question reminds the proposed method of production attosecond electron and photon pulses at linear SASE FEL[22].

REFERENCES

1. Synchrotron Radiation News, 18, N4, 2, 2005.
2. T. Pfeifer, C. Spielman and G. Gerber, Rep. Prog. Phys. 69, 443, 2006.
3. A.A. Zholents and M.S. Zolotorev. Phys. Rev. Lett. 76, 912, 1996.
4. R.W. Schoenlein et al, Appl. Phys. B71, 1, 2000.
5. R.W. Schoenlein et al, Science, 287, 2237, 2000.
6. C. Steier et al, Proc. PAC2005, Knoxville, Tenesse, p. 4096.
7. S. Khan et al, Proc. PAC2004, Lucerne, p. 2287.
8. K. Holldack, T. Kachel, S. Khan, T. Quast, Phys. Rev. STAB, 8, 040704, 2005.
9. S. Khan, H. Holdack, T. Kachel, T. Quast, Phys. Rev. Lett. 97, 074801, 2006.
10. H. Wiedemann, Particle Accelerator Physics, Springer, Berlin, 1993.
11. H. Holdack, S. Khan, R. Mitzner, T. Quast, Phys. Rev. Lett. 97, 074801, 2006.
12. K.A. Ispirian, M.K. Ispiryan, Femtosecond transversal deflection of electron beams with the help of laser beams and its possible applications, ArXiv:hep-ex/0303044, 2003.
13. E.D. Gazazyan et al, Proc. PAC 2005, Knoxville, Tenessee, pp. 2944 and 4054.
14. A. Zholents et al, Nucl. Instr. and Meth. A 425, 385, 1999.
15. R. Palmer, SLAC-PUB-4107, 1988.
16. K. Harkay et al, Proc. PAC2005, Knoxville, Tenessee, p. 668.
17. D. Robin et al, Proc. PAC2005, Knoxville, Tenessee, p. 3659.
18. R.O. Avakian, K.A. Ispirian, Proc. SPIE, v. 5974, p. 597409, 2005.
19. K.A. Ispirian, A.G. Oganesian, X-Ray Radiation of Ultrarelativistic Electrons in Magnetic Undulators and Its Applications, Preprint YerPhI-ME-4, 1971.
20. A.I. Alikhanian et al, Pisma Zh. Eksp. Teor. Fiz. 15, 142, 1972.
21. V.N. Litvinenko et al, AIP, 395, 1997; Proc. PAC2001, Chicago, 2002, p. 2614.
22. P. Emma et al, Phys. Rev. Lett. 92, 074801, 2004; SLAC-PUB-10712, 2004.

BUNCH TIME STRUCTURE DETECTOR WITH PICOSECOND RESOLUTION

A. Margaryan[a,1], R. Carlini[b], N. Grigoryan[a], K. Gyunashyan[c], O. Hashimoto[d], K. Hovater[b], M. Ispiryan[e], S. Knyazyan[a], B. Kross[b], S. Majewski[b], G. Marikyan[a], M. Mkrtchyan[a], L. Parlakyan[a], V. Popov[b], L. Tang[b], H. Vardanyan[a], C. Yan[b], S. Zhamkochyan[a], C. Zorn[b]

[a]*Yerevan Physics Institute, 2 Alikhanian Brothers Str., Yerevan, 375036, Armenia,* [b]*Thomas Jefferson National Accelerator Facility, Newport News, VA 23606 USA,* [c]*Yerevan State University of Architecture and Construction, Yerevan, Armenia,* [d]*Tohoku University, Sendai, 98-77, Japan,* [e]*University of Houston, 4800 Calhoun Rd, Houston, TX 77204 USA*

Abstract: We propose a device measuring bunch time structure of continuous wave beams, based on radio frequency (RF) analysis of low energy secondary electrons, (SEs). By using a currently developed 500 MHz RF deflector it is possible to scan circularly and detect the SEs, amplified in multi-channel plates (MCP). It is demonstrated that the noise induced by RF source is negligible and the signals, generated in MCP, can be processed event by event, without integration, by using available nanosecond-time electronics. Therefore, this new device can be operated alone as well as in combination with more complex setups, using common electronics and providing time resolution for single SE better than 20 ps.

Key words: Bunch shape; Secondary electrons; RF deflector; Picosecond techniques

1.   INTRODUCTION

It is well known that timing systems based on radio frequency (RF) fields can provide picosecond temporal precision. For instance, a resonant microwave cavity was used previously as a deflecting system to investigate the structure of electron bunches in a linear accelerator[1] and in a microtron[2] with ~2 ps temporal resolution. In a photo-cathode RF electron gun, 0.5 ps

[1] Corresponding author; e-mail address: mat@mail.yerphi.am

Vasili Tsakanov and Helmut Wiedemann (eds.), Brilliant Light in Life and Material Sciences, 455–464.
© 2007 *Springer.*

time resolutions have been reached[3]. In these investigations RF deflectors were used to analyze electron bunches directly.

Transverse RF modulation was successfully used in the fifties to measure the upper limit of time dispersion of the secondary emission[4].

The longitudinal profiles of the charged particle beams can be investigated by analyzing the produced low energy secondary electrons as well. In this case, the temporal structure of the beam under study is coherently transformed into that of low energy secondary electrons and then into a spatial structure via RF modulation. The first proposal to use low energy secondary electrons (SEs) for this purpose was made in the early sixties[5]. To obtain a phase-dependent separation of the electrons, it was suggested that their velocity be modulated transversally in two perpendicular directions, i.e. by a circular scan. The circular scan provides a measuring interval equal to a full period of the RF deflecting field. The first real device was proposed and created in the mid-seventies[6].

Improved bunch length detectors (BLDs) with transverse modulation of the velocity of the electrons have been developed and built at the INR[7-9]. In these devices, the velocity of the electrons is transversally modulated by an RF sweep field in one direction, providing a phase-dependent separation after a drift space. These BLDs have been installed in the INR linac and are being used successfully to tune the accelerator and to test the beam quality. Monitors of this type have found application in a number of laboratories: CERN, DESY, Fermilab, GSI, KEK, SSC and TRIUMF. Experience with commissioning these new beam diagnostic devices, their use for beam measurements, and studies of the machine performance were discussed elsewhere[10]. All these BLDs have been operated in pulsed mode. Recently, a similar BLD with capability to work in continuous wave, CW regime, has been created and tested in Argonne National Laboratory[11]. The device provides measuring of bunch phase spectrum integrated over a 33 ms period, therefore does not allow to recognize the precise value of the fluctuation frequencies.

Streak cameras based on similar principles are used routinely for measurements in the picosecond range and have found an increasing number of applications; in particular, in particle accelerators permitting both precise measurements and instructive visualizations of the beam characteristics and behavior that can not be obtained using other beam instrumentation[12]. With a streak camera operating in the repetitive mode known as "synchroscan", typically, a temporal resolution of 2 ps (FWHM) can be reached for a long time exposure (more than one hour) by means of proper calibration[12, 13].

The basic principle of the RF timing technique or streak camera's operation is the conversion of the information in the time domain to a spatial domain by means of ultra-high frequency RF deflector. Both techniques provided measuring of bunch phase spectrum mainly integrated over long time, tenths of ms, period.

This paper is devoted to the description of bunch length measuring technique, which is based on the recently developed 500 MHz RF circular sweep deflector[14,15]. Dedicated secondary electron detection system allows detect circularly scanned single electrons with time resolution better than 20 ps FWHM by using available nanosecond electronics. The results of investigations of device characteristics by using thermionically emitted electrons are presented.

2. PRINCIPLE OF OPERATION AND OBTAINED RESULTS WITH THERMO ELECTRON SOURCE

The primary beam hits the target wire (1) and produces SEs. These electrons are accelerated by a negative voltage V applied to the target wire and collimated by the narrow whole collimator-electrode (2). An electrostatic lens (3) then focuses the electrons onto the SEs detector plane at the far end of the device (7). Along the way, the electrons are deflected by the circular sweep RF deflection system, consisting of electrodes (4) and $\lambda/4$ coaxial RF cavity (6), which operates with a frequency v equal to multiple of the accelerator bunching frequency. Thus a SE image of the beam bunches on the screen is formed and is detected (5).

The general layout of the proposed device is shown in Fig. 1.

The distribution of the SEs on the circle is a function of the phase angle and is proportional to the time structure of the primary beam. Therefore the time structure of the primary beam bunch is coherently transferred to the SE distribution on the scanned circle. The termionically emitted electrons are distributed randomly and form a circle. Several factors determine the time resolution of such a device[16, 17]:

1. **Physical time resolution of the SE emission, i.e., the time dispersion or delay of the electron emission.** The value of time dispersion for metals was estimated theoretically to be 10^{-14} s [18] and was measured experimentally to be smaller than 6 ps[4].

2. **Physical time resolution of the electron tube** is mainly determined by the SEs' initial energy and angular spread. A Monte Carlo code has been developed for the accurate estimate of the spread, ΔT in electron transit times from the wire to the collimator. The energy distribution of SEs were generated using $P(E) = (E/E_0)\exp(-E/E_0)$, where E is the kinetic energy of the emitted electron and $E_0 = 1.8$ eV is the most probable electron

energy. Such energy dependence corresponds to the experimental energy spectrum of SEs knocked out of metallic foils[18]. In calculating the transit time fluctuations the actual structure of the target wire electrostatic field was considered. For 100 μm diameter target wire, V=2.5 kV applied negative voltage and 2 mm diameter collimator at L=1cm distance from target wire, one obtains $\Delta T = 2.5$ ps.

3. **Technical time resolution of the electron tube** is determined by the electron transit time dispersion and in a carefully designed system this time dispersion can be minimized to be in ps range.

4. **Technical time resolution of the RF deflector.** By definition the technical time resolution is $\Delta\tau_d = d/v$ where d is the size of the electron beam spot or the position resolution of the secondary electron detector if the electron beam spot is smaller, while v is the scanning speed: $v = 2\pi R/T$ here T is the period of the RF field, R is the radius of the circular sweep on the position-sensitive detector. For example, if $T = 2\times10^{-9}$ s (f = 500MHz), R = 2cm, and d = 1.0mm, we have $v \geq 0.5\times10^{10} cm/s$ and $\Delta\tau_d \leq 20\times10^{-12} s$. Consequently, the phase resolution $\Delta\phi/\phi = d/2\pi R \leq 0.008$.

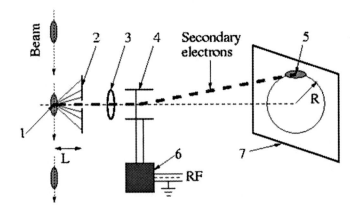

Figure 1. The schematic layout of the BLD (for notations see the text).

To scan circularly 2.5 keV electrons we have used a dedicated RF deflecting system developed by our group[14,15] at a frequency of $f = 500\,MHz$. This RF deflecting system consists of a $\lambda/4$ coaxial cavity and deflection electrodes. It is constructed in such a way that the deflection electrodes serve as a capacitance element which contains all the applied RF power localized inside. The deflection electrodes formed part of resonance circuit with a Q-factor of about 130. In addition the special design of the deflection electrodes allows avoiding transit time effects. The sensitivity of this new and compact RF deflector is about 1mm/V or $0.1\,rad/W^{1/2}$ and is an order of magnitude higher than the sensitivities of the RF deflectors used previously. The experimental setup has been tested by using thermo-electrons emitted from a heated wire. The wire was heated up by current flowing through it. For visual tuning of the experimental setup a phosphor screen is situated at the far end of the electron tube. About 1 W (on 50Ω), 500 MHz RF power have been used to scan circularly and reach 2 cm radius (see Fig. 2), or 20 ps resolution, for 2.5 keV electrons. For comparison, in the reference[11] to reach ~20 ps time resolution or 16 cm transverse deflection, 30 kW of 97 MHz RF power has been used for 10 keV electrons.

2.1 SE detector

The detection of the SE beam is accomplished with position-sensitive detector based on multi-channel plates, (MCPs)[15,19], the schematic of which is displayed on the Fig. 3. A dual, chevron type, MCP detection system is used to obtain a high gain. Both MCPs are identical, have 32 mm diameter and can sustain a maximum bias of 1000 VDC each. The dual MCP system has a combined maximum gain of about 10^7. The position sensitive resistive anode (Fig. 4) is situated about 3 mm behind the second MCP. It is biased at ~300V relative to the MCP output to allow electrons multiplied in the MCP reach the anode. In order to tune the system under visual control, the anode has a hole in the center and a phosphor screen is placed about 3 mm behind it. It is biased at ~2 kV relative to the MCP output to convert the accelerated electrons kinetic energy into visible photons. We use wire planes made of 2 mm thick G10 plates to feed bias voltages. Both planes use 20μm diameter gold-plated tungsten wires and have wire spacing 1

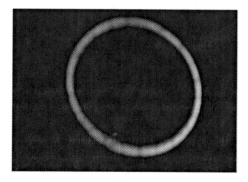

Figure 2. Photograph of the circularly scanned 2.5 keV electron beam on the phosphor screen.

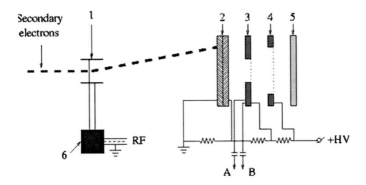

Figure 3. Schematic view of the position sensitive detector based on microchannel plates 1- RF deflector, 2- micro-channel plates, 3- position sensitive anode and accelerating electrode, 4- accelerating electrode, 5- phosphor screen, 6- quarter wavelength coaxial cavity, A and B are position signals.

The visually tuned image of circularly scanned 2.5 keV electron beam amplified in the MCP is also displayed on the Fig. 4.

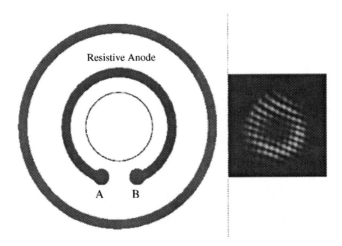

Figure 4. Schematic of the resistive anode (left side) and image of circularly scanned and amplified in the MCP 2.5 keV electron beam on the phosphor screen (right side).

The shadow of the wires of both electrodes is seen clearly and can be used to determine an absolute scale of the image. From Fig. 4 it follows that the beam size of amplified in the MCP SE is about mm, therefore, for R = 2 cm scanned circle, the time resolution of detection of a single SE is expected to be $\Delta\tau_d \leq 20 \times 10^{-12}$ s. By increasing the RF power we match the size of the circle to the size of the position sensitive anode. Then we decrease the intensity of SE flux by decreasing the heating current, to be able to read event by event the signals A and B on the oscilloscope. Typical signal, detected by digital 400 MHz scope, is displayed in Fig. 5. It consists of two parts: signal generated in MCP by circularly scanned 2.5 keV single electrons; and signal induced by the RF deflector's 500 MHz RF noise. One can see that the amplitude of the induced 500 MHz RF noise from 1 W RF power is about an order of magnitude smaller than the amplitude of amplified in MCP signals of single SE.

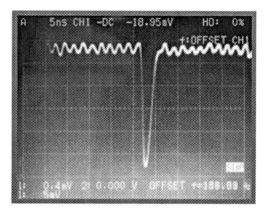

Figure 5. The oscilloscope screen showing the signal of circularly scanned single SE from position sensitive anode.

Thus, signals from such a device can be processed without integration, by using common nanosecond-time electronics (amplifiers, discriminators, ADCs, logic units), and time resolution better than 20 ps can be achieved for single SE. This device can be used for bunch time structure measurement. Detection of SEs, event by event, by using nanosecond fast electronics, allows observation of bunch center fluctuation frequencies and amplitudes with respect to the reference RF phase. For comparison in the case of BLDs in which integrated signals are used, only the amplitude of the energy jitter can be measured[11].

The time resolution can be improved, if necessary, by using higher harmonics of bunching frequencies, since it is possible to operate the developed RF deflector in the frequency range 500-1500 MHz and scan circularly keV energy electrons. Several picosecond intrinsic time resolutions are achievable, which is negligible to the pulse lengths to be measured in most accelerator applications (30-200ps FWHM).

By using direct readout scheme such as an array of small (~1mm^2) pixels with one readout channel per pixel, the device can be used as a single bunch length digitization tool in the sub nanosecond domain.

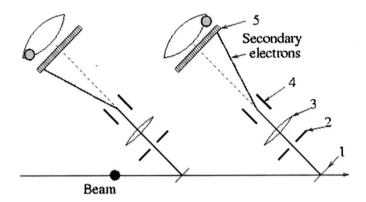

Figure 6. The schematic layout of the time-of-flight system for heavy ions. 1- Thin foil for producing SE, and accelerating electrode, 2- collimator-electrode, 3- electrostatic lens, 4- RF deflector, 5- SE detector.

Important additional feature of the technique is the ability of operation in coincidence with other technique or fast electronic gating. This helps extracting phenomena like partial instabilities that would otherwise be overshadowed. The fast gating is achieved by processing signals with common nanosecond electronics.

Bunch length measurement is important for study of beam dynamics and accelerator tuning. The performance of Storage Ring based FELs depends much on the bunch length and longitudinal phase stability of the electron beam. The proposed technique can be used to observe various effects of coupling, detuning and interaction between the electron and FEL light pulse, while assessing the effects of feed-back systems.

Two such kind identical devices placed at some distance from each other (Fig.6) can serve as a precise time-of-flight (TOF) measurement system in the heavy ion beams. Here again by using common nanosecond electronics, better than 20 ps time resolution can be achieved.

3.   CONCLUSIONS

A new 500 MHz RF timing device for the measurement of bunch time profile has been developed and tested with thermoelectrons. The device is capable of measuring the bunch shape of CW beams by detecting and analyzing, event by event, the secondary electrons with better than 20 ps resolution.

We intend to investigate possibilities of using this technique for detection of delayed fission with subnanosecond lifetimes in future hypernuclear experiments with CW electron beams[20, 21]. Beside this application, the technique can be used in heavy ion beam experiments[22] as a compact and high precision TOF measurement instrument.

ACKNOWLEDGMENTS

We thank H. Asatryan, R. Avakian, H. Bayatyan, H. Bhang, D. Bowman, L. Cardman, K. Egiyan, R. Ent, K. Ispiryan, E. Laziev, C. Leemann, Ed Hungerford, B. Mecking, H. Mkrtchyan, E. Oset, M. Petrosyan, V. Sahakyan, A. Sirunyan and S. Taroyan for their interest of this work.

This work was supported in part by International Science and Technology Center- ISTC, Project A-372.

REFERENCES

1. L. E. Tzopp, Oscillography of relativistic electron beams, Radio Engineering and Electron Physics 4 (1959) 1936.
2. V. P. Bykov, Investigation of Electron Bunches in a Microtron, Soviet Physics JETP 13, n6 (1961) 1169.
3. X. J. Wang, X. Qiu, and I. Ben-Zvi, Experimental observation of high-brightness microbunching in a photocathode rf electron gun, Phys. Rev. E 54, R3121 (1996).
4. E. W. Ernst and H. Von Foerster, Time Dispersion of Secondary Electron Emission, Appl. Phys. 26 (1955) 781.
5. I. A. Prudnikov et all, A Device to Measure a Bunch Phase Length of Accelerated Beam, USSR invention license, H05h7/00, No.174281 (in Russian).
6. R. L. Witkover, A Non-Destructive Bunch Length Monitor For a Proton Linear Accelerator, Nucl. Instr. and Meth. 137 (1976) 203.
7. A. V. Feschenko, Methods and Instrumentation for Bunch Shape Measurements, Proceedings of the 2001 Particle Accelerator Conference, Chicago, 2001, p. 517
8. A.V. Feschenko, Bunch Shape Monitors Using Low Energy Secondary Electron Emission. AIP Conf. Proc. No. 281, Particles and Fields, Series 52, Accelerator Instrumentation Forth Annual Workshop, Berkeley, Ca. 1992, p.185.
9. S. K. Esin, A.V. Feschenko, P. N. Ostroumov, INR Activity in Development and Production of Bunch Shape Monitors. Proc. of the 1995 Particle Acc& Conf. and Int. Conf. on High-Energy Accelerators, Dallas, May 1-5, 1995, p.2408.
10. P. N. Ostroumov, Review of beam diagnostics in ion Linacs, Proceedings of the 1998 Linac Conference, Chicago, IL, August 23–28, 1998, p. 724.

11. N. E. Vinogradov et al., A Detector of Bunch Time Structure for CW Heavy-Ion Beams. Nucl. Instr. and Meth. A526 (2004) 206.
12. K. Scheidt, Review of Streak Cameras for Accelerators: Features, Applications and Results, Proceedings of EPAC 2000, Vienna (2000) p. 182.
13. Wilfried Uhring et al., Very high long-term stability synchroscan streak camera, Rev. Sci. Instrum. 74 (2003) 2646.
14. R. Carlini, N. Grigoryan, O. Hashimoto et al., Proposal for Photon Detector with Picosecond Resolution, H. Wiedemann (ed), Advanced Radiation Sources and Applications, NATO Science Series, Vol. 199, 2006 Springer, p. 305.
15. A. Margaryan, R. Carlini, R. Ent et al., Radio frequency picosecond phototube, accepted for publication in Nucl. Instr. and Meth. A (2006).
16. E. K. Zavoisky and S. D. Fanchenko, Image converter high-speed photography with sec time resolution, Appl. Optics, 4, n.9 (1965) 1155.
17. R. Kalibjian et al., A circular streak camera tube, Rev. Sci. Instrum. 45, n.6 (1974) 776.
18. I. M. Bronstein, B. S. Fraiman, Secondary Electron Emission, Moscow, Nauka, 1969 (in Russian).
19. G. Pietri, Contribution of the channel electron multiplier to the race of vacuum tubes towards picosecond resolution time, IEEE Transactions of Nuclear Science, NS-24, No.1 (1977) 228.
20. A. Margaryan, L. Tang, S. Majewski, O. Hashimoto, V. Likhachev, Auger Neutron Spectroscopy of Nuclear Matter at CEBAF, LOI to JLAB PAC 18, LOI-00-101, 2000.
21. S. Majewski, L. Majling, A. Margaryan, L. Tang, Experimental Investigation of Weak Non-Mesonic Decay of Hypernuclei at CEBAF, e-Print Archive: nucl-ex/0508005.
22. Richard Pardo et al., RIA Diagnostics Development at Argonne. http://www.oro.doe.gov/riaseb/wrkshop2003/papers/p-2-4-4.pdf.

AUTO-CALIBRATING BASED DISPERSOMETER FOR HIGH-ACCURACY ALIGNMENT SYSTEMS

Vladimir Hovhannisyan[1], Davit Baghdasaryan[1], Ara Grigoryan[1], Hrant Gulkanyan[1], Ingesand Hilmar[2]
[1]*Yerevan Physics Institute, Yerevan Armenia,* [2]*Swiss Federal Institute of Technology, Zurich, Switzerland*

Abstract: Simple technique for determination of the dispersion parameters in two-wavelength dispersometer and subsequent correction for distortion based on the analysis of the standard deviations of measured coordinates of two color beams is suggested and realized experimentally.

Key words: Atmospheric turbulence, sight correction, dispersion

1. INTRODUCTION

Different optical measuring and monitoring systems are applied during accelerators construction as well as for alignment of detectors and magnets[1,5]. However the atmospheric turbulence, air density and temperature gradients and other optical effects caused, in particular, by heated components of large experimental, energy and industrial constructions produce contortions in indications of the instrument. Very often atmosphere induced errors exceed admissible errors more than an order[2,3]. For correction for atmosphere related effects a very elegant technique with using two-wave-lengths optical radiation has been suggested[4,5]. In this technique the spatial separation of two beams with different frequencies is used to obtain a correction factor to be applied in determining the true line of sight. The quantity of the atmospheric

Vasili Tsakanov and Helmut Wiedemann (eds.), Brilliant Light in Life and Material Sciences, 465–468.
© 2007 *Springer.*

refraction induced light bending is determined from relationship $L_1 = \Delta L /$ $(1-n_2/n_1)$, where L_1 is the absolute displacement from the straight line of sight of the first wavelength beam, n_1 and n_2 are refractive indices of a medium for the first and second components respectively, and ΔL is separation between the two wavelength beams at a plane of measurement. The optical systems (dispersometers) have been developed for precise measurements of angle[3] and transversal coordinates[6].

2.   THEORY AND METHOD OF DISPERSOMETRY

The light propagation through a medium with the refractive index n is described by the following differential equation

$$\frac{d}{ds}(n\frac{d\vec{r}}{ds}) = \vec{\nabla}n$$

where s is the length of the light path measured from the emitting point and $\vec{r}$ is its position radius-vector. For the air medium with the temperature 'profile' $t(\vec{r})$ and the pressure P (we ignore its $\vec{r}$ - dependence), the dependence of the refractive index on $\vec{r}$ and the wavelength λ can be described by the following empirical parameterization

$$n(\vec{r},\lambda) = 1 + \frac{P}{P_0}\frac{\delta(\lambda)}{1+kt(\vec{r})} \text{ with } \delta(\lambda) = 10^{-6}(272.6 + \frac{1.5294}{\lambda^2} + \frac{0.01367}{\lambda^4})$$

where λ is in μ m and $k = 1/273\cdot15°C$. Here we neglect 0.3% correction to $\delta(\lambda)$ caused by the atmospheric moisture.

It can be shown that, when two light beams with λ_1 and λ_2 pass a thermally inhomogeneous medium, the ratios $r_x = \Delta x(\lambda_1)/\Delta x(\lambda_2)$ and $r_y = \Delta y(\lambda_1)/\Delta y(\lambda_2)$ of their deflections from a straight line are practically independent of the temperature gradient along their path and are equal to

$$r_x = r_y = \delta(\lambda_1)/\delta(\lambda_2)$$

However, accurate experiments have shown that intrinsic dispersion of the registration system and inexact coincidence of two probe beam trajectories bring to result that the experimental quantity $\Delta x,y(\lambda_1)/\Delta x,y(\lambda_2)$ can essentially differ the theoretical value $r_{x,y}$.

In the present report we use simple technique for determination of the parameters r and subsequent correction for distortion based on the analysis of the standard deviations and the correlation coefficient of measured coordinates of two beams.

The technique is based on principle utilizing optical turbulence for the determination of the combined dispersion of the atmosphere and the detection system[3]. It was shown that parameters r is possible to determine solely based on the acquired data from relation: $r = 2\sigma^2_1/(\sigma^2_1+\sigma^2_2-\sigma^2_{1-2})$, where σ_1, σ_2 and σ_{1-2} are mathematical dispersions of Y_{uv}, Y_{ir} (or X_{uv} X_{ir}) and ΔL respectively (see Fig. 1).

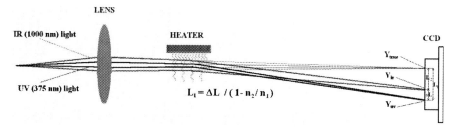

Figure 1. Optical scheme of the experiment.

Measurements were carried out on the setup described in[6] with using UV (375 nm) and IR (1000 nm) LEDs and CCD camera, as well as, in the slightly modified setup, with laser diodes and position sensitive detector (PSD). It was shown, that heat induced distortions, reaching up to 1.2 mm, were corrected by this technique so that transversal coordinates coincided with those in absence of air heating with an accuracy of 10 microns, i.e. in limits of admissibility during the construction alignment in industry and accelerator physics[1-3] (Fig. 2-3).

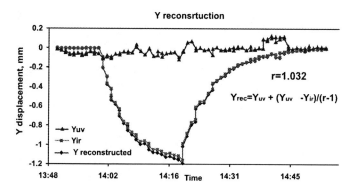

Figure 2. Light spot Y coordinate of UV and IR images and the reconstructed coordinate. Light deflection has been induced by heater switched on from 14:00 to 14:20 h.

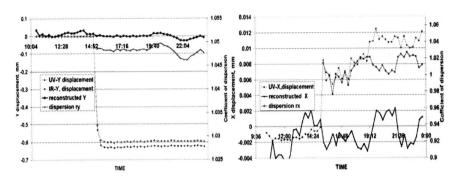

Figure 3. X and Y coordinates reconstruction in long -term measurement.

Single- ended version of the dispersometer when light sources, registration and synchronization systems and controlling computer are located on the same side of the test area, while a mirror is placed on the monitored object is developed and tested.

This system is especially effective for long-distance monitoring or when the object is temporarily inaccessible (at high radiation or temperature, harmful conditions etc.).

3.   REFERENCE

1.   Dekker H. et al. The RASNIK/CCD 3-dimensional Alignment System, Proc. IWAA93, 3-rd Int. Workshop on Accelerator Alignment, CERN, Geneva, 1993.
2.   Grigoryan A.A. and Teymurazyan. A. Light ray displacements due to   air temperature gradient. ALICE INT-2000-13, 2000.
3.   Bockem B.. Development of a Dispersometer for the Implementa-tion into Geodetic High-Accuracy Direction Measurement Systems. Diss. ETH, No14252, Zurich, 2001.
4.   Geodetic Refraction, Effect of Electromagnetic Wave Propagation Through the Atmosphere, New York.1984.
5.   US Patent 5233176: Precision Laser Surveying Instrument using Atmospheric Turbulence Compensation, 1993.
6.   Azizbekyan H. et al. Double-Lambda Sensor Prototype for Optical Alignment in Inhomogeneous Environment. IWAA 2004, CERN, Geneva, 2004.

THIN WIRE SECONDARY EMISSION TARGET FOR PICO-FEMTO SECOND RF TIMING TECHNIQUE

S. Zhamkochyan
Yerevan Physics Institute, Alikhanyan Bros. St. 2, Yerevan, Armenia

Abstract: A dedicated Monte-Carlo code is developed for the accurate estimation of the spread in electron transit times for thin wire secondary electron emission target. In calculating the transit time fluctuations the energy distribution of secondary electrons and the actual structure of the target wire electrostatic field was considered. It demonstrated that in the case of thin wire the femto second time spread for secondary electrons could be achieved with moderate applied voltages. For example, in the case of 10 kV applied voltage and 10 μm of wire radius, the time dispersion is in the rage of 70 fs.

Key words: Bunch length detector, Thin wire, Secondary electrons, Monte-Carlo, pico-femto second.

1. INTRODUCTION

Measurements of the longitudinal distribution of charge in a bunch and the longitudinal emittance are among the most important tools for a beam dynamics study or an accelerator tune. The main requirements of a bunch shape detector are high time resolution, low beam distortion and wide range of measurements in primary beam characteristics. One of the types of the bunch length detectors (BLD) is based on the secondary electron emission from the thin wire. The schematic picture of the concerned device is similar to the BLD developed for CW beams[1,2] and is shown on Fig. 1.

Vasili Tsakanov and Helmut Wiedemann (eds.), Brilliant Light in Life and Material Sciences, 469–473.
© 2007 *Springer.*

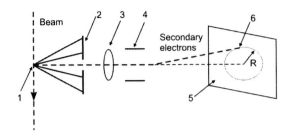

Figure 1. The schematic layout of the bunch length detector.

The primary beam hits the target wire (1) and produces SEs. These electrons are accelerated by a negative voltage V applied to the target wire and collimated by the narrow whole collimator (2). An electrostatic lens (3) then focuses the electrons onto the screen at the far end of the detector (5). Along the way electrons are deflected by the circular sweep RF deflector (4)[2], which oscillates with frequency equal to multiple of the accelerator bunching frequency. Thus a SE image of beam bunches on the screen and is detected (6). The time structure of the primary beam bunch is coherently transferred to the SE distribution on the circle image on the screen.

The time resolution limit[3] of the detector mainly depends on the time dispersion between the wire and the narrow whole collimator. As it is shown bellow it can be in a range of femto seconds. Such high resolution is achieved due to fast acceleration of the electron in the electric field of the thin wire (Fig. 2).

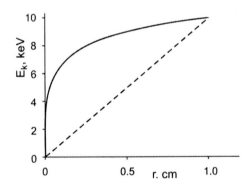

Figure 2. The rise of the electron kinetic energy with moving from the surface of the electrode for the cases of flat electrode (dash line) and the wire with 20 μm thickness (solid line).

2.   THE MONTE-CARLO SIMULATION

The FORTRAN code was written for Monte-Carlo simulation of electrons transit times and time dispersion.

The electric field was calculated according to the method of images. Since the electric potential of wire on distance r is $-2 \times \kappa \times \ln(r)$, where κ is the charge per unit length on wire, the field of the induced charges is replaced by the field of similar symmetrically placed opposite charged wire (such substitute keeps the plate equipotential).

For calculation of time dispersion the following expression was used:

$$\sigma = ((t_i - t_{av})^2/n)^{1/2},$$

where t_{av} is the average transit time of electrons, n is the number of generated electrons, t_i is the transit time of i-th electron. For the energy distribution of secondary electrons the formulae $P(E) = (E/E_0)\exp(-E/E_0)$ was used, where E is the kinetic energy of the emitted electron and E_0 is the most probable electron energy. The energy dependence given above corresponds approximately to the experimental energy spectrum of electrons knocked out from aluminum foils by 5.48 MeV alpha particles, if one takes[4] $E_0 = 1.8$ eV. For the polar and azimuthal angles isotropic distributions were used.

3.   RESULTS OF THE MONTE-CARLO SIMULATION

On Fig. 3 obtained time dispersions depending on the wire's radius are shown for 2.5, 5.0 and 10.0 kV applied voltages. One can see that the time dispersion may be notably reduced by using thinning the target wire and femto second time spread for secondary electrons can be achieved by using moderate voltages, e.g. for 10 μm radius and 10 kV applied voltage σ is around 70 fs. For comparison by using flat electrodes 70 fs resolution can be achieved with 170 kV applied voltage. The similar detector considered in[1] provided resolution in the range of 20 ps.

The dependence of electron transit time dispersion on the applied voltage for 10 μm wire's radius is shown on Fig. 4.

The calculations were also made for the case of two-plate collimator. The electric field was measured by replacing the field of induced charges with some great (theoretically infinite) number of alternately charged wires. The simulation has shown the similar result for the time resolution.

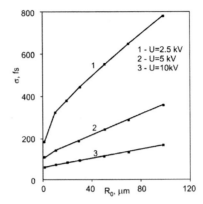

Figure 3. Rising of the electron's kinetic energy by the moving from the surface of the electrode for the cases of flat electrode (dash line) and the wire with 20 μm thickness (solid line).

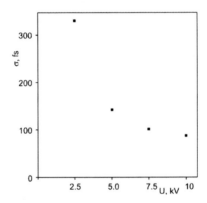

Figure 4. The dependence of electron transit time dispersion on the applied voltage for 10 m wire's radius.

4.   CONCLUSIONS

One can conclude that using the thin wire as a SE emitter allows to achieve several tens fs time resolution due to fast acceleration of electrons into the wire's electric field. Along with RF-timing technique it can be successfully used in BLDs for measurements of the longitudinal distribution of charge in a bunch.

REFERENCES

1. N. E. Vinogradov et al., A detector of bunch time structure for CW heavy-ion beams, Nucl. Instr. and Meth. A **526** (2004) 206.
2. A. Margaryan et al., Bunch time structure detector with picosecond resolution, these proceedings.
3. E. K. Zavoisky and S. D. Fanchenko, Image converter high-speed photography with $10^{-12} - 10^{-14}$sec time resolution, Appl. Optics, 4, n.9 (1965) 1155.
4. M. Bronstein, B. S. Fraiman, Secondary Electron Emission, Moscow, Nauka, 1969 (in Russian).

VIBRATION WIRE MONITOR FOR PHOTON BEAMS DIAGNOSTICS: PRELIMINARY TESTS ON LASER BEAMS

M.A. Aginian, S.G. Arutunian, M.R. Mailian
Yerevan Physics Institute, Alikhanian Brothers St 2., 375036 Yerevan, Armenia

Abstract: Developed vibrating wire scanner showed high sensitivity to the charged particles beam intensity (electron, proton, ions). Since the mechanism of response of frequency shift due to the interaction with deposited particles is thermal one, the vibrating wire scanner after some modification can be successfully used also for profiling and positioning of photon beams with wide range of spectrum and intensity. Some new results in this field are presented.

Key words: Beam diagnostics, Vibrating wire

1.   IRRADIATED WIRE FREQUENCY

One of the widespread methods of diagnostics in accelerators is the scanning of charged particles beams by thin wires (see e.g.[1,2]). The principle of operation is based on measurement of fluxes of secondary particles scattered off the wire and proportional to the quantity of primary particles colliding with the wire. We suggested[3] to use wire heating quantity as a source of information about the number of interacting particles. To do this the wire is strained and temperature shift is measured by wire natural oscillations frequency. The rigid fixation of the wire ends on the base allowed to obtain an unprecedented sensitivity of the frequency to the temperature and so to the flux of colliding photons. Such vibrating wire scanners were used for electron, ion, proton and photon beams profiling. Equation of transverse oscillations in one transverse direction of a stretched wire taking into account its elasticity is written as

Vasili Tsakanov and Helmut Wiedemann (eds.), Brilliant Light in Life and Material Sciences, 475–478.
© 2007 *Springer.*

$$I_M EX'''' - TX'' + \rho S\ddot{X} = 0 , \tag{1}$$

where X is the transverse deviation of the wire with length l, prime means derivative along the z axis of the wire, point means time derivative, I_M is the moment of inertia of the wire transverse cross-section (for a round wire of diameter d: $I_M = \pi d^4 / 64$), E is the wire modulus of elasticity, T is the wire tension, S is the wire cross-section, ρ is the material density.

Solving equation (1) by the method of partition at boundary conditions on solution of $X_1(0) = X_1'(0) = X_1(1) = X_1'(1)$ leads to the form

$$X_1 = c_1 \sin k_1 x + c_2 \cos k_1 x + c_3 \sinh k_3 x + c_4 \cosh k_3 x , \tag{2}$$

$$k_1 = \sqrt{(\sqrt{1 + 4\varepsilon\Lambda^2} - 1)/2\varepsilon}, \quad k_3 = \sqrt{(\sqrt{1 + 4\varepsilon\Lambda^2} + 1)/2\varepsilon}.$$

where $x = z/l$, $\varepsilon = d^2 E / 16 l^2 \sigma$, $\sigma = T/S$ and Λ is the root of equation:

$$2\frac{k_1}{k_3}\frac{1}{\cosh k_3} - 2\frac{k_1}{k_3}\cos k_1 + (1 - \frac{k_1^2}{k_3^2})\sin k_1 \tanh k_3 = 0 . \tag{3}$$

This degeneracy condition defines the dependence of the oscillations frequency on the task parameters. In our case of stretched vibrating wire $\varepsilon \ll 1$ and $\Lambda \approx \pi(1 + 2\sqrt{\varepsilon})$ so the wire oscillations first harmonics is:

$$f_1 = \frac{1}{2l}\sqrt{\sigma/\rho}(1 + \frac{d}{2l}\sqrt{E/\sigma}) . \tag{5}$$

The resonator can be represented as a base with wire with coefficients of thermal expansion α_B and α_S. If the electromechanical resonator is assembled at a certain temperature T_0 with unstrained wire length l_{S0} and the distance between wire ends fixation l_{B0}, the initial strain of the wire is $\sigma_0 = E(l_{B0} - l_{S0})/l_{S0}$. The relative wire frequency and strain change is then:

$$\frac{\Delta f}{f} \approx \frac{\Delta\sigma}{\sigma_0} = \frac{E}{\sigma_0}\frac{\alpha_B(T_B - T_0) - \alpha_S(T_S - T_0)}{l_{S0}} . \tag{6}$$

where T_S and T_B are average temperatures of the wire and base.

In Table 1 some characteristic values of $E/\sigma_{0.2}$ ($\sigma_{0.2}$, is a tension, for which the residual deformation after removal of the load is 0.2 %) for some materials are presented. We present also the temperature range $\Delta T_S = \sigma_{0.2}/\alpha_S E$ of the vibrating wire sensor.

Table 2. Temperature range of vibrating wire sensor.

Material	E, GPa	α_s, 1/K	$\sigma_{0.2}$, GPa	ΔT_S, K
Beryl Bronze hard	130	$1.90 \cdot 10^{-5}$	0.9	482
Wolfram Recrystal	400	$4.70 \cdot 10^{-6}$	0.5	433
Titan ($\sigma_{0.2}$ at 20^0 C)	110	$9.86 \cdot 10^{-6}$	0.66	909
Platinum, therm. Treated	160	$9.70 \cdot 10^{-6}$	0.07	341
SiC, fiber	400	$4.50 \cdot 10^{-6}$	2.0	856

The photon beam absorption in the wire causes heating of the wire. The average heating quantity q_1 from one photon passing though the wire is:

$$q_1 = K_{tr} \varepsilon \alpha_{abs} \frac{\pi}{2} r, \tag{8}$$

where α_{abs} is absorption coefficient, ε is photon energy, K_{tr} describe transformation ratio of photon energy losses into the wire heat and r is the wire radius. The power Q_S heating the wire is $Q_S = q_1 N_S$, where N_S is the number of photons deposited on the wire. Due to thermal losses mainly on thermoconductivity and convection (in air) Q_S causes the wire heating with respect to environment temperature by the value

$$T_{mean} = Q_S / (8\pi r^2 \lambda / l + 2\alpha_{conv}\pi r l), \tag{10}$$

where λ is the thermal conductivity coefficient of the wire, α_{conv} is the convection coefficient of the wire surface. Corresponding frequency shift is:

$$\Delta f = -0.25(fE/\sigma)\alpha_S T_{mean}. \tag{11}$$

2. LASER BEAM SCAN

The laser beam was scanned in air by a Vibrating Wire Monitor (VWM) with two wires located at distance 1 mm from each other. The operational aperture with diameter 5 mm is enough for narrow photon beams scanning. We used the feed system of a microscope with microfeed step accuracy 1.25 µm. Scan of the laser of about 1 mW power was done at different speeds. Typical scans at speed 66 µm/s and 7.6 µm/s forwards and backwards are presented at Fig. 1.

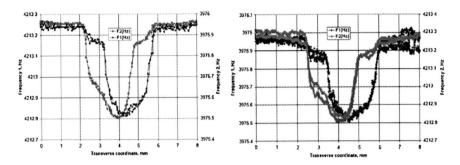

Figure 1. Laser beam scan at speed 66 μm/s (left) and 7.6 μm/s (right).

Shift of frequency is about 0.45 Hz corresponding to temperature increment about 0.02 K. So the slope of frequency dependence on wire temperature is about 20 Hz/K. The accuracy of frequency measurement is about 0.01 Hz, which corresponds to temperature resolution $5 \cdot 10^{-4}$ K. According to formula (9) the power deposited on the wire in the beam center about 17 μW.

In air experiments dominate the term of convectional losses 15 μW (at $\alpha_{conv} = 60$ W/m^2/K) compare with thermoconductivity losses about 0.6 μW.

The resolution of the VWM with respect to the density of laser radiation power is $\sim 1.3 \cdot 10^{-4}$ W/cm^2. When VWM is used for optical radiation it is preferable to set it in vacuum and thus increase the sensitivity up to $5 \cdot 10^{-6}$ W/cm^2.

ACKNOWLEDGEMENTS

We are very grateful to J. Bergoz and K. Unser for support in development of Vibrating Wire Monitor for photon beam diagnostics and to G.Decker for kindly consent to test Vibrating Wire Monitor at APS ANL.

REFERENCES

1. Elmfors P., Fasso A., Huhtinen M., Lindroos M., Olsfors J., Raich U. (1997) Wire scanners in low energy accelerators, *NIM (A)*, **396(1-2)**, 13-22.
2. Wittenburg K. (2004) Beam tail measurements by wire scanners, http://www.sns.gov/workshops/20021023_icfa/presentations/KayWireScanner.pdf.
3. Arutunian S.G., Dobrovolski N.M., Mailian M.R., Sinenko I.G., Vasiniuk I.E. , *Phys. Rev. Special Topics - Accelerators and Beams*, **2**, 122801, 1999.

USE OF MULTI-BEAM SCATTERING RADIATION FOR THE ABSOLUTE X-RAY ENERGY CALIBRATION

M.M. Aghasyan, A.H. Grigoryan, A.H. Toneyan, V.M Tsakanov
CANDLE, Yerevan, Armenia

Abstract: The calibration of the diffracted beam from SR monochromator working in the energy region of 5-30 keV by coplanar two-dimensional multi-beam diffraction is presented in this work. The simulation and numerical calculation of the intensity in the multi-beam channel for the silicon monocrystal monochromators were performed. For the given crystal and the energy sub range the optimal construction is suggested.

Key words: monochromator, multi-beam diffraction, energy calibration

1. INTRODUCTION

Resonant X-rays techniques such as absorption fine structure (XAFS), x-ray absorption near-edge absorption spectroscopy (XANES)[1], as well as the study of protein structure by means of x-ray anomalous dispersion (MAD)[2], that measure small changes in the position of spectral features around resonance, require accurate absolute calibration of the monochromatized energy spectra. If the energy is properly calibrated in the experiment, the energy resolution is defined by the monochromator crystal spectral purity and makes up $\Delta E/E \sim 10^{-5}$-10^{-6}. However, the energy of the prepared incident beam to several effects, such as the loss steps in the moving motor rotating the monochromator, the determination of the exact "zero" of the monochromator crystal Bragg angle may have an effect on the beam formation energy value established by the x-ray optical equipment. Such energy deviations require rapid recalibration of the energy reflected from the monochromator. The well-known energy calibration methods are

479

Vasili Tsakanov and Helmut Wiedemann (eds.), Brilliant Light in Life and Material Sciences, 479–483.
© 2007 *Springer.*

a) the usage of the geometry of considerable intensity changes of the radiation provoked by the Bragg secondary reflection in the monochromator crystal ("glitches"); b) EXAFS or XANS absorption determination using usual standard sample; c) the usage of crystal analyzer[3] or crystal energy calibrators[4] for measuring the Bragg angle. These methods bring to surmountable difficulties[3].

The aim of this paper is the development of such structural changes of well-known synchrotron monochromators[5], which will use the phenomena of coplanar n-beam diffraction in the monochromator internal calibration in conditions of the SR continuous spectrum incident on the monochromator crystal. The monochromator is intended to be built is a novelty and, in particular, it can be used in the following beam lines of the electron storage ring of the CANDLE synchrotron facility creation project: general diffraction and dispersion, x-ray absorption spectrum (EXAFS and XANS) studies, x-ray reflection and holography, protein crystallography studies.

2.   MONOCHROMATIZED RADIATION CALIBRATION BY ENERGY

It is well known that multi-beam coplanar diffraction occurs when several families (more than one) of the atomic planes of single crystal monochromator are in simultaneous diffraction conditions for a certain wavelength [6]. It means that at least three points (including the origin) of the crystal reciprocal lattice are on the great circle of the Ewald sphere (Figure 1).

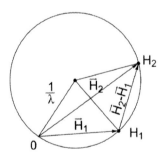

Figure 1. Reciprocal space representations of three-beam x-ray diffraction . -are reciprocal lattice vectors ,0 the origin of the reciprocal space, -reciprocal vector of coupling reflection i.e. H2-H1,K2-K1,L2-L1.

Therefore the radius of the sphere (1/l) is defined by the positional relationship of the reciprocal lattice points. For example, in cubic crystal it is define by the following formula:

$$\lambda = 2a\left(\frac{\left(H_1^2 + K_1^2 + L_1^2\right)\left(H_2^2 + K_2^2 + L_2^2\right) - \left(H_1H_2 + K_1K_2 + L_1L_2\right)^2}{\left(H_1^2 + K_1^2 + L_1^2\right)\left(H_2^2 + K_2^2 + L_2^2\right)\left[\left(H_1 - H_2\right)^2 + \left(K_1 - K_2\right)^2 + \left(L_1 - L_2\right)^2\right]}\right)^{\frac{1}{2}}$$

where H_i, K_i, L_i $(i = 1, 2)$ are the interference indexes of two reflections, a - is the constant of the cubic crystal lattice.

So the coplanar three-beam (n-beam, n ≥ 3) diffraction with certain configuration takes place only for strongly definable wavelength of a primary synchrotron radiation incident on the crystal monochromator. We discuss the diffraction for three-beam case.

Fig. 2 schematically show the monochromator working channel (for simplicity the first block of double-crystal monochromator is shown) and multi-beam-calibration channel. It is supposed that synchrotron radiation ribbon-type beam incidents on the crystal monochromator, a part of which (as small as possible) passing the slot of the monocrystal wall creates three-beam diffraction by $\vec{S}_0, \vec{S}_1, \vec{S}_2$ scattering vectors carrying the λ wavelength consequent three-beam reflections from monolith crystal walls.

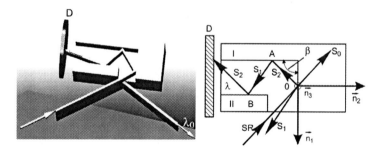

Figure 2. Schematic presentation of the working and multi-beam channels of the monochromator (a) and of the ray path in coplanar three-wave channel (b).

In points A and B consequently reflected three-beam diffraction is recorded by two-dimensional detector D. The working beam (the larger part of the ribbon-type beam) of the monochromator is reflected by λ_0 wavelength from the plains with h_o, k_o, l_o Miller indexes. λ_0 wavelength is not necessarily coincide with the λ wavelength of the three-beam diffraction. The following correlation exists between the scattering vectors

of the three-beam diffraction and the unit vectors of the coordinate system:
$\vec{s}_0 \cdot \bar{n}_2 < 0$, $\vec{s}_0 \cdot \bar{n}_1 > 0$, $\vec{s}_1 \cdot \bar{n}_2 > 0$, $\vec{s}_1 \cdot \bar{n}_1 < 0$, $\vec{s}_2 \cdot \bar{n}_2 < 0$, $\vec{s}_2 \cdot \bar{n}_1 < 0$ and
$\vec{s}_i \cdot \bar{n}_3 = 0$ (if $i = 0,1,2$). Vectors $\bar{n}_1, \bar{n}_2$ and $\bar{n}_3$ are shown in Fig. 2. The
work presents preliminary results of the monochromator energy scaling.
The following configuration of the monochromator is chosen: $\bar{n}_1 \| [111]$,
$\bar{n}_2 \| [1\bar{1}0]$ and $\bar{n}_3 \| [\bar{1}\bar{1}2]$, where [hkl] are the crystallographic directions
normal of the (hkl) plane, interference indexes of working planes. The
preliminary results of the computer simulation are presented in Table 1 and
in Fig. 3 for forth-beam configuration (-4,-4,-4; -2,-6,-4; 2,-2,0). DABAX
data files[6] and multibeam BRL online software[7] are used.

Table 1. Computer simulation results for Si monochromator with configuration of $h_0k_0l_0$ for
the working beam from 2-2 0. ... $\chi_{h1,h2}$ -. Fourier components of crystal susceptibility.

λ_0	θ_0	$H_2 K_2$ L_2	$H_1 K_1$ L_1	N	λ	$\|\chi_{h2}\|$ $\cdot 10^6$	$\|\chi_{h1}\|$ 10^6	$\|\chi_{h1-h2}\|$ $\cdot 10^6$ (coupling)
0.4249	6.3532	-4 0 -2	-1 -5 -3	3	1.6998	0.000	4.869	4.869
0.4771	7.1361	-7 -3 -5	1 -7 -3	3	1.0786	0.973	1.334	0.000
0.5178	7.7494	-3 -3 -3	0 -4 -2	4	2.0713	8.314	0.000	11.069
0.6409	9.6076	-7 -3 -5	0 -8 -4	3	1.0557	0.931	1.366	1.029
0.7111	10.6707	-5 -1 -3	-2 -6 -4	6	1.4221	3.378	3.476	3.378
0.8575	12.9032	-6 -2 -4	-3 -7 -5	3	1.1874	2.398	1.186	2.336
0.9086	13.6861	-6 -2 -4	1 -7 -3	3	1.1758	2.351	1.593	1.285
1.0552	15.9485	-5 -5 -5	1 -7 -3	4	1.2059	1.354	1.678	0.000
1.1243	17.0239	-6 -2 -4	0 -8 -4	4	1.1243	2.144	1.555	1.722
1.1922	18.0870	-5 -5 -5	-3 -7 -5	4	1.1922	1.322	1.197	5.461
1.2420	18.8699	-5 -1 -3	1 -7 -3	4	1.2420	2.561	1.782	2.115
1.3767	21.0083	-5 -1 -3	-3 -7 -5	3	1.1907	2.349	1.193	0.000
1.4515	22.2077	-4 -4 -4	-2 -6 -4	4	1.4515	4.073	3.625	8.144
1.4515	22.2077	-5 -5 -5	0 -8 -4	6	1.1612	1.252	1.662	2.232
1.4515	22.2077	-5 -1 -3	0 -8 -4	6	1.1612	2.232	1.662	1.252
1.7795	27.6062	-6 -2 -4	3 -9 -3	3	0.9366	1.475	0.603	0.436
1.8360	28.5608	-3 -3 -3	-1 -5 -3	8	1.8359	6.497	5.703	13.136
1.9176	29.9570	-5 -1 -3	3 -9 -3	8	0.9588	1.511	0.633	0.666
1.9437	30.4068	-4 -4 -4	1 -7 -3	4	1.3521	3.521	2.123	3.047
2.2740	36.3102	-5 -5 -5	3 -9 -3	4	1.0107	0.941	0.706	0.000
2.4288	39.2315	-4 -4 -4	0 -8 -4	4	1.2144	2.824	1.824	3.632
0.4249	6.3532	-4 0 -2	-1 -5 -3	3	1.6998	0.000	4.869	4.869

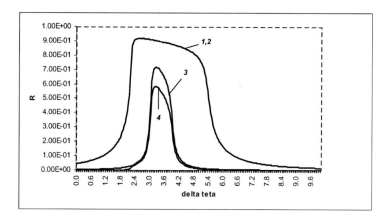

Figure 3. Reflection curves of crystal monochromator in dependence on the Braggs angle. 1-2 for reflection of 2-2 0, 2-2 0 accordingly. 3 - reflection for 4-4-4. 4- 1234 product.

3.   SUMMARY

The calibration principles of Si monochromator based on multi-wave diffraction and the preliminary results of computer simulation are presented. Three-wave diffraction large asymmetry-dependent limitations of energetic sub-region reflected from the monochromator, as well as monochromator calibration simulation for various configurations will be presented in the nearest future.

REFERENCES

1. S. Kraft, J. Stu¨mpel, P. Becker, and U. Kuetgens, Rev. Sci. Instrum. 67,681,(1996)
2. W. Hendrikson, C. M. Ogata, Meth. Enzymol, 276, 494 (1997).
3. J. O. Cross, A.J. Frenkel, Rev. Sci. Instrum, 70, 38 (1999).
4. M. Hart, J. Phys E: Sci Instrum, 12, 911 (1979).
5. M. Hart and Rodrigues ARD J. Appl. Cristallogr 11, 248 (1978).
6. S.-L. Chang, X-ray Multiple-wave Diffraction: Theory and Application, series in Solid-State Science, Vol. 143, Springer-Verlag, 7. Berlin-Heidelberg-New York-Tokyo (2004), 431 pages.
7. http://www.esrf.fr/computing/scientific/dabax/tmp_file/FileDesc.html
8. http://sergey.gmca.aps.anl.gov

GRAVITATIONAL WAVE DETECTORS - CANDLE CASE

Gravitational Wave Detectors, how CANDLE can be used

Avetis A. Sadoyan, Tatevik Sh. Navasardyan, Levon R. Sedrakyan

Affiliation Department of Physics, Yerevan State University, Armenia

Abstract: A proposal is made to establish a low budget Gravitational Wave (GW) detector at CANDLE site that will use as a source synchrotron radiation. The GW detector will work in MHz domain, with a possibility, to be improved, to work in kHz domain and lower. The advantage of using CANDLE as light source is motivated with wide frequency range of light source at CANDLE that will be necessary for "tuning" the detector in different frequencies of GWs.

Key words: Gravitational wave detectors, Synchrotron radiation sources.

Gravitational waves are ripples in space-time, in other words small perturbations of metric tensor that are propagating in space according to wave equation. Gravitational waves were predicted theoretically in the beginning of last century[1], but for a long time, up to mid fifties they were regarded as "coordinate waves" that are disappearing when one changes the coordinate system. Only after works of Piranee F.A.E.[2] who used tetrad method to describe what is happening to test masses in laboratory when a gravitational wave is passing through; it becomes clear that gravitational waves have physical meaning. From textbooks one can learn that there is no gravitational radiation of dipole origin, due to the third law of Newton, dipole component for gravitational radiation is zero[3]. Because of quadrupole origin of gravitational waves they have a very small intensity and are interacting with matter very weekly.

Gravitational waves can be generated if and only if third time derivative of quadrupol moment of source is different then zero. Gravitational waves are characterized by strain amplitude h, which is the ratio of displacement of two experimental particles over the distance of these particles. It is

485

Vasili Tsakanov and Helmut Wiedemann (eds.), Brilliant Light in Life and Material Sciences, 485–489.
© 2007 *Springer.*

proved[4], that a "Hertz type" experiment: an experiment when waves are generated by a source in laboratory, is impossible for gravitational waves. Most optimistic calculations are showing that the maximum amplitude of hand-maid gravitational wave sources on a distance of one wavelength from the source shall be 10^{-33} that is far away from being recordable. In contrary, gravitational waves of astrophysical origin can be detected. There are a lot of celestial sources that can generate gravitational waves with higher amplitudes: black hole- black hole collisions, anisotropic supernova explosions, neutron star- neutron star or neutron star -white dwarf or neutron star normal star mergers, white dwarf binaries, oscillating neutron stars or white dwarfs and so on.

There are two polarizations of gravitational waves "x" polarization and "+" polarization.

Detection of gravitational waves will give not only enough experimental confirmation for General Relativity (GR) Theory, but also will make possible to open a "new window" to the universe, to register information from those parts of universe that were unreachable in radio or optical measurements. While, if existence of gravitational waves will be questioned experimentally, that will be a major upset leading to establishment of new understanding of basic principles of general relativity.

Detectors of gravitational waves: There are three types-generations of gravitational wave detectors: resonant mass detectors, laser interferometers and space base detectors.

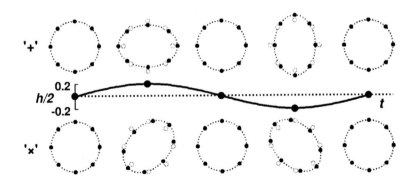

Figure 1. How experimental particles placed on a circle will be displaced if a gravitational wave with strain amplitude h and two polarizations "+" and "x" , will pass perpendicular to the plane of the circle. "t" is the period of the wave.

In late sixties Weber had proposed[4] and constructed first GW detector of resonant type, to register gravitational waves. Resonant mass antenna consists of a cylinder that is very well isolated from seismic, thermal, and other perturbations. Gravitational waves excite those vibrational modes of a resonant body that are close to frequency of falling gravitational wave, such as the fundamental longitudinal mode of a cylindrical antenna[5]. Resonant modes are strengthened by an amplifier and sent for registration. Amplifier introduce its own noise that is usually a white noise which is limiting the detector's bandwidth. Nowadays there are five working GW antennas in the world. Allegro is situated in Luisiana, Auriga in Lengaro, Explorer in CERN Geneva, Nautilus in Frascati, Italy and Niobe in Perth. All these detectors[6] are sensitive near 1 KHz. Due to the cylindrical form of resonant bar, it is "blind" in direction of axe of the cylinder. There are two ongoing projects Minigrail[6] at Leaden and Mario Schenberg in San-Paulo to build resonant detectors with spherical-shaped resonant masses that will be sensitive in all directions.

The second type of Gravitational Wave detectors are ground based interferometers that are consisting of two perpendicular interferometer arms from 300m to 4 km long. Laser beam traveling along arms will "feel" the difference when gravitational wave will pass through the detector. There are four detectors, already gravitational wave observatories, working online today: LIGO that consists from two identical detectors in Hanford and Louisiana, VIRGO near Pisa, TAMA at Tokyo, GEO at Hanover[7]. These detectors are sensitive in a wide range from 10 Hz to 1 KHz.

The third generation gravitational wave antenna is LISA (the Laser Interferometry Space Antenna) which is a joint ESA and NASA mission that will detect gravitational waves from massive black-hole mergers in the centers of galaxies, from the ultra-compact binary systems in our own Galaxy, and from many other sources, creating revolutionary research opportunities in astrophysics and fundamental physics[8]. LISA will be lunched in 2008 and will be ready to take data in 2012. It will consist of three spacecrafts interchanging coherent laser beams to find small displacements of test particles onboard for registration of gravitational waves. Space based mission will work in frequency band 10^{-1}-10^{-4} Hz.

The forth generation of GW detectors are accumulator type antenas. There is a nice idea by Kawamura et all to construct a new type of gravitational wave detectors, named by him "Displacement- and Timing-Noise Free" Gravitational-Wave Detector[9]. It's a low budget detector that will work, according to proposal, in a MHz domain, but in our understanding, if synchrotron radiation will be used as a source, it may work in kHz and even near 1 hz domain. The main difference from interferometers is that this detector is designed (Fig 2) to keep the light beam in one of the arms of interferometer some time that is equal to half of expected gravitational wave period, then send the beam to perpendicular arm to keep it there another half period. With this strategy the beam will remain all the time either on maximum or minimum of the wave and will accumulate displacement data as much times as it will pass between mirrors of arms.

The current status[10] of accumulator type detector is already reported with finesse around 100 (using available mirrors). A theoretical prediction for upper limit of cosmic gravitational wave background at 100MHz is $h < 10^{-33} \text{ Hz}^{1/2}$. This experiment correlates the outputs of two Interferometers for a few months at $h < 10^{-26} \text{ Hz}^{1/2}$.

Nowadays theory do not predict any significant source of gravitational waves in MHz frequency domain, may be there are some celestial bodies with ultra-superdense cores that can give something around 1 MHz or less, but 100 MHz should be free of sources. That's why its extremely interesting to build Gravitational wave detectors working in KHz and lower frequencies.

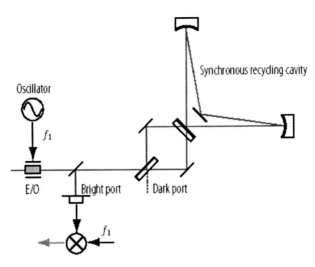

Figure 2. Design of Accumulator type GW detector or "Displacement- and Timing-Noise Free" Gravitational-Wave Detector. Synchrotron light will enter from the left, then will be splited into two parts to go into synchronous recycling cavity in two directions clockwise and anticlockwise. Difference of these two at bright port will indicate a GW at cavity.

CONCLUSIONS

We suggest construction of low cost accumulator type Gravitational Wave antenna of forth generation at CANDLE site. As positive aspects to support the project we can underline smooth radiation of source, wide spectrum that is necessary to operate the detector in different frequency ranges, low cost of detector, professional personnel at CANDLE site that will make construction and day to day maintenance of detector easier. As a negative aspect we are mentioning weakness of light source in a given frequency domain in comparison to laser sources. For the future, one can state that to work with high sensitivity, GW detector needs to be well isolated from seismic noises, especially to work in lower frequencies, near 1 Hz it is needed to put the detector underground, in mines, so seismic noises will be vanished. From that viewpoint CANDLE is the unique place, because in near vicinity there are old salt mines that are now used in scientific purposes, that will be a wonderful place to place the detector.

REFERENCES

1. Einstein, "Naherungsweise Integration der Feldgleichungen der Gravitation", Sitzungsberichte, Koeniglich Preussiche Akademie der Wissenshaften, Berlin, Erster Halbband, 688-696, 1916.
2. Pirani F.A.E. "On the Physical significance of the Riemann Tensor", Acta Phys. Pol. 15, 389, 1956.
3. Landau L.D., Lifshitz E.M., "Field Theory", Moscow, 1989.
4. Douglas, D. H., & Braginsky, V. G., " Experimental Gravitational wave Physics", in General Relativity: An Einstein Centenary Survey, ed. S. W. Hawking & W. Israel (Cambridge: Cambridge Univ. Press), 30, 1979.
5. Weber J.,"Gravitational radiation experiments", Phys.Rev.Lett.24 , pp.276 –279,1970
6. for further information consult web pages of detectors http://sam.phys.lsu.edu/ALLEGRO/allegro.html,http://www.auriga.lnl.infn.it/auriga/dete ctor/overview.html,http://www.roma1.infn.it/rog/explorer/explorer.html,http://www.roma 1.infn.it/rog/nautilus/nautilus.html,http://www.gravity.uwa.edu.au/bar/bar.html, http://www.minigrail.nl/
7. for further information on interferometer detectors see http://www.ligo-wa.caltech.edu/ , http://www.virgo.infn.it/, http://tamago.mtk.nao.ac.jp/,
8. Lisa project . http://lisa.esa.int ; see also http://www.lisascience.org.
9. Seiji Kawamura and Yanbei Chen, "Displacement- and Timing-Noise Free Gravitational-Wave Detection", arXiv:gr-qc/0504108 v3 19 Feb 2006.
10. Akutsu Tomotada, S. Kawamura, K. Arai, D. Tatsumi, S. Nagano, N. Sugiyama, T. Chiba et all, Development of a laser interferometer for MHz gravitational-wave detection, in presentations of Gravitational Wave Advanced Detectors Workshop – VESF meeting La Biodola, Isola d'Elba (Italy) – May 27th -June 2nd, 2006,http://131.215.114. 135:8081/talks.asp

RUBY BASED DETECTOR IN UV AND VUV SPECTRAL REGION

Z. Aslyan[a], K. Madatyan[†a], J. Vardanyan[a], V. Tsakanov[c], R.Mikaelyan[c]

[a]Yerevan State University, [b]Institute of Applied Problems of Physics, Yerevan, Armenia, [c]CANDLE, Yerevan, Armenia

Abstract: Recently major scientific centers (APS, ESRF, SRS etc.) systematically hold meetings on development and application of ionizing radiation detectors. Wide spectral range, long time frame and big power range characteristics of the synchrotron radiation, along with various used technologies, make the optimal good choice (from the point of view of the time-space characteristics) of the detector very actual. Really, each technology, each experiment–beamline, demand much and very unique detecting system.

Key words: converter, ruby, UV

Some of main requirements for a detector in UV are:
1. The detectors must be solar blind.
2. The detectors must be photon counting.
3. The detector should have high detective quantum efficiency (QE).
4. The detector should have high local dynamic range.

One of the earliest techniques employed for recording system (detecting, imaging) ionizing radiation scintillator-based system (imagers) remain one of the most flexible and successful techniques.

In a UV and X-Ray range was suggested[1,2] and is suggested Al_2O_3 as a converter[3]. Specific characteristics and advances technologies of the crystal growth make α-Al_2O_3 more competitive. Aging of the ruby under the influence of the irradiation is small, and the influence on optical characteristics is negligible[4].

According to absorbing (blue line) and emission (red line) abilities[1,2] of the ruby (Fig 1.) a 200-400nm detector was developed[5].

Vasili Tsakanov and Helmut Wiedemann (eds.), Brilliant Light in Life and Material Sciences, 491–494.
© 2007 *Springer.*

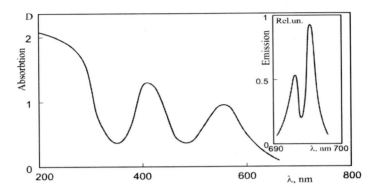

Figure 1. Absorbing and Luminescent properties of the ruby.

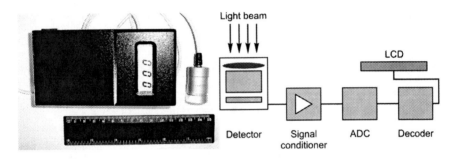

Figure 2. Photo and Block Diagram Measurements of the detector.

DC –dark current $\leq 1 \cdot 10^{-10}$A: it coincides with a threshold of sensitivity of the apparatus. Virtually the dark current is missing.

The bottom threshold of sensitivity of the device $\sim 10^{-4}$W/m^2.

Sensibility of the detector is coupled with QE of the ruby. QE is determined analogues to the method described in[6].

Samples with a concentration of Cr in the range 0.02-0.05% per weight were studied.

QE in the range of 200-400nm is about 0.9.

Also the following parameters and characteristics were been measured: dark current, total photocurrent, integral sensitivity, energetic characteristics of the relative spectral sensitivity, dependence of the photocurrent on irradiation intensity for various wave lengths.

Measurements were been carried according to the scheme (Fig. 2).

Luminescent intensity of the ruby linearly depends on intensity of the incident flux in a rather wide range of the intensity (Fig. 3), i.e. value of the output current is proportional to intensity of the incident flux and of the wholly absorbed irradiation.

Measurements were been carried on the samples with different concentrations of the Cr (1- 0. 05 %, 2-0.03% и 3-0.02 %) . Light source - Sun light imitator.

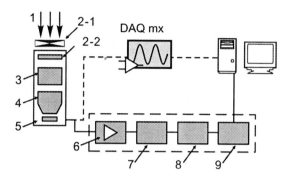

Figure 3. Measurements scheme. 1-light (SR, mercury or xenon lamp, imitator Sun light), 2-1 chopper, 2-2 filter, 3- converter, 4- optical amplifier, 5-CCD, 6- amplifier, 7-ADC, 8-signal processing module, 9-interface and LCD, as an alternative to 6-9 can be used PC and National Instruments-LabVIEW, DAQmx software packages.

Sensitivity ruby as well as the range of the sensitivity can be raised by means of additional activator, in particular titanium – non-radiant transition in ruby-titanium system[7] and also by means of radiation treatment[8].

On the basis of the developed UV detector in the spectral range 200-400nm of the wave lengths and the projected detector in the range 160-700nm[9], as well as[7,8], it is necessary to study, develop and manufacture a detector for the intensity measurements in range the 50-700nm.

Changes in the registering system will allow to study also time characteristics of the detector, to carry out investigations for the whole range of the excitation wave lengths (using ruby monocrystals with different concentrations of various imperfections) due to alignment of the main optical axes of the monocrystal.

Different modifications (particularly different geometry and sizes) of the suggested detector can be used in various spheres of the scientific and applied investigations, different wave length irradiation intensity measurements, especially for the ozone layer of the atmosphere, in medicine, in private life, etc.

REFERENCES:

1. V.I. Baryshnikov, T.F. Martynovich, Transformation of the Colouring Centers in Al_2O_3 Monocrystals, FTT, 1986. v. 28, N 4, pp. 1258-1260.
2. T.F. Martynovich et. al., Luminescence of colouring centers in a- Al_2O_3 crystals. Novosibirsk, Nayka, 1985, pp. 132-136.
3. Marianne C Aznar, Real time in-vivo luminescence dosimetry in radoitherapy and mammograpy using Al_2O_3:C, Riso-PhD-12 (EN), June 2005
4. J. H. Crawford, Jr., J. Nucl. Mater., 108-109, 644-654 (1982).
5. K.A Madatjan et.al., Sun UV Measurements by means of Fluorescent Materials and Phototransformers, Renewable Energy Sources, Part II, Yerevan, 1985.
6. V.A.Gevorkjan, K.A.Madatjan, et.al., Quantum Exit of a Ruby Luminescence, Crystals Spectroscopy, Nauka, 1970.
7. S.V. Grum-Grzhimajlo et. al., The Main Properties of a Ruby Monocrystal, Ruby and Sapphire, Nauka, 1974.
8. V.I. Baryshnikov et.al., Mechanism of Transformation and Fracture of Colouring Centers in a- Al_2O_3 Monocrystals. FTT, v.32, issue 1, pp. 291-293, 1990.
9. K. Madatyan, J. Vardanyan, S. Arzumanyan, Z. Asliyan, UV Detector Development for CANDLE Synchrotron Radiation Facility, Design Report Support Materials, ASLS-CANDLE R-002-02, Yerevan, 2002.

TWO-LAYER ULTRA-HIGH DENSITY X-RAY OPTICAL MEMORY

Hakob (Akop) P. Bezirganyan[1], Siranush E. Bezirganyan[2], Hayk H. Bezirganyan Jr[1] and Petros H. Bezirganyan Jr[3]

[1]Yerevan State University, Yerevan Armenia; [2]Department of Medical & Biological Physics, Yerevan State Medical University Yerevan Armenia; [4]Department of Computer Science, State Engineering University of Armenia, Yerevan Armenia

Abstract: Data reading procedure from nanostructured semiconductor X-ray optical memory (X-ROM) system detects data by measuring the changes in x-ray micro beam intensity reflected from the various surface points of data storage media. Two different mechanisms of the digital information read-out procedure, which are utilizing grazing-angle incidence X-ray backscattering diffraction (GIXB) and grazing-angle incidence X-ray reflection (GIX) techniques respectively, enable, in principle, the fabrication and exploitation of two-layer X-ROM. Angle of incidence of the x-ray micro beam is different for each storage layer of the proposed two-layer X-ROM.

Key words: two-layer X-ray optical memory; data ultra-high density; grazing-angle incidence X-ray backscattering diffraction; grazing-angle incidence X-ray reflection.

1.   INTRODUCTION

Optical data storage systems were introduced in the 1970s. Storage capacity was increased using the same resolution improvement as in classical microscopy and optical lithography, namely, a reduction in the source wavelength and an increase in the numerical aperture of the imaging optics. Even using an advanced multiplexing method based on the detection of optical angular momentum of a focused light beam, it can be expected to achieve a capacity of 125 gigabyte (GB) per one side.[1] The storage density limit in optical systems e.g. are considered in the paper of Kraemer et al.[2]

Vasili Tsakanov and Helmut Wiedemann (eds.), Brilliant Light in Life and Material Sciences, 495–498.
© 2007 *Springer.*

Nanofocusing probe limitations for ultrahigh-density optical memory connected with the wave-optics diffraction limit e.g. are discussed in the papers of Nikolov[3] and of Zhang et al.[4] Alternatively, a new scientific research direction devoted to development of the x-ray wavelength operating data handling system is proposed.[5-7]

2.   X-RAY OPTICAL MEMORY (X-ROM)

Most important aspect of nanotechnology applications in the information ultrahigh storage is the miniaturization of data carrier elements of the storage media with *emphasis on the long-term stability*. In the presented theoretical paper we consider the X-ROM as a crystalline semiconductor layer, in which the nanosized mirrors (reflecting speckles) are embedded. Data are encoded due to certain positions of these nanosized domains.

The data recording procedure of the cap layer can e.g. be performed by the in-focus femtosecond high-fluence laser pulses of a free electron laser at extreme ultraviolet or X-ray wavelengths. This method makes amorphous mirrors by melting and quenching nanosized subsurface domains in the cap silicon layer. If this two-phase cap layer is heated without quenching up to the activation internal energy, then the atoms in the amorphous phase change their location into the lower-energy crystalline structure phase. This can be used to erase digital information from the proposed storage layer, by the recrystallization of nanosized amorphous domains. Another way to make nanosized domains is the zone-plate-array lithography (ZPAL) technique.[8]

Digital data read-out procedure from X-ROM is performed via grazing-angle incident well-collimated and monochromatic x-ray micro beam. X-rays reflection from the proposed two-layer X-ROM, takes place either by the grazing-angle incidence X-ray backscattering diffraction (GIXB)[9,10] in first silicon layer or because of the difference between the refractive indexes of second silicon layer and of germanium domains embedded in it.[7] So, these two different techniques of the digital information read-out procedure enable, in principle, the fabrication and exploitation of two-layer X-ROM. For example, if the single-bit linear size is 10 nm and the bit spacing also is 10 nm, then an ultrahigh density two-sided digital data storage X-ROM with two layers per each side and 12 cm of edge size of the square shaped substrate has the following capacity:

$$4 \times 12\,cm \times 12\,cm \times (10\,nm + 10\,nm)^{-2} \times 1\,bit \approx 17000\,GB. \qquad (1)$$

3. X-ROM CHARACTERISTIC FUNCTION

Let introduce a characteristic function $\Delta R = R_2 - R$, where R_2 is the reflectivity coefficient of the X-ray plane wave reflected from the X-ROM wafer, R is the reflectivity coefficient of the x-ray plane wave reflected from the nanosized mirrors. Images of the introduced characteristic function are presented in Figure 1, which correspond to first and second layer of the X-ROM. Images are computed using the values of the incidence angle θ^i of the primary x-ray plane wave taken from the following angular region:

$$89{,}5^\circ \le \theta^i = \theta_B \le 89{,}9^\circ. \tag{2}$$

where the incidence angle θ^i of the primary x-ray plane wave is set equal to Bragg angle θ_B:

$$\lambda^i = 2 d_{hkl} \sin\left(\theta_B = \theta^i\right), \tag{3}$$

d_{hkl} is the space period of the wafer diffracting lattice planes (*hkl*).

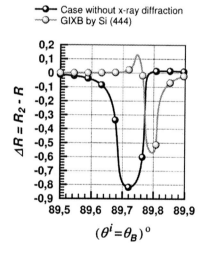

Figure 1. Images of characteristic function in particular case of two-layer X-ROM, cap layer of which consists of crystalline silicon wafer with regularly located amorphous nanosized mirrors, second storage layer consists of crystalline silicon layer with regularly embedded sub-surface germanium domains - a reflecting speckles produced by the advanced nano-lithography technology. Particular case of GIXB by *Si* lattice planes (*444*) is considered.

The minimum values of X-ROM characteristic function are most appropriate for the proposed digital data read-out procedure, so the mentioned minimum values give the possibility to choose optimally the angle of incidence θ^i of the X-ray micro beam and, consequently, the incident radiation wavelength λ^i, according the Bragg condition (3).

4.   CONCLUSION

Next generation ultra-high density optical storage media will operate on x-ray optics and X-ray diffraction optics, which allow to read-out the data from storage layer with the single-bit linear size reduced up to 10 nm or less.

REFERENCES

1. A. S. van de Nes, J. J. M. Braat, and S. F. Pereira, High-density optical data storage, *Rep. Prog. Phys.* **69,** 2323–2363 (2006).
2. D. Kraemer, B. J. Siwick, and R. J. D. Miller, Ultra high-density optical data storage: information retrieval an order of magnitude beyond the Rayleigh limit, *Chem. Phys.* **285,** 73-83 (2002).
3. I. D. Nikolov, Nanofocusing probe limitations for a ultra-high density optical memory, *Nanotechnol.* **15,** 1076-1083 (2004).
4. F. Zhang, Y. Wang, X.-D. Xu, H.-R. Shi, J.-S. Wei, and F.-X. Gan, High-Density Read-Only Memory Disc with $Ag_{11}In_{12}Sb_{51}Te_{26}$ Super-Resolution Mask Layer, *Chin. Phys. Lett.* **21**(10), 1973-1975 (2004).
5. H. P. Bezirganyan, H. H. Bezirganyan Jr., S. E. Bezirganyan, and P. H. Bezirganyan Jr., Specular beam suppression and enhancement phenomena in the case of grazing-angle incidence x-rays backdiffraction by the crystal with stacking fault, *Opt. Commun.* **238**(1-3), 13-28 (2004).
6. H. P. Bezirganyan, H. H. Bezirganyan Jr., S. E. Bezirganyan, P. H. Bezirganyan Jr., and Y. G. Mossikyan, An ultrahigh-density digital data read-out method based on grazing-angle incidence x-ray backscattering diffraction, *J. Opt. A: Pure Appl. Opt.* **7**(10), 604-612 (2005).
7. H. P. Bezirganyan, S. E. Bezirganyan, H. H. Bezirganyan Jr., and P. H. Bezirganyan Jr., Two-dimensional ultrahigh-density x-ray optical memory, *J. Nanosci. Nanotech.*, to be published (2006).
8. R. Menon, A. Patel, D. Gil, and H. I. Smith, Maskless lithography, *Mater. Today* **2,** 26–33 (2005).
9. H. P. Bezirganyan, and P. H. Bezirganyan, Solution of the two-dimensional stationary Schroedinger equation with cosine-like coefficient (in view of x-ray diffraction), *Phys. Status Solidi (a)* **105,** 345-355 (1988).
10. H. P. Bezirganyan, X-ray reflection from and transmission through a plane-parallel dielectric plate with cosine-like polarizability (symmetrical Laue case; $\theta_B \approx \pi/2$), *Phys. Status Solidi (a)* **109,** 101-110 (1988).

INDEX

Printed in the United States
73895LV00001B/19

9 781402 057229